Introduction to MECHATRONICS

APPUU KUTTAN K.K.
Professor
Department of Mechanical Engineering
National Institute of Technology Karnataka
Surathkal

OXFORD
UNIVERSITY PRESS

OXFORD
UNIVERSITY PRESS

YMCA Library Building, Jai Singh Road, New Delhi 110001

Oxford University Press is a department of the University of Oxford.
It furthers the University's objective of excellence in research, scholarship,
and education by publishing worldwide in

Oxford   New York

Auckland   Cape Town   Dar es Salaam   Hong Kong   Karachi
Kuala Lumpur   Madrid   Melbourne   Mexico City   Nairobi
New Delhi   Shanghai   Taipei   Toronto

With offices in

Argentina   Austria   Brazil   Chile   Czech Republic   France   Greece
Guatemala   Hungary   Italy   Japan   Poland   Portugal   Singapore
South Korea   Switzerland   Thailand   Turkey   Ukraine   Vietnam

Oxford is a registered trade mark of Oxford University Press
in the UK and in certain other countries.

Published in India
by Oxford University Press

© Oxford University Press 2007

The moral rights of the authors have been asserted.

Database right Oxford University Press (maker)

First published 2007
Second impression 2008

All rights reserved. No part of this publication may be reproduced,
stored in a retrieval system, or transmitted, in any form or by any means,
without the prior permission in writing of Oxford University Press,
or as expressly permitted by law, or under terms agreed with the appropriate
reprographics rights organization. Enquiries concerning reproduction
outside the scope of the above should be sent to the Rights Department,
Oxford University Press, at the address above.

You must not circulate this book in any other binding or cover
and you must impose this same condition on any acquirer.

ISBN-13: 978-0-19-568781-1
ISBN-10: 0-19-568781-7

Typeset in Times Roman
by Text-o-Graphics, Noida 201301
Printed in India by Ram Book Binding House, New Delhi 110020
and published by Oxford University Press
YMCA Library Building, Jai Singh Road, New Delhi 110001

Preface

Mechatronics refers to the multidisciplinary approach comprising mechanics, electronics, and computer technology that aims at improving the performance and quality of engineering products. Mechatronics today has found numerous applications in designing, manufacturing, and maintenance of a wide range of engineering products. A lot of devices we see around, such as DVD players, washing machines, and microwave ovens, are products of mechatronic engineering.

In the past, only one form of technology was used in designing and manufacturing of mechanical devices. But now a synergistic integration of mechanical, electronics and computer engineering disciplines is being used to produce efficient devices. The application of mechatronic concepts enhances the productivity and quality of the products. Apart from automobiles, consumer electronics, telecommunications and robotics, mechatronic systems are today also being used in biomedicine and aerospace industries. Mechatronics is fast developing as the core of all the activities in production technology.

The subject of mechatronics is also becoming increasingly popular in engineering colleges for research as well as education. Research topics related to mechatronics are diverse and include actuators/sensors, microelectromechanical systems (MEMS), mechatronic devices/machines, etc. Engineering colleges are also updating their courses in mechatronics to equip the students with the essential tools to successfully face today's industrial challenges. Mechatronics, thus, has a key role to play in future technological advancements.

About the Book

Introduction to Mechatronics provides a complete coverage of the basic principles of mechatronics and their applications. Beginning with the basic concepts, the book moves on to cover the topics such as system modelling and analysis, microprocessors, microcontrollers, sensors, actuators, need of intelligent systems for accurate operation of mechatronic systems, and applications of mechatronic systems in autotronics, bionics, and avionics. Case studies on slip casting for ceramic products and pick-and-place robots enhance the value of the text.

Content and Coverage

The book comprises eleven chapters and two appendices. Exercises are provided at the end of each chapter to help the readers assess their comprehension of the subject matter studied in the chapter. Ample number of examples are interspersed throughout the text to illustrate the concepts.

Chapter 1 contains an introduction to mechatronics and its applications and objectives. The advantages and disadvantages of mechatronics are also discussed.

Chapter 2 begins with a discussion on the challenges faced by today's manufacturing industries and then deals with computer integrated manufacturing and just-in-time production systems.

Chapter 3 discusses force, friction, and lubrication in detail as also stress–strain behaviour, bending of beams and torsion. The chapter also includes fits and tolerance, surface texture and scraping, and machine structure.

Chapter 4 introduces the readers to the applications of electronics in mechatronics. Conductors, insulators, and semiconductors along with digital, passive, and active electrical components are discussed comprehensively in this chapter.

Chapter 5 covers various analog and digital computers as well as analog to digital and digital to analog conversions. Microprocessor, microcontroller, and programmable logic controller (PLC) are also described in this chapter.

Chapter 6 commences with a description of the control system concept. It further discusses time response of a system, frequency domain analysis, and the modern control theory. Sequential and digital control systems are also included in this chapter.

Chapter 7 deals with the motion control devices, i.e. the elements of mechatronic systems responsible for transforming the output of a microprocessor or a control system into a controlling action on a machine or device. Hydraulic, pneumatic and electrical actuators as well as DC servomotor, AC servomotor, and stepper motor are described at length in this chapter along with brushless permanent magnet DC motors and microactuators.

Chapter 8 focusses on various types of internal and external sensors. The working principles of some microsensors are also presented in this chapter.

Chapter 9 deals with computer numerical control (CNC) and direct numerical control (DNC) machines. It discusses at length the CNC machine operation.

Chapter 10 provides a detailed description of the applications of intelligent systems in designing of mechatronic systems. Designing principles of some consumer mechatronic products—such as washing machine, automatic camera, and alarm indicator—are dealt with to make the subject matter practice-oriented. This chapter also includes case studies on the slip casting process and pick-and-place robots.

Chapter 11 discusses three major specialized study areas of mechatronics, namely, autotronics, bionics, and avionics.

There are two appendices at the end of the book. Appendix A lists Laplace transforms of different time domain functions. Appendix B has a table comparing Laplace and Z transforms.

The book is designed as a textbook for undergraduate students of mechanical, electronics, and electrical engineering disciplines. In addition, it will also prove to be useful for postgraduate students as well as practising engineers.

Acknowledgements

I sincerely thank the editorial team of Oxford University Press for the help provided to me in bringing out this book. I am also grateful to the reviewers who reviewed my manuscript and offered comments that greatly enhanced the quality of the book.

APPUU KUTTAN K.K.

Contents

Preface — *iii*

Chapter 1 Mechatronic Systems — **1**
1.1 Synergy of Systems — 1
1.2 Definition of Mechatronics — 3
1.3 Applications of Mechatronics — 5
1.4 Objectives, Advantages, and Disadvantages of mechatronics — 9
  Illustrative Examples — 10
  Exercises — 11

Chapter 2 Mechatronics in Manufacturing — **12**
2.1 Production Unit — 12
2.2 Input/Output and Challenges in Mechatronic Production Units — 13
2.3 Knowledge Required for Mechatronics in Manufacturing — 16
2.4 Main Features of Mechatronics in Manufacturing — 17
2.5 Computer Integrated Manufacturing — 21
2.6 Just-in-Time Production Systems — 22
2.7 Mechatronics and Allied Subjects — 23
  Illustrative Examples — 23
  Exercises — 26

Chapter 3 Mechanical Engineering and Machines in Mechatronics — **27**
3.1 Force, Friction, and Lubrication — 27
3.2 Behaviour of Materials Under Load — 32
3.3 Materials — 38
3.4 Heat Treatment — 38

3.5	Electroplating	39
3.6	Fits and Tolerance	39
3.7	Surface Texture and Scraping	40
3.8	Machine Structure	41
3.9	Guideways	42
3.10	Assembly Techniques	44
3.11	Mechanisms used in Mechatronics	44
	Illustrative Examples	52
	Exercises	56

Chapter 4 Electronics in Mechatronics — 57

4.1	Conductors, Insulators, and Semiconductors	57
4.2	Passive Electrical Components	58
4.3	Active Elements	65
4.4	Digital Electronic Components	72
	Illustrative Examples	78
	Exercises	81

Chapter 5 Computing Elements in Mechatronics — 82

5.1	Analog Computer	83
5.2	Timer 555	85
5.3	Analog to Digital Conversion	86
5.4	Digital to Analog Conversion	88
5.5	Digital Computer	89
5.6	Architecture of a Microprocessor	91
5.7	Microcontroller	94
5.8	Programmable Logic Controller	95
5.9	Computer Peripherals	97
	Illustrative Examples	105
	Exercises	109

Chapter 6 Systems Modelling and Analysis — 110

6.1	Control System Concept	112
6.2	Standard Test Signals	118
6.3	Time Response of A System	120
6.4	Block Diagram Manipulation	130
6.5	Automatic Controllers	132

6.6	Frequency Domain Analysis	134
6.7	Modern Control Theory	135
6.8	Sequential Control System	141
6.9	Digital Control System	141
	Illustrative Examples	147
	Exercises	149

Chapter 7 Motion Control Devices — 151
7.1	Hydraulic and Pneumatic Actuators	152
7.2	Electrical Actuators	160
7.3	DC Servomotor	166
7.4	Brushless Permanent Magnet DC Motor	169
7.5	AC Servomotor	170
7.6	Stepper Motor	171
7.7	Microctuators	172
7.8	Drive Selection and Applications	176
	Illustrative Examples	179
	Exercises	183

Chapter 8 Sensors and Transducers — 185
8.1	Static Performance Characteristics	186
8.2	Dynamic Performance Characteristics	187
8.3	Internal Sensors	189
8.4	External Sensors	195
8.5	Microsensors	207
	Illustrative Examples	213
	Exercises	215

Chapter 9 CNC Machines — 216
9.1	Adaptive Control Machine System	220
9.2	CNC Machine Operations	221
	Exercises	246

Chapter 10 Intelligent Systems and Their Applications — 248
10.1	Artificial Neural Network	249
10.2	Genetic Algorithm	257
10.3	Fuzzy Logic Control	260
10.4	Nonverbal Teaching	263

10.5	Design of Mechatronic Systems	264
10.6	Integrated Systems	268
	Case Study 10.1: Slip Casting Process	277
	Case Study 10.2: Pick-and-Place Robot	282
	Exercises	285

Chapter 11 Autotronics, Bionics, and Avionics   **286**

11.1	Autotronics	286
11.2	Bionics	300
11.3	Avionics	309
	Exercises	316

Appendix A: Laplace Transform	317
Appendix B: Laplace and Z Transforms	319
Bibliography	320
Index	327

Chapter 1

Mechatronic Systems

1.1 Synergy of Systems

The quest for making life better, easier, and comfortable has been at the heart of all the endeavours by human beings. All inventions and innovations have one aim, that of making life comfortable. In this pursuit, human beings have gone beyond exploiting only the basic sciences for producing technology products that are in tune with their aim. The technological advancement that we see today is a product of putting together some or all the basic sciences. This has resulted in as diverse and useful products as human beings can go in exploiting the knowledge base. This technological advancement has made it possible to produce products that can substitute for human effort. There are two forms of human effort: physical effort and mental effort. Technology is, accordingly, classified as machine technology and fine technology. Machine technology aims at relieving physical stress and strain of human beings. Some examples of the products of this category are machine tools, power generators, pumps, etc. Fine technology, on the other hand, is concerned with relieving mental strain of human beings. Measuring systems, computers, communication systems, etc., come under this category.

Communication and information transfer is vital to put together machine technology and fine technology. A combination of these can offer both physical and mental prowess. A combination of these two technologies has produced products that can substitute for human element in both its facets, physical effort and mental effort.

Typewriter has the distinction of being the first communication system ever used. Typewriter is purely a mechanical system. The first patent of a typewriter was made by Henry Mill in the year 1714. Until 1830 patents were made for typewriters working with different constructions. In 1864 Peter Mitter Hofer, a carpenter, made the first typewriter which resembled the modern type writer. In 1874 onwards, industrial production of typewriter started. Even today, most of the data transfer systems use the typewriter keyboard as the basic input device.

French philosopher Rene Descarts (1596–1650) found the relation between algebra and geometry. He illustrated the theory in a graph an algebraic function. It was the basis for analytical geometry. Later another French mathematician Cocagne prepared a monogram. Solutions to different mathematical calculations could be obtained with the help of the monogram.

But the accuracy of such calculations depended upon the accuracy of the drawing of monogram and the behaviour of the paper with respect to atmospheric humidity and temperature. In the course of time, mechanical linkages were developed to make calculations with greater accuracy, ease, and speed. Computer is the most advanced and accurate system for making large and repetitive calculations today. The focus now is on developing an interface capable of feeding data at a rate that is compatible with the computer's ability to process data.

The human can communicate directly or indirectly with the computer. Direct communication means that information is entered into the computer by means of switches on a console, or by using the enquiry typewriter. The information is received from the computer in the form of visual displays or audible alarms or printed matter. Indirect communication involves an intermediate medium such as magnetic type, punched card, or magnetic disk or compact disk. Indirect communication is necessary because of the great mismatch in speed between human input and computer consumption.

Traditional mechanical systems were built using mechanical components only. In the steam engine, James Watt used steam energy to convert it into kinetic energy by the rotational motion of shaft using a reciprocating mechanism. The speed of steam flow was controlled by a fly-ball governor. This is an example of a mechanical feedback system. In the twentieth century, electrical energy and electric signals were made available for industrial applications. These proved more efficient in the conversion of electrical energy into mechanical energy because electrical energy is easier to process as signals for measurement and control. Consider a mechanical system such as a shaper machine which removes material during the forward stroke only. In this system, the backward stroke is made faster since metal removal does not take place during a backward stroke. The time required for the backward stroke can be minimized using computer and electronic gadgets. This provides minimum cycle time and better production rate. The last few decades have seen digital computers taking place of most of the analog devices. Faster and more accurate results are obtained using a digital computer. As a result, most of the production processes and products are a mix of mechanical, electrical, and digital components. The design and manufacture of mixed systems requires the knowledge from all these disciplines. Mechatronics is the science that deals with mechanical, electrical, and digital components needed for the mixed systems. An inter-disciplinary knowledge is must for specialized engineers in mechanical, electrical, or digital systems.

1.2 Definition of Mechatronics

The integration of mechanical engineering, electronics engineering and computer technology is increasingly forming a crucial part in the design, manufacture and maintenance of a wide range of engineering products and processes. As a consequence of the synergy of systems in industry, it is becoming increasingly important for engineers and technicians to adopt an interdisciplinary and integrated approach towards engineering problems. The term 'mechatronics' is used to describe this integrated approach. In the design of cars, robots, machine tools, washing machines, cameras, microwave ovens, and many other machines, an integrated and interdisciplinary approach to engineering design is increasingly being adopted.

The term 'mechatronics' was first coined by the Japanese scientist Yoshikaza in 1969. The trademark was accepted in 1972. Mechatronics is a subject which includes mechanics, electronics, and informatics (Fig. 1.1).

Mechanics involves knowledge of mechanical engineering subjects, mechanical devices, and engineering mechanics. Basic subjects such as lubricants, heat transfer, vibration, fluid mechanics, and all other subjects studied under mechanical engineering directly or indirectly find application in mechatronic systems. Mechanical devices include simple latches, locks, ratchets, gear drives, and wedge devices to complicated devices such as harmonic and Norton drives, crank mechanisms, and six bar mechanisms used for car bonnets.

Engineering mechanics discusses the kinematics and dynamics of machine elements. Kinematics determines the position, velocity, and acceleration of machine links. Kinematic analysis helps to find the impact and jerk on a machine element. Change in momentum, causes an *impact,* whereas change in acceleration causes a *jerk.* Dynamic analysis gives the torque and force required for the motion of link in a mechanism. In dynamic analysis, friction and inertia play an important role.

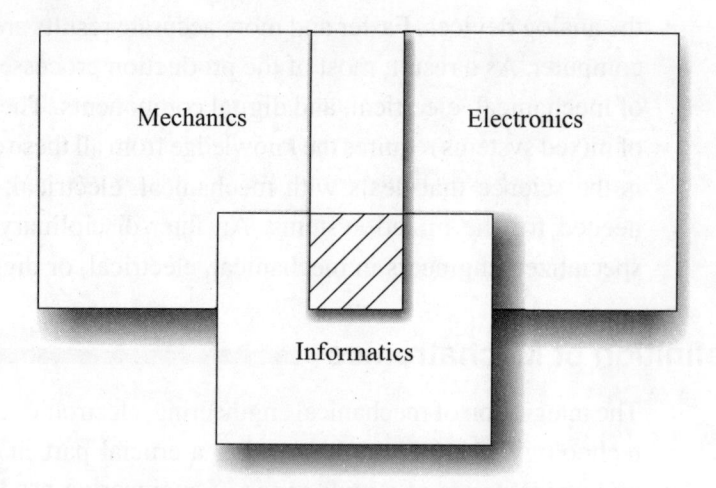

Fig. 1.1 Concept of mechatronics

Electronics involves measurement systems, actuators, power electronics, and microelectronics. Measurement systems, in general, are made of three elements, namely, the sensor, signal conditioner, and display unit. A sensor responds to the quantity being measured, giving an electrical output signal that is related to the input quantity. The signal conditioner takes the signal from the sensor and manipulates it into conditions which is suitable for either display or control any other systems. In a display system, the output from the signal conditioner is displayed. Actuation systems comprise the elements which are responsible for transforming the output from the control system into the controlling action of a machine or device. Power electronic devices are important in the control of power-operated devices to actuate through a small gate power of the order milliwatts. The silicon controlled rectifier (thyristor) is an example of a power electronic device which is used to control dc motor drives. The technology of manufacturing microelectronic devices through very large scale integrated (VLSI) circuit designs is also gathering momentum. Microsensors and microactuators are subdomains of the mechatronic system, which are used in many applications.

Informatics includes automation, software design, and artificial intelligence. The programmable logic controller (PLC) or microcontroller, or even personal computers, are widely used as informatic devices. A completely automated plant reduces the burden on human beings in respect of decision-making and plant maintenance, among other things. Software is used not only for solving complex engineering problems but also in finance systems, communication systems, or virtual modelling. Wide area networks, such as internet facilities, have large data storage facilities and the data can be retrieved from anywhere in the world. Informatics systems can make decisions using artificial intelligence. Artificial neural networks, genetic systems, fuzzy logic, hierarchical control systems, and knowledge-base systems are effective tools used in artificial intelligence.

1.3 Applications of Mechatronics

Mechatronics has a wide range of applications, as discussed in the following subsections.

1.3.1 Design and Modelling

Design and modelling are simplified to a large extent by the use of mechatronic systems. Basically, design involves drawing, analysis, and documentation. In earlier days, the processes of design were performed manually and it took weeks or months together. Now, the computer is used to complete processes of design faster. There are many designing tools such as AUTOCAD, IDEAS, and PROENGG, through which 2D or 3D drawings can be made. There are a number of tools to edit drawings at a faster rate. Analysis of the design involves working out the stress distribution, temperature distribution, weight analysis, and animations.

The virtual modelling of a manufacturing plant gives an idea of the time taken for a particular component to be manufactured and also shows virtually how the operations will be performed. The drum plotter, x-y plotter, printer, etc. give complete documentation of design drawings. Important parameters such as surface roughness and tolerance value can be incorporated in the drawing. Digitizers, plotters, CD drives, and many such devices are mechatronic systems.

1.3.2 Software Integration

Different kinds of software are used in manufacturing, design, testing, monitoring, and control of the manufacturing process. Examples of such software include computer aided design (CAD), computer aided testing (CAT), computer aided engineering (CAE), and computer aided processing planning (CAPP). The integration of the packets of software leads to computer integrated manufacturing (CIM) or just-in-time (JIT) manufacturing. Software integration is not only used for manufacturing but also for communication networks, economic analysis, etc.

1.3.3 Actuators and Sensors

Mechanical, electrical, hydraulic, and pneumatic actuators are widely used in the industry. Toggle linkage and quick return mechanics are typical examples of mechanical actuators. Switching devices, solenoid-type devices, and drives such as alternative current (ac) and direct current (dc) motors can be used as electrical actuators. Hydraulic and pneumatic drives use linear cylinders and rotary motors as actuators.

The term sensor is used for an element which produces a signal relating to the quantity being measured. For example, an electrical resistance temperature device transforms the input of temperature into change in resistance. The term transducer is often used in place of sensor. Transducers are defined as devices which when subject to some physical change experience a related change. In the displacement transducer, force is not an error. Addition of extra force into the system reduces backlash and play. For example, in the dial gauge, an additional tension spring is provided on the rack so that the play between the set of gear trains is minimized. Similarly, in a force-transmitting transducer, the provision of more displacement is not an error. Reduction in the play in force-transmitting devices produces a loss in power due to friction.

1.3.4 Intelligent Control

Feedback control systems are widespread not only in nature and the home but also in industry. There are many industrial processes and machines which control many variables automatically. Temperature, liquid level, fluid flow, pressure, speed, etc. are maintained constant by process controllers. Adaptive control and intelligent manufacturing are the areas where mechatronic systems are used for decision making and controlling the manufacturing environment.

1.3.5 Robotics

Robot technology uses mechanical, electronic, and computer systems. A robot is a multifunctional reprogrammable machine used to handle materials, tools, or any special items to perform a particular task. Manipulation robots are capable of performing operations, assembly, spot welding, spray painting, etc. Service robots such as mail service robots, household servant robots, nursing robots in hospitals are being used nowadays.

1.3.6 Manufacturing

In the domain of factory automation, mechatronics has had far-reaching effects in manufacturing. Major constituents of factory automation include computer numerically controlled (CNC) machines, robots, automation systems, and computer integration of all functions of manufacturing. Low volume, more variety, higher levels of flexibility, reduced lead time in manufacture, and automation in manufacturing and assembly are likely to be the future needs of customers, and mechatronic systems will play an important role in this context.

1.3.7 Motion control

A rigid body can have a very complex motion which might seem difficult to describe. However, the motion of any rigid body can be considered to be combinations of translational and rotational motions. By considering a three-dimensional space, a translational movement can be considered to be one which can be resolved into components along one or more of three axes. The rotation of a rigid body has rotating components about one or more of the axes. A complex motion may be a combination of translational and rotational motion. Motion control is important in many industrial applications such as robots, automated guided vehicles, NC machines, etc. If the robot arm cannot reach a particular location, then the movements of workpiece have to be analysed further. Any body has six degrees of freedom, three translations and three rotations. A point has only three translations. In a machine tool, the workpiece has six degrees of freedom and the tools also have six degrees of freedom. Thus, a machine tool with twelve degrees of freedom can be manufactured. Such a tool can perform a complicated machining operation.

1.3.8 Vibration and Noise Control

When a machine member is subjected to a periodic dynamic force, it will vibrate. If the vibration level ranges from a frequency of 20 Hz to 20,000 Hz, it produces noise. Vibration and noise isolation are important in industry. Vibration isolation can be achieved by passive, semi-active, or active dampers. In passive dampers the structure is mounted on damping materials with initial spring loading. In semi-active dampers, both passive and active damping elements are used. In active damping, extra energy is used to damp the structure. When a structure is subjected to a pulse input, a shock is produced. Different types of shock absorbers are used to reduce the shock amplitude.

Noise isolation is equally important in industry since noise is harmful to human beings. Adaptive control techniques are used for noise isolation. In this method, the system predicts the noise level in each interval of time and noise is introduced through the speaker in phase opposition. This adaptive control system reduces the noise level.

1.3.9 Microsystems

It is fair to say that microsystems are a major step towards the ultimate miniaturization of machines and devices such as dust-size computers and needle-type robots. The advancement of nanotechnology will certainly result in the realization of superminiaturized machinery. The need for miniaturization has increased manifold in recent years, and engineering systems and devices have become more and more complex and sophisticated. Picosatellites, spacecrafts, table-top manufacturing units, and microelectromechanical systems will become a reality in the future. The knowledge of mechatronics is very useful for microsystems.

1.3.10 Optics

All slip gauge blocks are calibrated against light wavelength as a standard. Angle gauges can be calibrated to an accuracy of 0.1 sec using a light wave standard—the angstrom unit. A combination of optical and electronic principles has led to the development of instruments such as the midarm which measures angular displacement with an accuracy of 0.05 sec. Optical angle measurement systems for inertial guidance with an accuracy of 0.02 sec have been in use since 1961. Opto-electronic systems use a lens or telescope to form an optical image of an object under study on a photocathode image detector tube. The motion of the object causes the motion of the photocathode optical image and the corresponding motion of the electron image. The optical image is obtained by a conventional videcon camera or a coupled charge device. The camera converts an array of analog signals, in 236×236 pixels in a square centimetre. The analog signals are then converted into digital signals for each pixel and transmitted to an electron image grabber to produce an electron image. As the image starts deviating from the neutral position, the photo multiple layer output tends to drive back by means of a deflection coil. Thus, any main object can be brought to the aperture continuously.

The application of still and motion picture photography often allows qualitative and quantitative analyses of complex motion. The photoelastic method is convenient to determine the stress distribution in a machine element. The basic phenomenon of double refraction under load is used in photoelasticity. Double refraction takes place when light travels at a different speed in a transparent material depending on the direction of travel relative to the direction of the principle stress and also depending on the magnitude of the difference between principle stresses for two-dimensional fields. Due to double refraction, light waves form an interference pattern of fringes on a photograph. The photograph is then used to determine the principal stresses. By the use of the frozen stress technique, the method can be extended to three-dimensional problems.

The cathode ray tube provides display devices for computers and other entertainment devices such as the television, projector, etc. Electron guns with basic columns can be obtained in a pixel. Cathode ray tubes for picture displays usually have 256×256 pixels/cm^2. As the number of pixels increases per square centimetre, the clarity of the picture becomes better. Systems are available which permit each pixel in grey levels (256 levels) in a black-and-white display. Grey levels (light intensity levels) are called grey scaling. With the basic three colours 256^3 colour combinations can be obtained with grey scaling.

In the ordinary film, only the magnitude of intensity is recorded, which in turn gives two-dimensional images. By recording the amplitude and phase of the reflected light from an object, a hologram can be obtained. A hologram gives three-dimensional ghost images of three-dimensional objects. An optical computer with a hologram will give faster computation in future. Coding and decoding is not required as in conventional computer operation. A ghost image from the hologram gives a grey-scaled image on each voxel. 256^3 voxels can be accommodated in a cubic centimetre of laser hologram. Sintering in each voxel can be obtained by packing the metal particles in the ghost image. Thus in future any complicated article can be manufactured in seconds using the laser hologram technique.

1.4 Objectives, Advantages, and Disadvantages of Mechatronics

The objectives of mechatronics are the following.

1. To improve products and processes
2. To develop novel mechanisms
3. To design new products
4. To create new technology using novel concepts

Earlier the domestic washing machine used cam-operated switches in order to control the washing cycle. Such mechanical switches have now been replaced by microprocessors. A microprocessor is a collection of logic gates and memory elements whose logical functions are implemented by means of software. The application of mechatronics has helped to improve many mass-produced products such as the domestic washing machine, dishwasher, microwave oven, cameras, watches, and so on. Mechatronic systems are also used in cars for active suspension, antiskid brakes, engine control, speedometers, etc. Large-scale improvements have been made using mechatronic systems in flexible manufacturing engineering systems (FMS) involving computer controlled machines, robots, automatic material conveying and, overall supervisory control.

There are many advantages of mechatronic systems. Mechatronic systems have made it very easy to design processes and products. Application of mechatronics facilitates rapid setting up and cost effective operation of manufacturing facilities. Mechatronic systems help in optimizing performance and quality. These can be adopted to changing needs.

Mechatronic systems are not without their disadvantages. One disadvantage is that the field of mechatronics requires a knowledge of different disciplines. Also, the design cannot be finalized and safety issues are complicated in mechatronic systems. Such systems also require more parts than others, and involve a greater risk of component failure.

Illustrative Examples

Example 1.1 An electrical switch is a man-made mechatronic system, used to control the flow of electricity. The toggle of the switch is a mechanical system and the human brain or a control system used to actuate the switch acts as an informatics system. The brain or informatics system decides whether we need to turn on the switch. If we do, the brain controls the movement of our limbs and we turn on the switch. When the switch is on, the resistance of contact is nearly zero and energy flow takes place. When the switch is off, the resistance is infinity and no current flows.

Example 1.2 A thermostatically controlled heater or furnace is a mechatronic system. The input to the system is the reference temperature. The output is the actual temperature. When the thermostat detects that the output is less than the input, the furnace provides heat until the temperature of the enclosure becomes equal to the reference temperature. Then the furnace is automatically turned off. Here, the bimetallic strip of the thermostat acts as informatics since it automatically turns the switch on or off. The lever-type switch is mechanical system whereas the heater acts as an electrical system.

Example 1.3 Most washing machines are operated in the following manner. After the clothes to be washed have been put in the machine, the soap, detergent, bleach, and water are put in required amounts. The washing and wringing cycle time is then set on a timer and the washer is energized. When the cycle is completed, the machine switches itself off.

When the required amount of detergent, bleach, water, and appropriate temperature are predetermined and poured automatically by the machine itself, then the machine is a mechatronic system. The microprocessor used for this purpose acts as the informatics system. The electrical motor actuated for wriggling is an electrical system. The agitator and timer are mechanical systems. The washing machine is an ideal example of a mechatronic system.

Example 1.4 The automatic bread toaster is a mechatronic system, in which two heating elements supply the same amount of heat to both sides of the bread. The quality of the toast can be determined by its surface colours. When the bread is toasted, the colour detector sees the desired colours, and the switch automatically opens and a mechanical lever makes the bread pop up. Mechanical, electrical, and informatics systems are involved in the operation of the bread toaster.

Exercises

1.1 Identify mechanics, informatics, and electronic systems of the (a) washing machine, (b) jet engine, (c) bread toaster, and (d) automatic camera.

1.2 What is meant by the kinematic and dynamic analyses of a machine element?

1.3 What are the tools for artificial intelligence?

1.4 Define impact and jerk.

1.5 Identify the areas where mechatronic systems can be used.

1.6 Explain the objectives of mechatronics.

1.7 What are the advantages and disadvantages of a mechatronic system?

Chapter 2

Mechatronics in Manufacturing

2.1 Production Unit

Industrial production is a transformation process that converts raw material into finished products. The products are made by a combination of manual labour, machine tools, and energy. This transformation process involves a sequence of steps, each step bringing the material closer to the desired final state. Individual steps form part of what is called the product operation. Depending on the nature of the product operation, there are two types of industries.

1. Manufacturing industries—examples are car, computer, machine tool, and aircraft industries.
2. Process industries—examples are petroleum refineries, food-processing, chemical, and plastic industries.

In the nineteenth century and before, a nation was considered powerful on the basis of its human resources. The industrial revolution made a change in this. And, in the twentieth century, which was the era of machine power, mechanical advancement determined a country's power. Productivity and growth were the need of the hour. In the twenty-first century, knowledge is considered to be the indicator of growth. The success of a business is now determined by how efficiently, creatively, and accurately it uses the knowledge available to it. Application of the knowledge available is essential for all manufacturing units.

The term 'knowledge worker' was first used in 1960. The growth of information technology and electronics brought about great changes in almost all walks of life. Knowledge is different from all other kinds of resources. This resource constantly makes itself obsolete. Knowledge is of two types:

1. Explicit knowledge—which is codable, and in which communication through data, formulae, etc. is possible, and

2. Tacit knowledge—highly personal, hard to formulate, and difficult to communicate or share with others. Welding and painting are examples of tacit knowledge.

Knowledge creation is a process of interaction between explicit and tacit knowledge. For the transformation of tacit knowledge into explicit knowledge, a company should create an environment to enhance knowledge.

New knowledge is created by an individual, but the organization has to facilitate the adoption of such knowledge. The acquisition of knowledge can create a significant competitive advantage over other companies in the market. In the domain of production unit automation, mechatronics has had far-reaching effects in manufacturing and will become even more important in the future.

Major constituents of mechatronic automation involve CNC machines, robots, automation systems, and computer integration of all functions of manufacturing. Basically, these advanced manufacturing solutions consist of mechatronic systems.

2.2 Input/Output and Challenges in Mechatronic Production Units

Figure 2.1 illustrates the block diagram of a production unit with inputs, challenges, and product output. The inputs are men, machines, material, money, market, and methods. The challenges are competition, communication, commitment, compatibility, and cost. Productivity can be improved by overcoming the challenges, and by technology-based utilization of the six inputs. Globalization has led to increase in competition. The success of an enterprise depends on the integration of management technology and human-resource management. Along with effective use of information technology, this would enhance productivity of technology and management.

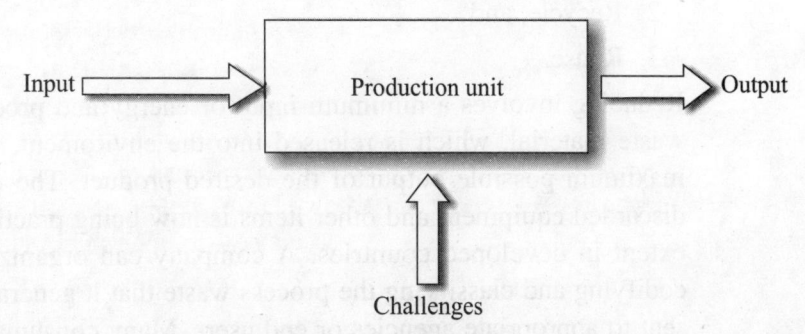

Fig. 2.1 Production unit

An intelligent manufacturing system based on machines equipped with artificial intelligence can handle, compare, and vary conditions without interrupting the automatic operation. Small and medium-size enterprises require a low-cost module or computer integrated manufacturing design and system.

Availability of the material plays an important role in production. Usually material cost per product cannot be reduced unless an alternative material is made. But a major reduction in cost can be achieved by reduction in transportation time.

An organization's economic power is dependent upon the level and growth rate of its productivity. Economists use indexes to measure productivity. The various measures for productivity are labour productivity index, capital productivity index, energy productivity index, and material productivity index.

In the last few decades, market and consumer needs have changed significantly. The priorities in decision making have also changed. In the 1970s a competitive price was the topmost priority in decision making. But quality took centre stage in the 1980s and 1990s when cost and quality became essential parameters for any product to come to the market and delivery and speed assumed a pivotal role for decision making. The new consumers of this century expect uniqueness and personalized features in products.

Technology should provide a service or product ensuring that the consumer expends a minimum amount of energy, material, and other resources. At the same time, the process of production and the product itself should cause minimum environment pollution. The product and services are the main outputs from a production unit. However, unwanted scrap and waste are also produced in this unit. These should also be converted into useful by-products. The technology involved in such conversion is known as green technology. All green technology essentially improves the environment with the following concepts:

1. Reduce,
2. Recycle, and
3. Reuse.

Reducing involves a minimum input of energy and production of minimum waste material, which is released into the enviroment, while producing the maximum possible output of the desired product. The recycling of used or discarded equipment and other items is now being practised to an increasing extent in developed countries. A company can organize itself by properly codifying and classifying the process waste that it generates, so that it can be sent to appropriate agencies or end users. Many consumable items which get contaminated or degraded due to use can be recycled. It is not uncommon to find a whole lot of equipment of all sizes and complexities and spare being consigned to the scrapyard in industries because they are either worn out or beyond repair.

Reconditioning is currently in vogue. It involves the use of vintage equipment rather than new equipment of the same design and productivity.

The new competitive era forced organizations to meet the challenges of staying technologically competitive in a fully globalized environment. It is evident that harnessing the technology has become of paramount importance for the success of any enterprise. As the process of globalization has spread, leading to an increasing need for competitiveness, management of technology has assumed greater importance. Gradually, the success or even survival of an enterprise will predominantly depend upon how much technology-driven the organization becomes. To meet the demands of global competitiveness, an enterprise has to develop the ability to respond to market demands for changes in production specification and product mix as rapidly as possible.

An accountant essentially takes a costing approach to productivity. Engineers generally seek costing as a measure of the physical asset and other resources, such as production per hour, work hours per unit, material requirement per unit, machine utilization, and spare utilization. Cost reduction is the most important priority for the management. Managers frequently use the accountancy ratio for achieving the objective of the general management. However, these are not basic standards according to which production may be measured in different situations.

Communication in a manufacturing environment refers to the adequate and timely flow of information with a feedback mechanism. The purpose of effective communication is to achieve a better understanding between the employees and management. Such communication helps to motivate the employees to improve productivity. The communication technique may not have a short-term impact on total productivity, but certainly has a positive effect in the long run. Modern methods of communication are paving the way for tele-service and will lead to increased productivity. The distinct feature of many Japanese companies is that they keep even the lowest level employee well informed about their financial status, realizing that all the employees and management are part of one company family whose objective is to produce products or services at the most competitive price and best quality possible. The employee knows that if this is not done, there will be no company and no job.

Commitment from employees can be achieved by training and education, and by creating and maintaining an environment in which people can accomplish goals efficiently and effectively. Training seeks to achieve required human productivity by increasing the ability level of the workforce. Some of the common forms of training are on-the-job training, outside courses, and visitation training. From childhood onwards, all of us have been educated about our surroundings. The industrial employees—workers, engineers, and managers—should have a certain level of formal or informal education. Different solutions to a given problem are proposed by different people because of their education levels. Education is the levels of knowledge one should possesses from visual observations, thinking, and perception of surrounding world. Improving working conditions also helps improve productivity. This involves a detailed analysis of working conditions. The factors that must be analysed are temperature, light, humidity, noise level, colour of the surroundings, how hazardous material is handled, etc. With the advancement of robotics, working conditions can be improved to a great extent.

Environment compatibility, national compatibility, and production compatibility play an important role in any enterprise. National policy and political influence play an important role in production and production compatibility. The International Product Standards are important for both hardware and software so that better compatibility can be achieved.

The integration of culture plays an important role in improving productivity. Religious, social, and cultural integration plays an important role in this context. There should be some established process for transforming the skills, ideas, and knowledge among various cultures.

2.3 Knowledge Required for Mechatronics in Manufacturing

Until the 1970s, machine tools were largely mechanical systems with very limited electrical or electronic content. However, since then there has been a dramatic change in technology with an increase in the content of electrical and electronic systems integrated with mechanical parts through mechatronics. Machine tools incorporating computer numerical control, electric servodrives, electronic measuring systems, and precision mechanical parts such as ball screws and antifriction guideways are examples of the application of mechatronic technology. The Japanese machine tool industry has flourished because Japanese have mastered electronics and have been able to combine precision mechanics and informatics in design. Computer and control system units are widely used in machine tools. Figure 2.2 shows how the Japanese have incorporated mechatronics in manufacturing. Computer aided design (CAD), electro-mechanical devices, control circuitry, and digital control systems are widely used in mechatronic systems in the manufacturing environment.

The development of computer numerically controlled (CNC) machines is an outstanding contribution to manufacturing industries. It has made possible the automation of the machining processes with flexibility to handle small to medium batch qualities in part production. Initially, the CNC technology was applied on basic metal cutting machines such as lathes and milling machines. Later, to increase the flexibility of machines handling a variety of components and to incorporate them in a single set-up in the same machine, CNC machines with material handling systems were developed. CNC machines are capable of performing multiple operations such as milling, drilling, boring, and tapping.

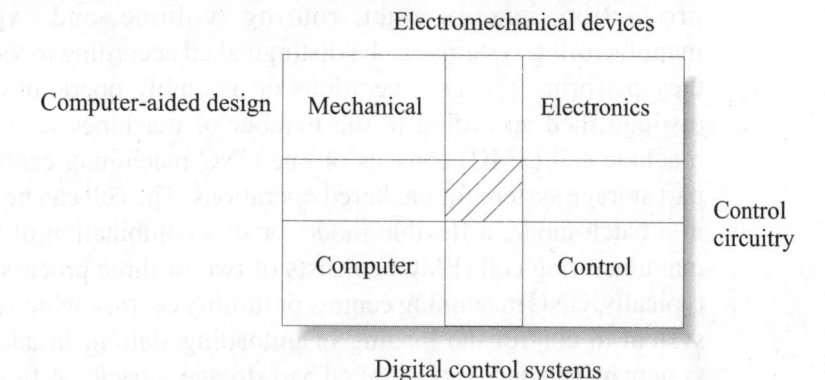

Fig. 2.2 Mechatronics in manufacturing

2.4 Main Features of Mechatronics in Manufacturing

We will discuss the main features of mechatronics in manufacturing in this section.

2.4.1 Flexible Manufacturing System

The flexible manufacturing system was first conceptualized for machines and it required the prior development of numerical control. The concept is credited to Parid Williamson, a British engineer employed by Moliars during the mid 1960s. Moliars patented the invention. A flexible manufacturing system (FMS) is a highly automated group-technology machine cell consisting of a group of processing workstations—usually CNC machine tools, interconnected by an automated material handling and storage system, and controlled by a distributed computer system. The reason the FMS is called flexible is that it is capable of processing a variety of different part styles simultaneously at various workstations. To qualify as being flexible, a manufacturing system should satisfy several criteria.

1. Part variety test—the system should be able to process different part styles in a non-batch mode.
2. Schedule change test—the system should readily adopt to changes in product schedule.
3. Error recovery test—the system should be able to recover in the case of equipment malfunctioning and breakdown.

4. New part test—the system should be able to introduce new part designs into the existing product.

The types of flexibility in manufacturing are those relating to the machine, production, mix, product, routing, volume, and expansion. Flexible manufacturing systems can be distinguished according to the kind of operations they perform—process operations or assembly operations. They can also be distinguished according to the number of machines in the system. A single machine cell (SMC) consists of one CNC machining centre combined with a part storage system for unaltered operations. The cell can be designed to operate in a batch mode, a flexible mode, or in a combination of the two. A flexible manufacturing cell (FMC) consists of two or three processing workstations—typically, CNC machining centres or turning centres, which plan a part handling system to control the loading or unloading station. In addition, the handling system usually includes a limited part storage capacity. A flexible manufacturing system has four or more processing workstations connected mechanically by a common part-handling system and electronically by a distributed computer system. The benefits that can be expected from the FMS include:

1. increased machine utilization,
2. fewer machine requirements,
3. reduction in factory shopfloor requirement,
4. reduced inventory requirements,
5. greater responsiveness to change,
6. lower manufacturing load time,
7. reduced direct labour requirements and high labour productivity, and
8. opportunity for unattended production.

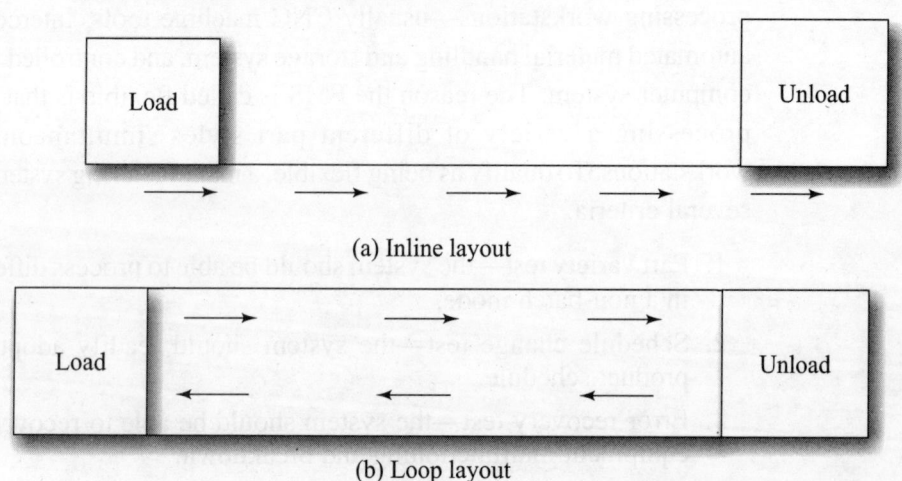

(a) Inline layout

(b) Loop layout

Machine station

(c) Robot-centred layout

Load/unload

(d) Ladder layout (e) Open-field layout

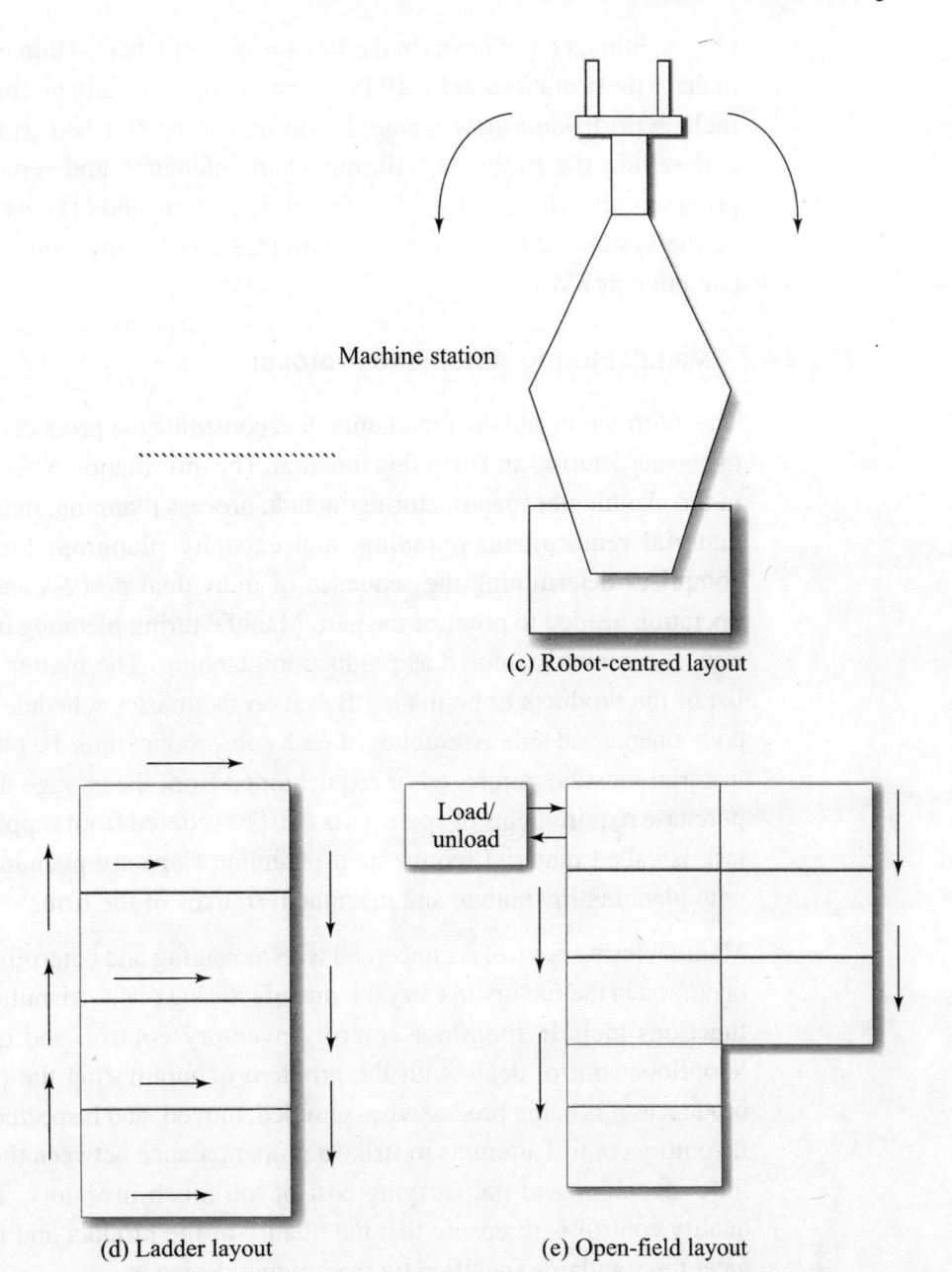

Fig. 2.3 FMS layouts

The material handling system establishes the FMS system. Most layout configurations found in today's FMS's can be divided into five categories—(a) inline layout, (b) loop layout, (c) robot-centred layout, (d) ladder layout, (e) open-field layout, which are shown in Fig. 2.3.

One additional component in the FMS is human labour. Humans are needed to manage the operations of the FMS. The functions typically performed by humans include (a) loading new materials, (b) unloading finished parts, (c) changing and setting the tools, (d) equipment maintenance and repair, (e) NC part programming, (f) operating the computing system, and (g) overall management of the system. The status report provides an instantaneous snapshot of the condition of FMS.

2.4.2 Manufacturing Automatic Protocol

The information and documentation that constitute the product design flow into the manufacturing and planning function. The information processing activators in the planning of manufacturing include process planning, master scheduling, material requirement planning, and capacity planning. Process planning comprises determining the sequence of individual process and the assembly operation needed to produce the part. Manufacturing planning includes logistic issues, commonly known as production planning. The master schedule is the list of the products to be made. Based on the master schedule, the individual components and sub-assemblies of each sub-product must be planned. The raw material must be purchased or requisitioned from the storage department, and purchase requirements of spare parts must be ordered from suppliers. The entire task is called material requirement planning. Capacity planning is concerned with planning the human and machine resources of the firm.

Manufacturing control is concerned with managing and controlling the physical operation in the factory to carry out manufacturing plans. Manufacturing control functions include shopfloor control, inventory control, and quality control. Shopfloor control deals with the problem of monitoring the progress of the product as it is being processed, assembled, moved, and inspected in the factory. Inventory control attempts to strike a proper balance between the danger of too little inventory and the carrying cost of too much inventory. The mission of quality control is to ensure that the quality of the product and its components meet the standards specified by the product designer.

The elements discussed above can be automated. Automation can be defined as a technology concerned with the application of mechanics, electronics, computers, and control systems to operate and control production. The automated elements of a production system are said to comprise manufacturing automatic protocol.

2.4.3 Technical Office Protocol

Business functions are the principal means of communicating with the customer. They form the beginning and end of the information processing cycle. Business functions include sales and marketing, sales forecasting, order entry, cost accounting, customer billing, and payroll. It is most important for any business enterprise to cater to the customer's needs. This can be done in the following ways.

1. Through an interview—either by in-person or over telephone.
2. Through comment cards—these allow the customer to give a feedback regarding the level of sophistication of the product and service.
3. Through a formal survey—this is accomplished by mass mailing.
4. Through focus groups—several customers or potential customers serve on the panel.
5. Through study of complaints—this is a statistical review of data on customer complaints.
6. By seeking information about the reason why a customer has returned a product.
7. Through the Internet—this is a relatively new way of gathering customer opinion.
8. Through field intelligence—this involves the collection of second-hand information from employees who deal directly with the customer.

Quality function deployment is a systematic procedure for defining customer needs and interpreting them in terms of product features and process characteristics. The automation of a business function is called technical office protocol.

2.5 Computer Integrated Manufacturing

The ideal computer integrated manufacturing (CIM) system applies computer and communication technology to all of the operational functions and information processing functions in manufacturing from order receipt, through design and production, to packaging and shipment. The scope of CIM is compared with the more limited scope of CAD/CAM in Fig. 2.4. CIM involves an integration of all the functions in one system that operates throughout the enterprise. Computer aided design (CAD) involves the use of computer systems in product design. Computer aided manufacturing (CAM) involves the use of computer systems in functions related to manufacturing engineering, such as process planning and numerical control part programming.

Design	Manufacturing
Business function	Manufacturing control

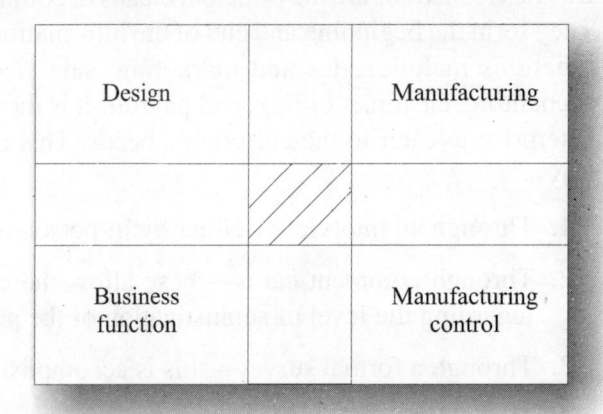

Fig. 2.4 Scope of computer integrated manufacturing

Some computer systems perform both CAD and CAM. So the term CAD/CAM is used to indicate the integration of design and manufacturing. Computer integrated manufacturing includes CAD/CAM, and also those business functions of the company that are related to manufacturing.

2.6 Just-in-Time Production Systems

Just-in-time (JIT) production systems were developed in Japan to minimize inventories—work-in-progress inventories and those of other types are seen by the Japanese as a waste that should be minimized or eliminated. The ideal JIT production system produces and delivers exactly the required number of each component to the downstream operation in the manufacturing sequence just at the time when that component is needed. Each component is delivered just in time. The delivery discipline minimizes work-in-progress and manufacturing lead time as well as the space and money invested in work in progress. JIT discipline can be applied not only to the production operation but to the supplier delivery operation as well.

The philosophy of JIT has been adopted by many US manufacturing firms. Continuous flow manufacturing is a widely used term in the US that denotes a JIT style of production operation. Certain requisites for the JIT production system to operate successfully are (a) a pull system of production control, (b) small batch size and reduced set-up times, and (c) stable and reliable production.

In the pull system of production control, parts are ordered for and delivered at each workstation when they are required. In a push system of production control, parts at each workstation are produced irrespective of when the parts are needed. To reduce the average inventory level, the batch size must be reduced. And to do so, the set-up cost must be reduced, which means minimizing set-up times. The reduced set-up time permits a smaller batch and lower work in progress. JIT production requires near perfection as regards timely delivery, quality of the parts, and equipment reliability. JIT also requires zero defects, which means that the production unit should be highly reliable.

2.7 Mechatronics and Allied Subjects

Mechatronics involves the synergetic integration of various engineering disciplines and is applied to the designing and manufacturing of products. The mechatronics engineer does not have to study in full mechanical engineering, electronics/electrical engineering, computer engineering, and control engineering. It is enough to study the selected topics as far as the designing and manufacturing of the product are concerned. Chapters 3 to 8 in this book deal with the topics relevant to mechatronics engineers. Mechatronics cannot be defined in a specific way. The scope of the subject is vast and it can be enclosed within the domain of designing, manufacturing, and marketing. Chapters 9 to 11 deal with the application and modelling of mechatronic systems.

Illustrative Examples

Example 2.1 A company manufacturing computers produces 10000 computers by employing 50 people at 8 hr/day for 25 days. What is the production and labour productivity? If the company produces 12000 computers by hiring 10 additional workers at 8 hr/day for 25 days, what is the production and labour productivity?

Solution

In the first case,

Production = 10000 computers

$$\text{Labour productivity} = \frac{\text{output}}{\text{human power input}}$$

$$= \frac{10000}{50 \times 8 \times 25} = 1 \text{ computer per work hour}$$

In the second case,

Production = 12000 computers

$$\text{Labour productivity} = \frac{12000}{60 \times 8 \times 25} = 1 \text{ computer per work hour}$$

The analysis clearly shows that the production of computers has increased by 20%, but that the labour productivity is the same.

Example 2.2 The output from a machine is 120 pieces/hr, while the standard rate is 180 pieces/hr. What is the operator efficiency? What is the effectiveness, if the productivity index is 2.0?

Solution

Efficiency is the ratio of the actual output to the standard output. Hence

$$\text{Efficiency} = \frac{120}{180} = 66.67\%$$

Effectiveness is the degree of accomplishment of the objective.

$$\text{Productivity index} = \frac{\text{effectiveness}}{\text{efficiency}}$$

Hence,

Effectiveness = $2 \times 66.67 = 133$ pieces/hr

Example 2.3 A company manufactues 100 items in 1 day. Of these, 10 are rejected. The processing cost is Rs 100/1000 items and the error correction cost of the rejected items is Rs 1000/1000 items. Calculate the quality productivity ratio. If the company manufactures the same number of items in two days, what is the percentage of improvement?

Solution

$$\text{Quality productivity ratio} = \frac{\text{number of perfect items}}{\text{cost for manufacturing} + \text{cost for error correction}}$$

In the first case,

$$\text{Quality productivity ratio} = \frac{100-10}{100 \times 0.1 + 10 \times 1} = 4.5 \text{ items per rupee}$$

In the second case,

$$\text{Quality productivity ratio} = \frac{100-5}{100 \times 0.1 + 5 \times 1} = 6.33 \text{ items per rupees}$$

Percentage of improvement = 40.67 %

Example 2.4 A turret lathe in a section has six machines, all devoted to the production of the same part. The section operates 10 shifts/week. Each shift is for 8 hr. The average production rate of each machine is 17 units/hr. Determine the weekly production capacity.

Solution

Production capacity $= nSHR_p$

where n is the number of work centres $= 6$

S is the number of shifts per week $= 10$

H is the number of hours per shifts $= 8$

R_p is the hourly production rate $= 17$ units/hr

$\therefore$ Production capacity $= 6 \times 10 \times 8 \times 17$

$= 8160$ units/week

Example 2.5 A production machine operates 80 hr/week, two shifts for 5 days at full capacity. Its production rate is 20 units/hr. The machine produced 1000 parts and was idle during the remaining time.

1. Determine the production capacity of the machine.
2. What is the utilization of the machine during the week under consideration?

Solution

Machine operation $= 80$ hr/week

Production rate $= 20$ units/hr

Capacity of the machine $= 80 \times 20$

$= 1600$ units/week

Utilization is equal to number of parts made/capacity of the machine.

Utilization of the plant $= \dfrac{1000}{1600} = 0.625$

$= 62.5\%$

Exercises

2.1 Distinguish between the manufacturing industry and process industry.
2.2 Compare SMC, FMC, and FMS.
2.3 Explain different types of FMS layouts.
2.4 Distinguish between manufacturing automatic protocol and technical office protocol.
2.5 Explain the concept of CIM
2.6 Explain the concept of JIT.

Chapter 3

Mechanical Engineering and Machines in Mechatronics

3.1 Force, Friction, and Lubrication

In this section, we will study force, friction, and lubrication.

The subject 'machines in machatronics' is that branch of engineering and science which deals with the study of relative motion between the various parts of a machine as well as the forces acting on them. The knowledge of this subject is very essential for an engineer to design the various parts of mechatronic systems. Force is an important factor as an agent that produces or tends to produce, destroys or tends to destroy motion. When a body does not move or tend to move, the body does not have any friction force. Whenever a body moves or tends to move tangentially with respect to the surface on which it rests, the interlocking properties of the minutely projected particles due to the surface roughness oppose the motion. This opposition force, which acts in the opposite direction of the movement of the body, is called force of friction or simply friction. Force and friction play an important role in engineering, especially in mechatronic systems. Force and friction, together with lubrication, are the topics of our discussion in this section.

3.1.1 Forces

Force is that physical quantity which causes or tends to cause a change in the state of rest or motion of a body. The line of action of a force is a line drawn through the point of application of the force and along the direction along which the force acts. If a change in motion is prevented, force will cause a deformation or change in the shape of the body. In statics, it is often convenient to consider the effect of a force which acts on a rigid body. The perfect rigid body will not suffer deformation under the action of any force. A force is completely defined by its magnitude, point of application, and direction. A body is said to be in equilibrium under the action of a system of forces if all forces acting on it are in balance. In statics all forces come in pairs—an action and a reaction. A body is in equilibrium under the action of two forces provided the forces are equal in magnitude and have the same line of action but act in opposite directions.

The action of a force on a rigid body tends to move or rotate the body. The turning effect or tendency of a force to cause rotation about any point equals the product of the force and the perpendicular distance of the line of action of the force from at the point. This turning effect is called moment of a force at the point, and the distance is called moment arm. If a body is at rest under the action of a number of coplanar forces, the moments must be balanced; otherwise the unbalanced resultant moment will cause a rotation or linear motion of the body. Figure 3.1 illustrates the static and dynamic positions of a body, and the turning moment.

45 N ←————————→ 45 N
(a) Static position

60 N ←————————— 45 N
(b) Dynamic position

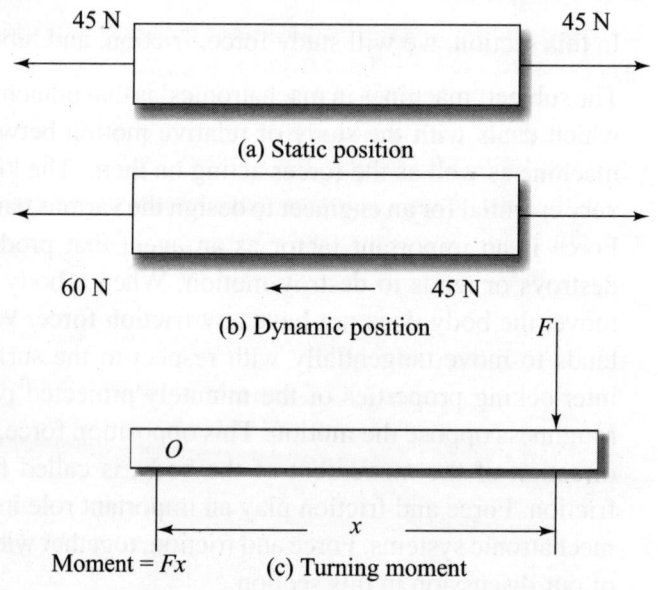

O
x
F
Moment = Fx (c) Turning moment

Fig. 3.1 Static and dynamic positions and the turning moment

Dynamics deals with linear forces or moments acting on a body in motion, linear or rotational. In linear motion, the force acting on the body is equal to the mass of the body multiplied by acceleration. For rotational motion, it is called torque, which is equal to the mass moment of inertia multiplied by angular acceleration. In Fig. 3.1(a) two equal forces act in opposite direction and hence the body is in euilibrium. In this case, the body is strained. In Fig. 3.1(b) an additional 15 N force acts on one side and hence the body moves in the direction of the additional force. In this case, the body is subjected to strain as well as motion. When a body is in motion it is said to be dynamic.

A couple is formed by two equal, parallel forces which are not collinear and act in opposite directions. The moment of a couple is the product of one of the forces and the perpendicular distance between the lines of action of the forces.

3.1.2 Friction

Suppose a body of weight W rests on a surface. It then exerts a normal force in a direction opposite to that in which the weight is acting. To move the body, a force F tangential to the surface has to be overcome by applying an external force P. This force F is known as friction. Figure 3.2 illustrates the various forces acting on a body at rest. If the magnitude of the external force, P, is increased, the frictional force also increases until its magnitude reaches a certain maximum value F_M. If P is increased further, the force of friction cannot balance the external force any more and the body starts moving. The ratio of this maximum frictional force F_M to the normal reaction R is constant and depends only on the nature of the pair of surfaces in contact. The ratio of F_M to R is called coefficient of static friction, which is denoted by μ_s. When sliding is just about to start, the pulling force $P = \mu_s R$. As soon as the body starts moving, the magnitude of the frictional force drops from F_M to F_R. This frictional force F_R is given by $\mu_R R$, where μ_R is called the coefficient of sliding friction and is less than the coefficient of static friction. Sliding friction depends on the nature and roughness of the surface, but is independent of the area of contact and the speed of the slide.

W = weight of the body (in newton)
R = normal reaction (in newton)
P = external force (in newton)
F = frictional force (in newton)

(a) Sliding friction

F_M = limiting applied force to slide
F_R = limiting applied force to roll

(b) Friction characteristic

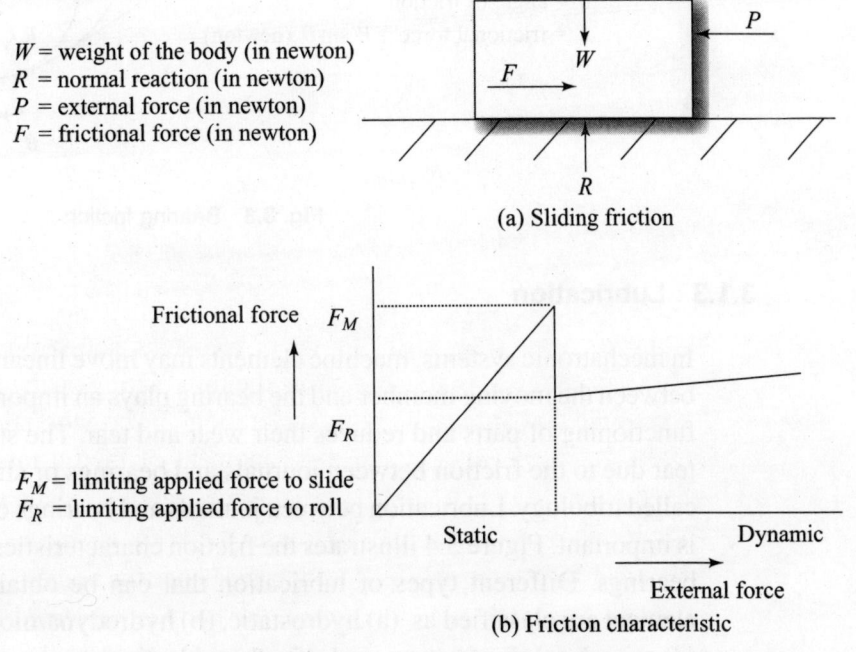

Fig. 3.2 Friction on a rigid body

For a wheel to roll on the ground without slipping, there must be a force of friction at the contact between the wheel and the track. There is a resistance to the rolling motion of the wheel due to deformation of the wheel and the track under the load. The resistance is called the rolling resistance. It acts opposite to the linear motion of the axle. The ratio of the maximum rolling resistance to the normal reaction between the wheel and the track is known as the coefficient of rolling resistance.

Journal bearings are used to provide lateral support to rotating shafts and axles. If they are fully lubricated, the frictional resistance depends upon the speed of rotation, the clearance between the axle and the bearing, and the viscosity of the lubricant. The coefficient of viscous friction due to fluid flow is less than the coefficient of rolling resistance. Hence, for skating, wheels are not used—the skate moves by sliding. Figure 3.3 illustrates the bearing and journal assembly which produces a viscous friction.

W = weight acting on the journal
R = normal reaction (newton)
I = torque in (newton metre)

θ = angle of friction
F = frictional force = $W \sin \theta$ (newton)

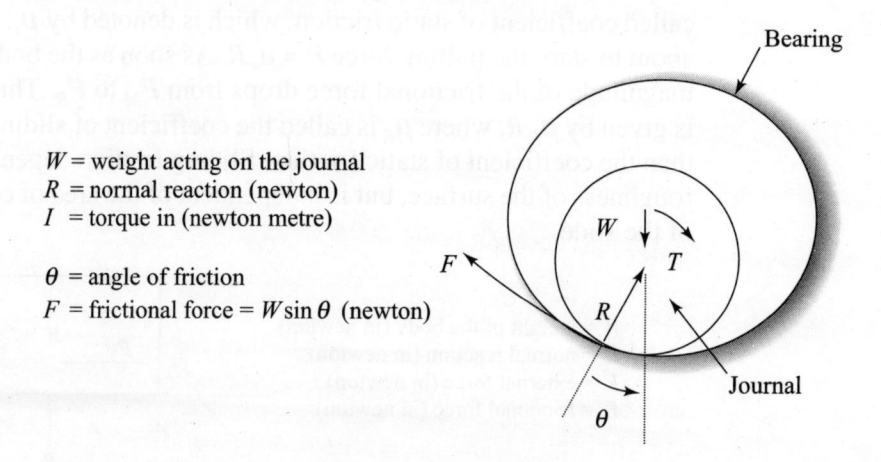

Fig. 3.3 Bearing friction

3.1.3 Lubrication

In mechatronic systems, machine elements may move linearly or rotate. Friction between the moving member and the bearing plays an important role in a smooth functioning of parts and reduces their wear and tear. The study of the wear and tear due to the friction between journals and bearings or sliders and bearings is called tribology. Lubrication between journals and bearings or sliders and bearing is important. Figure 3.4 illustrates the friction characteristics of linear and rotary bearings. Different types of lubrication that can be obtained in any moving element are classified as (a) hydrostatic, (b) hydrodynamic, (c) boundary layer, (d) mixed (e) dry friction, and (f) elastoplastic.

(a) Linear   (b) Rotary

Dry | Boundary | Mixed | Hydrodynamic

Friction coefficient

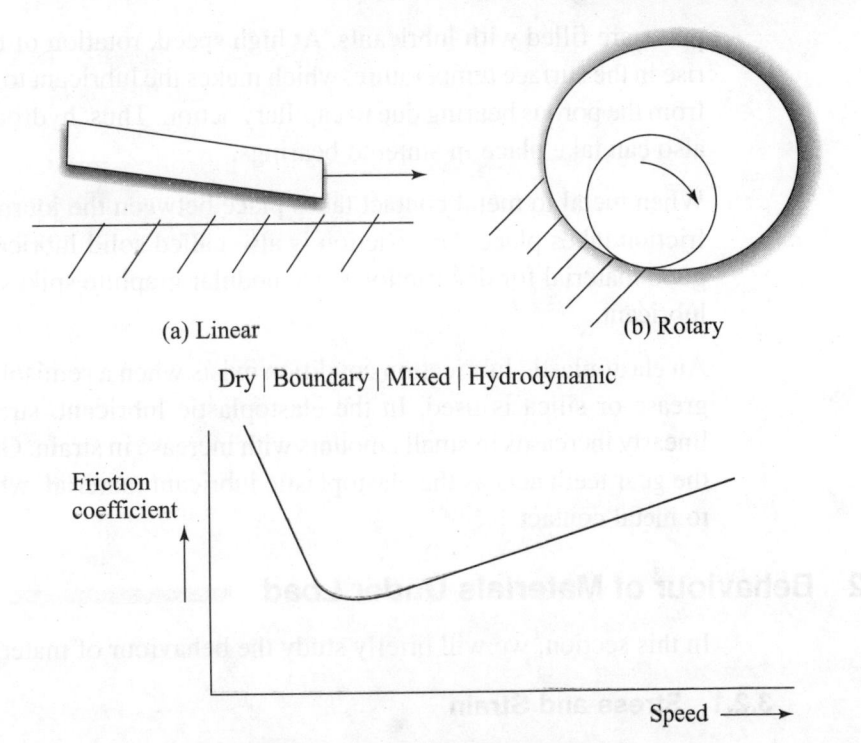

Speed →

Fig. 3.4 Linear and rotary bearings

When a lubricant is supplied between a journal and a bearing or a slider and a bearing at pressure, hydrostatic lubrication takes place. The load-carrying capacity of the journal under hydrostatic lubrication can be obtained by multiplying the projected area with the supply lubricant pressure.

When the journal rotates, the lubricant squeezes between the bearing and the journal due to the eccentricity between them. Due to the squeezing effect, pressure builds up, which prevents metal to metal contact between them. Such lubrication is called hydrodynamic lubrication.

At higher loads and low speeds, hydrodynamic lubrication fails. In such conditions, lubricants stick to the surface of the bearing like a brush since the hydrocarbon molecules of the lubricant have a long, open-chain structure. This property is called oiliness of the lubricant. This is called boundary layer lubrication. Here, the lubricant molecules separate the journal and bearing to prevent metal to metal contact.

When lubrication is partially boundary layer and partially hydrodynamic, we call it mixed lubrication. In a self-lubricated bearing, there is mixed lubrication. Sintered bearings are called self-lubricated bearings. Such bearings are manufactured using the powder metallurgy technique. They are porous, and the pores are filled with lubricants. At high speed, rotation of the journal causes a rise in the surface temperature, which makes the lubricant to come to the surface from the porous bearing due to capillary action. Thus, hydrodynamic lubrication also can take place in sintered bearings.

When metal to metal contact takes place between the journal and bearing, dry friction takes place. Dry friction is also called solid lubrication. Cast iron is a good material for dry friction since nodular graphite spikes act as a good solid lubricant.

An elastoplastic lubrication condition exists when a semisolid lubricant such as grease or silica is used. In the elastoplastic lubricant, stress in the lubricant linearly increases in small amounts with increase in strain. Grease used between the gear teeth acts as the elastoplastic lubricant material, which prevents metal to metal contact.

3.2 Behaviour of Materials Under Load

In this section, we will briefly study the behaviour of materials under load.

3.2.1 Stress and Strain

When an external force acts on a body, an internal resistance is set up within the body to balance the external load. The resistance carried per unit area is called the stress. When the applied force tends to compress the material or crush it, the material is said to be in compression and the stress is referred to as compressive stress. When the force tends to expand the material or tear it apart, the material is said to be in tension and the stress is referred to as tensile stress. When the force tends to cause the particles of the material to slide over one another, the material is said to be in shear stress. In the case of tensile force and compressive force, the area carrying the force is the area of the cross section in the plane perpendicular to the direction of the force. In the case of a shear force, the area carrying the force is the area to be sheared in the direction of the line of action of the force.

Figure 3.5 illustrates the different loading conditions on a rigid body. If a change of motion of the rigid body is prevented, the force applied will cause a deformation or change in the shape of the body. Strain is the change in dimension that takes place in the material due to an externally applied force. Linear strain is the ratio of change in length when a tensile or compressive force is applied. Shear strain is measured by the angular distortion caused by an external force. The load per unit deflection in a body is called stiffness. Deflection per unit load is called compliance. If deformation per unit load at a point on the body is different from that at the point of application of the load then compliance at that point is called cross compliance. In a machine structure cross compliance is an important parameter for stability analysis during machining.

Stress = P/A, where A is the cross-sectional area.

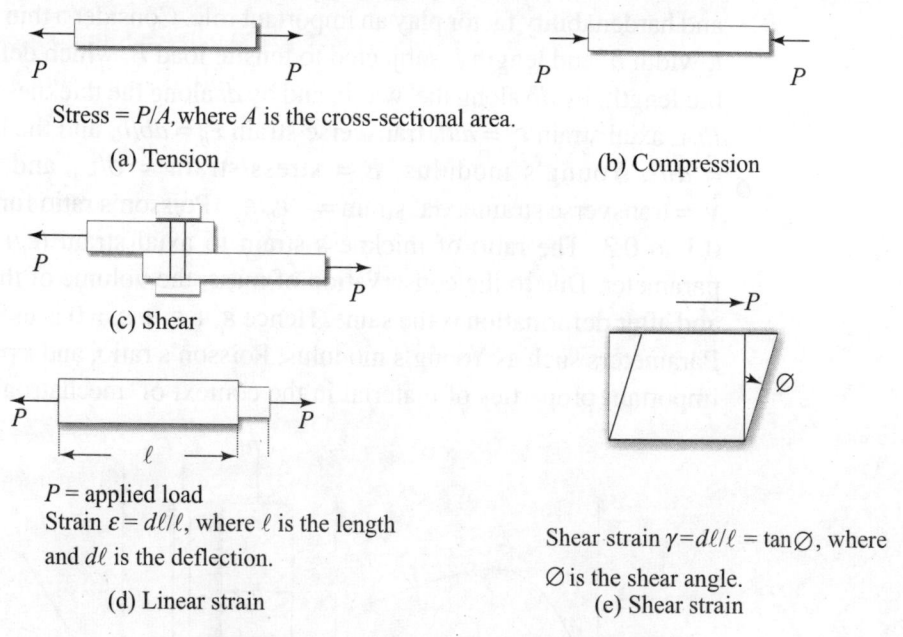

(a) Tension

(b) Compression

(c) Shear

P = applied load
Strain $\varepsilon = d\ell/\ell$, where ℓ is the length and $d\ell$ is the deflection.

(d) Linear strain

Shear strain $\gamma = d\ell/\ell = \tan\emptyset$, where $\emptyset$ is the shear angle.

(e) Shear strain

Fig. 3.5 A body in different loading conditions

3.2.2 Stress–Strain Behaviour

The strength of a material is expressed as the stress required to cause it to fracture. The maximum force required to break a material divided by the original cross-sectional area at the point of fracture is called the ultimate tensile strength of the material in tension.

All materials are elastic to a certain extent. In the elastic region, the material stretches if a tensile force is applied to it and returns to its original length on the removal of the force. There is a limit to this elastic property in every material, which is known as the elastic limit. If the force exceeds the elastic limit, the body deforms. The stress corresponding to the elastic limit is called yield stress. The stress–strain curves for ductile and brittle materials are shown in Fig. 3.6. When the material is loaded within the elastic limit, the stress is proportional to strain. The constant of proportionality is called the modulus of elasticity or Young's modulus. Similarly, when shear stress is divided by shear strain, the constant obtained is termed the modulus of rigidity.

It is obvious that the stress allowed in any component of a machine must be less than the stress that would cause permanent deformation. A safe working stress is chosen with regard to the conditions under which the material is to work. The ratio of the yield stress to allowable stress is termed the factor of safety.

In the material machining operation, a number of parameters such as strain rate and hardenability factor play an important role. Consider a thin plate of thickness t, width b, and length l, subjected to tensile load P, which deforms by dl along the length, by db along the width, and by dt along the thickness. Then, stress $\sigma = P/A$, axial strain $\varepsilon_a = dl/l$, transverse strain $\varepsilon_b = db/b$, and the thickness strain $\varepsilon_t = dt/t$. Young's modulus, E = stress/strain = σ/ε_a, and Poisson's ratio, v = transverse strain/axial strain = $-\varepsilon_b/\varepsilon_a$. Poisson's ratio for steel varies from 0.1 to 0.3. The ratio of thickness strain to axial strain ($\varepsilon_t/\varepsilon_a$) is called the r parameter. Due to the conservation of mass, the volume of the material before and after deformation is the same. Hence $\varepsilon_a + \varepsilon_b + \varepsilon_c = 0$ is used to determine ε_t. Parameters such as Young's modulus, Poisson's ratio, and r parameter are very important properties of material in the context of mechatronic systems.

Elastic | plastic

Cup and cone

Fracture

l = length of the specimen
b = width of the specimen
t = thickness of the specimen
P = applied load
A = cross-sectional area

(a)

(b)

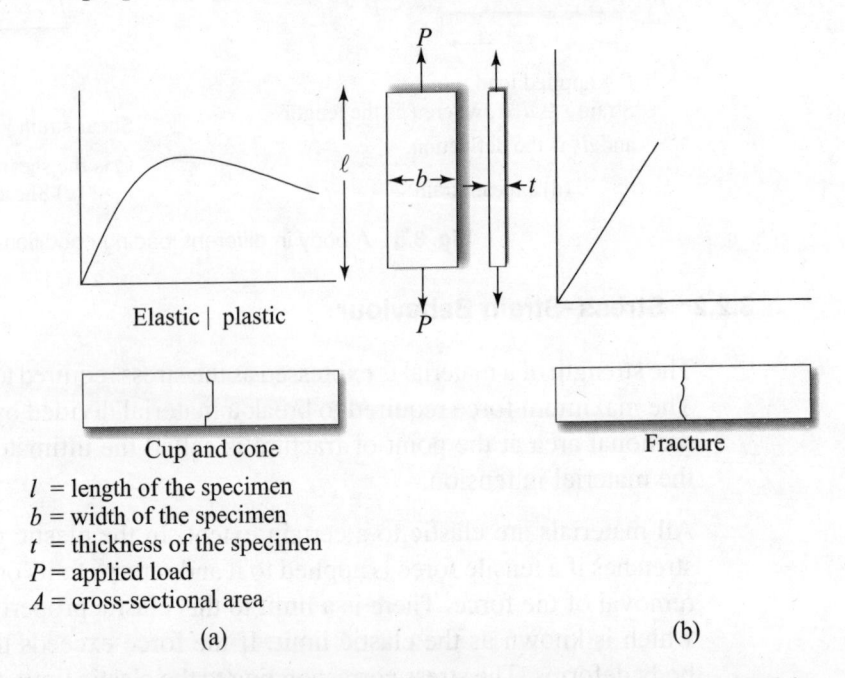

Fig. 3.6 Stress–strain curves (a) ductile material and (b) brittle material

For isotropic materials, Young's modulus E and Poisson's ratio v are constants along the x, y, and z directions. Steel, alloys, and non-ferrous metals behave like isotropic materials. If Young's modulus and Poisson's ratio are different along the x, y, and z directions, then the material is said to be anisotropic. Resins, composites, metal matrix composites, and fibre reinforced plastic (FRP) are examples of anisotropic materials. Metals and alloys have unique physical properties. Tailored physical properties can be obtained in composite form by different orientations of fibres or reinforcement.

Figure 3.7 illustrates the stress–strain relationship of different materials under different loading conditions. Materials may be ideal plastic or ideal elastic. After the elastic limit, materials may behave as elastoplastic or visco-elastoplastic materials. Adhesive materials or brazed materials behave like elastoplastic materials whereas materials at high temperature deform under visco-elastoplastic material model. Visco-elastoplastic material behaviour is a time-dependent problem.

Strictly speaking, for elastic, elastoplastic, visco-elastoplastic, or brittle materials in the elastic range, the stress–strain relationship is not linear. The non-linear stress–strain relationship can be expressed as $\sigma = k\varepsilon^n$, where k is called the hardenability factor and n is the hardenability index. The hardenability factor and hardenability index play an important role in the deformation behaviour of elastoplastic or visco-elastoplastic materials.

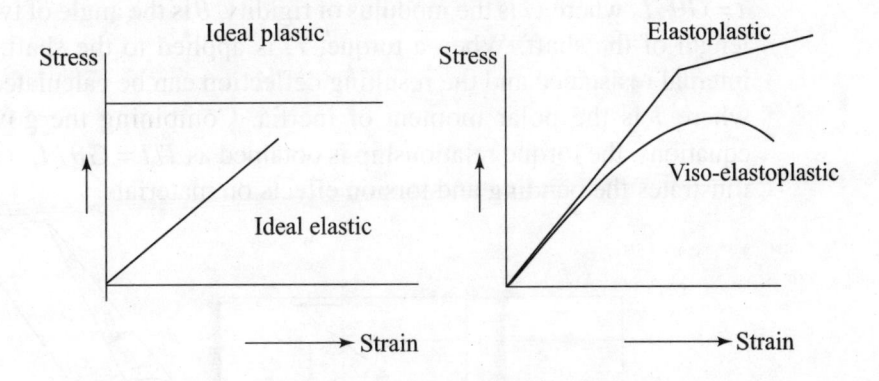

(a) Ideal elastic and ideal plastic   (b) Elastoplastic and visco-elastoplastic meterials

Fig. 3.7   Stress–strain behaviour

3.2.3   Bending of Beams and Torsion on Shaft

The term 'beam' is used for a single rigid length of the material that can support a system of external forces at right angles to its axis. A beam may be loaded with a concentrated load or a distributed load, or both. A rigid, long material that can carry axial loads only is called a strut.

A beam which is fixed at one end, leaving the other end free, is called a cantilever. A beam with its ends resting freely on a support is called a simply supported beam, and a beam with both ends fixed is called a fixed beam.

The loads on a beam tend to shear the beam and bend it as well. These effects on the beam are measured by measuring the shearing force and bending moment, respectively. The effect of a bending beam is to cause tensile and compressive stresses in it. Tensile stresses are set up in the lower half of the section and compressive stresses in the upper half. The axis with a stress value of zero is called the neutral axis. The stress at any distance y from the neutral axis is given by $\sigma_t = E(y/R)$, where E is Young's modulus and R is the radius of curvature of the neutral axis. The deflection of a beam at any point at a distance x is calculated from the equation

$$d^2y/dx^2 = M/I = I/R$$

where M is the bending moment and I is the area inertia of the section. Combining the governing equations, one can get the famous beam relation

$$M/I = E/R = \sigma_t/y$$

The moment of a force applied to a shaft which tends to twist or turn it is termed as turning moment or torque. The shaft suffers a shear stress. The shear stress at any point in a cross section at a distance r from the axis of the shaft is given by $\tau = G\theta r/L$, where G is the modulus of rigidity, θ is the angle of twist, and L is the length of the shaft. When a torque, T, is applied to the shaft, it develops an internal resistance and the resulting deflection can be calculated as $\theta = TL/GJ$, where J is the polar moment of inertia. Combining the governing torsion equations, the torque relationship is obtained as $T/J = G\theta/L = \tau/r$. Figure 3.8 illustrates the bending and torsion effects on materials.

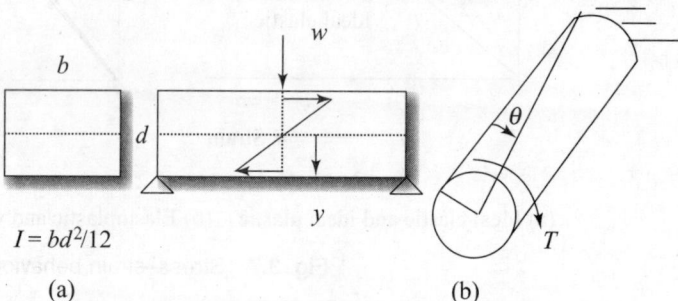

Fig. 3.8 (a) Bending of beams and (b) torsion of shaft

3.2.4 Trusses and Frames

Structures composed of bar elements and those composed of beam elements are classified as trusses and frames, respectively. Bars can only carry axial loads and deform axially, whereas beams can take transverse loads and bending moments about an axis perpendicular to the plane of the member. All the members of a truss are subjected to only axial loads and a truss cannot carry transverse shearing forces and bending moments.

In a truss all the members are connected to each other through pins that allow free rotation about the pin axis. On the other hand, in welded or riveted joints, a number of frame elements are connected by rigid connectors, so that axial and transverse forces and bending moments can be developed in the member. In many truss-and-frame structures, the bar and beam elements are found in many different orientations. Quantities such as displacement and forces for bar and beam elements are shown in Fig. 3.9.

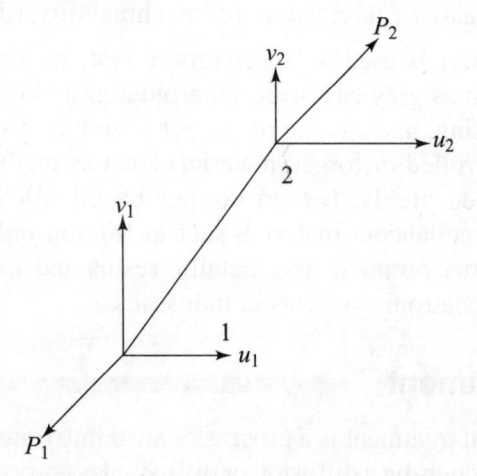

u = deflection along the x direction, v = deflection along the y direction, and P = axial load.

The axial deflection is resolved into components along the x and y directions.

(a) Bar element

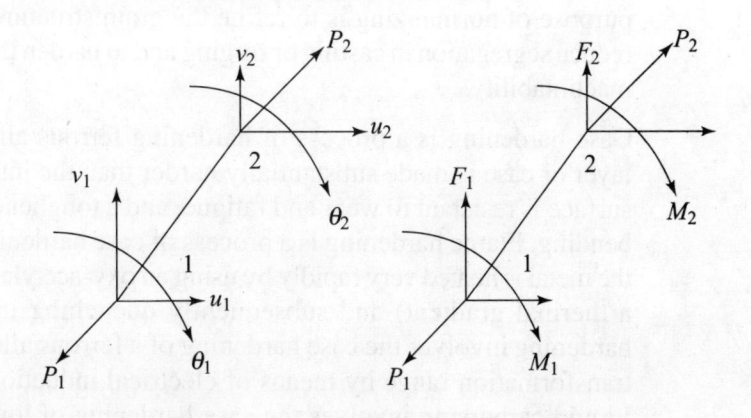

Displacement vector   Force vector

F = shear force, M = bending moment, and θ = slope.

(b) Beam element

Fig. 3.9 Displacement and forces in bar and beam elements

3.3 Materials

The modern mechatronic systems are designed to operate at higher speeds and feeds. They should possess improved accuracy and higher rigidity, and should operate at reduced noise levels. This calls for optimizing the design of mechatronic system elements and selection of the right type of materials. Effective heat treatment and fabrication are of the utmost importance. The factors to be considered while selecting the materials are: (a) functional requirement, (b) ease of fabrication, (c) machinability, (d) cost, and (e) availability.

Materials used for mechatronic systems are classified as (a) casting materials such as grey cast iron, spheroidal graphite cast iron, malleable cast iron, steel casting, and casting of copper-based alloys and aluminium-based alloys, and (b) rolled or forged materials such as medium carbon-steel, alloy steel, high-speed steels, forged copper-based alloys and aluminium-based alloys. Miscellaneous materials such as friction-reducing materials turcite, plastic, and rubber products and metallic resins and fibre composites are widely used in mechatronic systems in industries.

3.4 Heat Treatment

Heat treatment is a process of austenite transformation into slow or fast cooling products by a diffusion or diffusionless process. Annealing refers to any heating, soaking, and cooling operation which is usually done to induce softening. Annealing may also be carried out to relieve internal stress induced by a cold or hot working process. Normalizing is a process of heating steel to 50°C above the upper critical temperature followed by cooling in still air. The primary purpose of normalizing is to refine the grain structure prior to hardening or to reduce segregation in casting or forging and to harden the steel slightly to improve machinability.

Case hardening is a process of hardening ferrous alloys in which the surface layer or case is made substantially harder than the interior or core. The harder surface is resistant to wear and fatigue, and a toughened core resists impact and bending. Flame hardening is a process of case hardening wherein the surface of the metal is heated very rapidly by using an oxy-acetylene flame (thereby creating a thermal gradient) and subsequently quenching in water or oil. Induction hardening involves the case hardening of a ferrous alloy by heating it above the transformation range by means of electrical induction and cooling as desired. Liquid carbursing involves the case hardening of low-carbon and alloy steels by heating above the critical range in a molten cyanide salt, which induces both carbon and nitrogen into the case.

Nitriding is a process of case hardening of a special alloy steel in an atmosphere of ammonia or in contact with nitrogen material at a temperature below the transformation range. Surface hardness is caused by the absorption of nitrogen without quenching.

3.5 Electroplating

Electroplating or electrodeposition is a branch of electrometallurgy which deals with the art of depositing metals by means of electric current. The mechanism of electroplating is based on the theory of ionic dissociation and electrolysis. Copper can be electroplated on both ferrous and non-ferrous metals by using solutions of copper cyanide or acids. Commercial electroplating generally involves the use of copper cyanide, since more copper can be electroplated by this method. Most nickel plating salts are a mixture of nickel sulphate and nickel chloride, basic acid and stabilizers to control the quantity of nickel deposits. Chromium plating is done for decorative purposes, to improve corrosion resistance or to produce hard, abrasion-resistant surfaces. Tin, cadmium, silver, and gold plating are commonly used in mechatronic systems in industries.

3.6 Fits and Tolerance

The relationship resulting from the difference between the sizes of two features, the hole and the shaft, is called fit. Fits have a common basic size. They are broadly classified as clearance fit, transition fit, and interference fit. A clearance fit is one that always provides a clearance between the hole and shaft when they are assembled. A transition fit is one that provides either a clearance or an interference between the hole and the shaft when they are assembled. An interference fit is one that provides an interference all along between the hole and the shaft when they are assembled. There are two types of systems of fits—(a) the hole-based system and (b) the shaft-based system.

In the hole-base system, the size of the hole is the basic size, denoted by H. The size of the shaft is varied to get the required fit. Letters a, b, c, d, e, f, and g indicate loose fit, h, i, j, k and l indicate transition fit, and the rest $(m, n, \cdots, z)$ indicate interference fit. Similarly, in the shaft-based system, the design size of the shaft is the basic size, denoted by h. The size of the hole is allowed to vary according to the type of fit. A, B, C, D, E, F, and G indicate loose fit, H, I, J, K, and L indicate transition fit, and the rest indicate interference fit. The hole-based system is generally used in industries. This is because a shaft is easy to manufacture with a variety of sizes. The shaft-based system should only be used where there are definite economic advantages of doing so. For instance, in the assembly of a piston inside a bored cylinder, boring of the cylinder is more economical than changing the piston to get the correct fit.

The production of a part with exact dimensions repetitively is often difficult. Hence, it is sufficient to produce parts with dimensions accurate within two permissible limits of size. The difference in the limits of the size is called tolerance. Tolerance can be provided on both sides of the basic size (bilateral tolerance) or on one side of the basic size (unilateral tolerance). The ISO system of tolerance provides for a total of 20 standard tolerance grades, which are indicated as IT0, IT01, and IT1 to IT18.

In any engineering industry the components manufactured should also satisfy the geometrical tolerances in addition to the common dimensional tolerances. Geometrical tolerances are classified as:

(a) Form tolerance—straightness, flatness, circularity, cylindricity, profile of any line, and profile of any surface.

(b) Orientation tolerance—parallelism, perpendicularity, and angularity.

(c) Location tolerance—position, concentricity, co-axiality, and symmetry.

(d) Runout tolerance—circular runout and axial runout.

Geometrical tolerance should be specified only where it is essential. Geometrical tolerance features are covered by ISO: 1101 – 1983.

3.7 Surface Texture and Scraping

Any surface, cylindrical or flat, has to be defined for the requirements of surface roughness depending on its functional application to guarantee quality of products. Surface roughness should not be specified if it is not required. It can be measured and specified with different parameters such as the centre line average (CLA) value R_a, ten-point average value R_z, and maximum peak value R_{max}. The R_a value is the most commonly used international parameter for roughness. It is defined as the arithmetic mean of the absolute values of the profile departure within the sampling length. The roughness symbols, equivalent roughness values, roughness grade numbers, and machining processes are given in Table 3.1. Each quality class is established by the number of contact pairs between the scraped surface and the inspection surface plate. The contact points may be almost uniformly distributed in an area of 25 mm × 25 mm and also over the entire area of the scraped surface and checked by applying blue paste. In the normal case, blue paste is applied to the masterpiece and the workpiece is checked against it. For finer scraping, blue paste is used for the masterpiece and a thin layer of red oxide mixed with linseed oil is used for the workpiece. The red oxide gives a contrasting background and also prevents the blue paste from spreading when the masterpiece is rubbed against the workpiece grades and the various required parameters associate with them. Table 3.2 shows the various scraped quality grades.

Table 3.1 Roughness symbols and grades

Roughness symbol	Roughness value R_a (mm)	Roughness grade number	Machining process
~∇	50	N12	Casting, un-machined surface
∇	25	N11	
	12.5	N10	Rough turning, planing, etc.
	6.3	N9	
∇∇	3.2	N8	Finish turning, milling, and drilling
	1.6	N7	
∇∇∇	0.8	N6	
	0.4	N5	Grinding, boring, and reaming
	0.2	N4	
∇∇∇∇	0.1	N3	
	0.05	N2	Lapping, honing, etc.
	0.025	N1	

Table 3.2 Scraped quality grades and other parameters

Scraped quality grade	Number of contact points	Surface roughness (μm)	Application
1	22–32	0.1	Precision measuring machines, instruments
2	15–21	0.2	Surface plates with straight edges, guideways (machine tool slides)
3	10–14	0.4	Jigs and fixtures, guideways (machine tools)
4	6–9	0.8	Surface tables, clamps
5	4–5	1.6	Rotary tables, covers of gear boxes

3.8 Machine Structure

The machine structure is the load carrying and supporting member of a machine tool. Motors that drive mechanisms and other functional assemblies of the machine tool are aligned to each other and rigidly fixed to the machine structure. Machine structures are subjected to static and dynamic forces. It is therefore essential that they do not deform or vibrate beyond the permissible limit under the action of forces. The machine structure configuration is also influenced by considerations of manufacture, assembly, and operation. The basic design factors involved in the design of a machine tool structure are static load, dynamic load, and thermal load. The static load of machine tools results from weights of slides and the workpiece, and from the force due to cutting. The structure should have adequate stiffness to keep the deformation of the structure within the permissible value. Dynamic loads are often used for the constantly changing forces acting on the structure while movement takes place. These forces may cause the whole machine system to vibrate. The origins of such vibration are (a) unbalanced rotating parts, (b) improper machine gears, (c) bearing irregularities, and (d) interrupted cuts while machining.

In a machine tool, there are a number of local heat sources which set up thermal gradients within the structure of the machine. Some of the heat sources are (a) the electric motor, (b) friction in mechanical drives, (c) friction in bearings, (d) the machining process, and (e) the temperature of the surrounding objects.

The heat sources cause localized deformation, resulting in considerable inaccuracies in machine performance.

3.9 Guideways

Guideways are used in machine tools to control the direction or line of action of the carriage or the table on which a tool or workpiece is held and to absorb all the static and dynamic forces. The shapes and sizes of the workpieces produced depend on the geometric and kinematic accuracy of the guideways. The geometric relationship of the slide and guideways with the machine base determines the geometric accuracy of the machine. Kinematic accuracy depends on the straightness, flatness, and parallelism error in the guideways. Moreover, any kind of wear in the guideways reduces the accuracy of the guide motion. The points that must be considered while designing guideways are (a) rigidity, (b) dampening capability, (c) geometric and kinematic accuracy, (d) velocity of slide, (e) friction characteristic, (f) provision for adjustment of ply, (g) position relative to work area, and (h) protection against swarf and damage.

Guideways are generally of two types—friction guideways and antifriction linear motion guideways. Figure 3.10 illustrates different friction guideways.

Antifriction linear motion guideways are used in CNC machines and tools because they (a) reduce the amount of wear, (b) improve smoothness of movement, (c) reduce friction, (d) reduce heat generation.

(a) Flat guideways   (b) Dovetail guideways

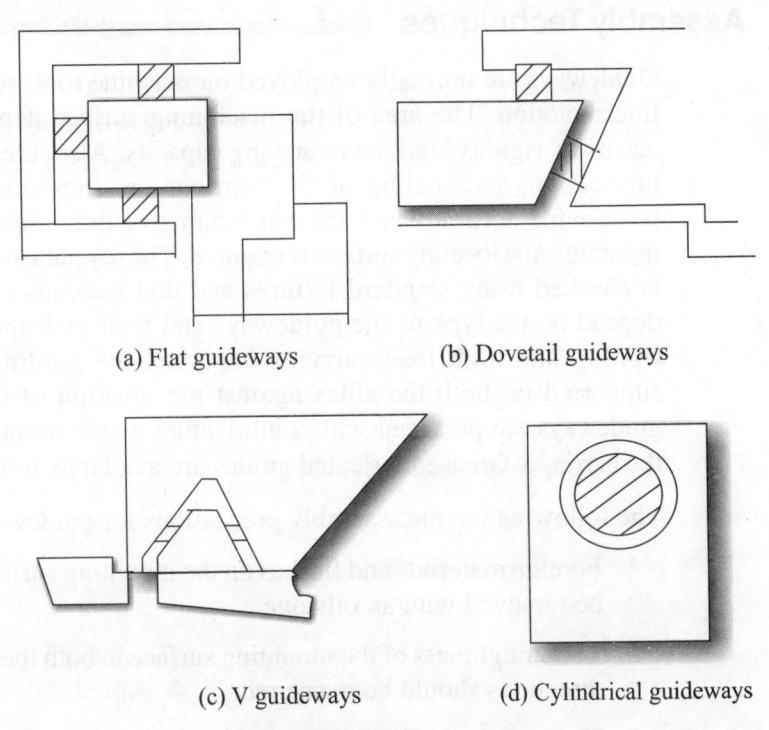

(c) V guideways   (d) Cylindrical guideways

Fig. 3.10 Friction guideways

Recirculation ball bearings or ball bushings are an example of antifriction linear motion guideways. Figure 3.11 illustrates the inner construction of a recirculating ball bush. A number of machines use rollers to provide for rolling motion rather than sliding motion. Hydrostatic and aerostatic guideways are used for precision machine tools. In hydrostatic guideways, the surface of the slide is separated from the guideways by a very thin film of fluid supplied at high pressure (about 300 bars). In aerostatic guideways, the slide is raised on a cushion of compressed air, which entirely separates the slide and the surface of the guideways. The major limitation of aerostatic guideways is their low stiffness, which limits their use to position application such as in the coordinate measuring machine (CMM).

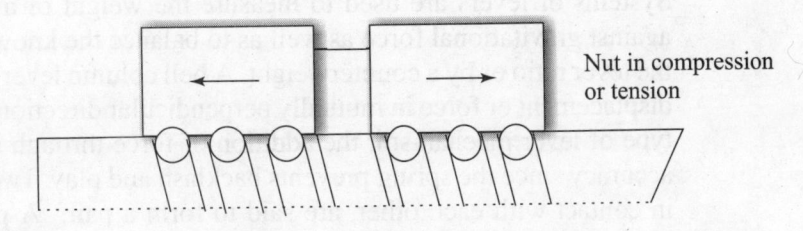

Nut in compression or tension

Fig. 3.11 Recalculating lead screw

3.10 Assembly Techniques

Guideways are normally employed on machine tools to obtain a rotating or a linear motion. The area of the machining surface depends upon the desired accuracy, rigidity, and load-carrying capacity. Also, the location of the guides, lubrication, and sealing of the bearings are important. To obtain a higher positioning accuracy and smooth running of the guiding system, an accurate mounting and locating surface is required. The parallelism of mounted guideways is checked using standard fixtures and dial indicators. The allowable values depend on the type of the guideways and their clearance. In order to achieve rigidity and high load-carrying capacity, the guiding elements should be supported on both the sides against the location of the surface. Normally, guideways are provided with a lubricating nipple mounted on the end face of the carriage. Grease lubricated guides are available to increase longevity.

The following are the assembly precautions for guideways.

1. Foreign materials and flashes on the mounting surfaces of the guide should be removed with an oilstone.

2. The straightness of the mounting surface in both the vertical and horizontal directions should be measured .

3. The rust-preventive coating applied on guides should be thoroughly wiped after the coat of oil is applied.

4. The presence of dust or foreign matter inside the guideways drastically increases wear, resulting in reduced life.

5. Care should be taken to prevent damage of the lip seal during assembly.

3.11 Mechanisms used in Mechatronics

In this section we will discuss some mechanisms used in mechatronics.

3.11.1 Lever and Four-Bar Mechanisms

Lever mechanisms are widely used in mechatronic systems. For even a small force, a lever should react without any friction or addition of any other force. Systems of levers are used to measure the weight of a body by balancing it against gravitational force as well as to balance the known weights by varying the lever ratio or by a counterweight. A bell column lever is used to transmit the displacement or force in mutually perpendicular directions. In the displacement type of lever mechanism, the addition of force through a spring improves the accuracy since the spring prevents backlash and play. Two levers of a machine, in contact with each other, are said to form a pair. A pair may be a sliding, turning, rolling, screw, or spherical pair. When the kinematic pairs are coupled in such a way that the last link is joined to the first link to transmit definite motion, they are said to form a kinematic chain. When any of the levers of the kinematic chain is fixed, the chain is known as a mechanism. A mechanism with four levers is known as a four-bar mechanism. When a mechanism is required to transmit power or to do some particular type of work, it becomes a machine. Obtaining different mechanisms by fixing different links in a kinematic chain is known as inversion of mechanism. Inversions of a four-bar chain mechanism are (a) crank-and-lever mechanism, (b) double-crank mechanism, and (c) double-lever mechanism.

Figure 3.12 illustrates the working of a simple lever and bell crank lever and Fig. 3.13 shows the four-bar mechanism and its inversion.

Lever ratio = B/A

(a) Simple lever   (b) Bell column lever

Fig. 3.12 Simple and bell column levers

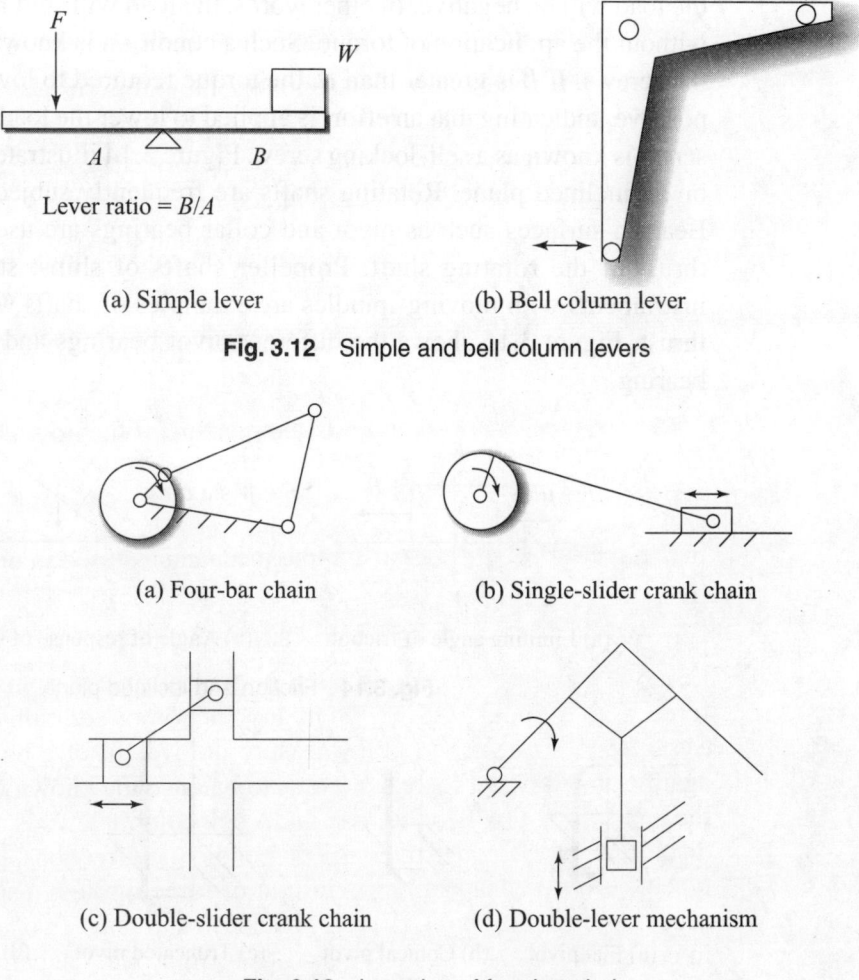

(a) Four-bar chain   (b) Single-slider crank chain

(c) Double-slider crank chain   (d) Double-lever mechanism

Fig. 3.13 Inversion of four-bar chain

3.11.2 Bearings

Mechanisms working on friction include the plane power screw, pivot and collar beams, and friction clutches. When a body rests and is pulled by a traction force, the limiting angle of friction, $\beta = \tan^{-1}\mu$, which is defined as the angle that the resultant reaction force makes with the normal reaction force. Here μ is the friction coefficient. An inclined plane is used for many applications working under friction. The body resting on the inclined plane may move down the plane when α is greater than β, where α is the inclination of the inclined plane. The screws used are power screws which work under the principle of the inclined plane. The torque required to overcome the friction between the screw and nut is $T = w \tan\alpha\, d/2$, where w is the load on the bolt, α is the helix angle, and d is the mean diameter of the screw. If β is less than α, the torque required to lower the load will be negative. In other words, the load will start moving downwards without the application of torque. Such a condition is known as overhauling of the screws. If β is greater than α, the torque required to lower the load will be positive, indicating that an effort is applied to lower the load. In such a case, the screw is known as a self-locking screw. Figure 3.14 illustrates the friction effect on an inclined plane. Rotating shafts are frequently subjected to axial thrust. Bearing surfaces such as pivot and collar bearings are used to take the axial thrust of the rotating shaft. Propeller shafts of ships, steam turbines, and instruments with moving spindles are examples of shafts which carry an axial thrust. Figure 3.15 shows the different pivot bearings and a single flat collar bearing.

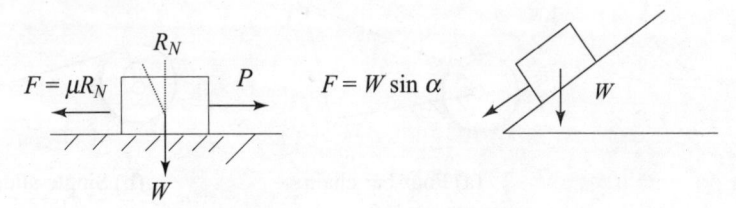

(a) Limiting angle of friction   (b) Angle of response of inclined plane

Fig. 3.14 Friction and inclined plane

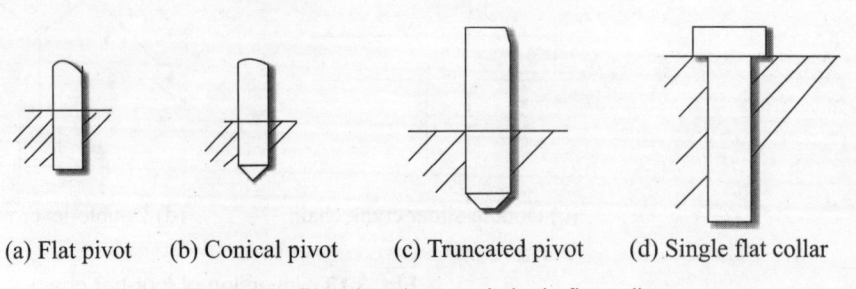

(a) Flat pivot   (b) Conical pivot   (c) Truncated pivot   (d) Single flat collar

Fig. 3.15 Pivot bearings and single flat collar

A rolling element bearing is considered a very important mechanical component as it is used for almost all rotary applications. Such a bearing consists of two circular metal rings and a set of rolling elements. One of the rings is larger than the other. The smaller of the two is called the inner ring and the larger one the outer ring. The inner ring fits well within the perimeter of the outer ring. Figure 3.16 shows a typical roller bearing. All the bearing parts are in transition fit since the bearing should allow a free rotation of the rolling element and the outer and inner rings should not get cold welded with the housing or body. Bearings have to deal with radial and thrust loads. For this purpose, different rolling elements such as balls, cylindrical rollers, and needle type or tapered rollers are used.

(a) Typical roller bearing (b) Types of rolling element

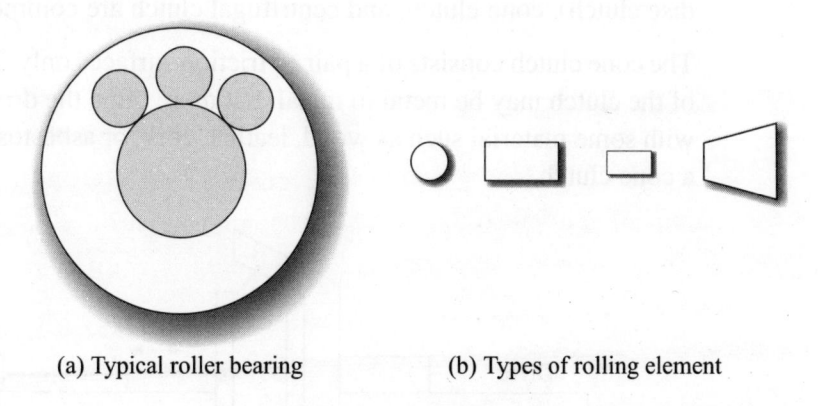

Fig. 3.16 Roller bearings

Mechanical friction losses are eliminated in magnetic bearings. Figure 3.17 shows a typical magnetic bearing. Electromagnetic bearings are used for journal suspension in magnetic bearings. Magnetic bearings are used for high-speed application and also where more accurate positioning is essential.

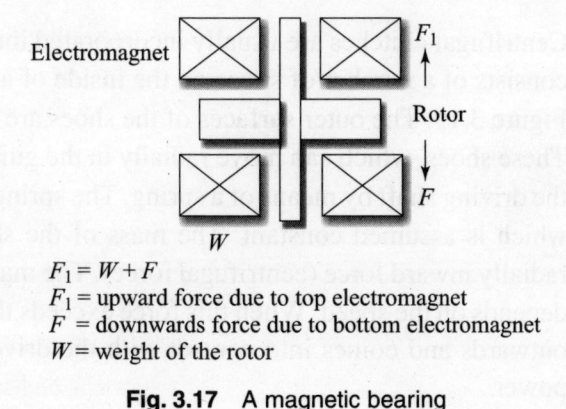

$F_1 = W + F$
F_1 = upward force due to top electromagnet
F = downwards force due to bottom electromagnet
W = weight of the rotor

Fig. 3.17 A magnetic bearing

Bearings are an integral part of nanomechanisms. Molecular bearings are likely to be used in the function of molecular machines. This type of bearing does not require lubrication.

3.11.3 Clutches

The principal application of the friction clutch is in the transmission of power of shafts and machines which must be started and stopped frequently. The force of friction is used to start the shaft from rest and gradually bring it up to the proper speed without excessive slippage of the friction surface. In an automobile, the friction clutch is used to connect the engine to the driven shaft. Friction clutches of different types such as disc or plate clutch (single disc or multiple disc clutch), cone clutch, and centrifugal clutch are commonly used.

The cone clutch consists of a pair of friction surfaces only. The contact surfaces of the clutch may be metal to metal, but more often the driven member is lined with some material such as wood, leather, cork, or asbestos. Figure 3.18 shows a cone clutch.

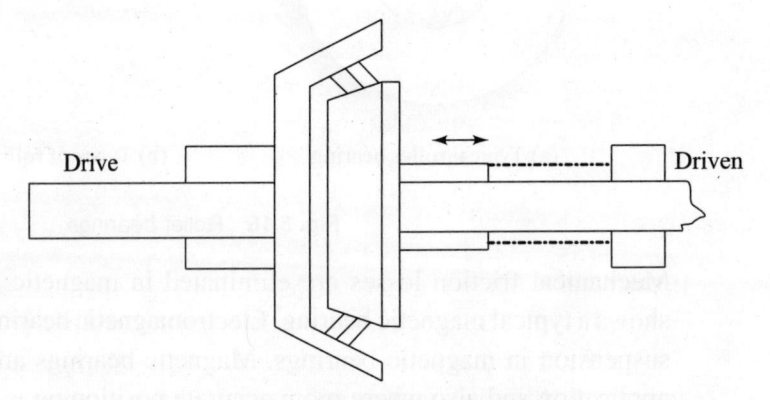

Fig. 3.18 A cone clutch

Centrifugal clutches are usually incorporated into motor pulleys. Such a clutch consists of a number of shoes on the inside of a rim of the pulley, as shown in Figure 3.19. The outer surfaces of the shoes are covered with friction material. These shoes, which can move radially in the guide, are held against a spider on the driving shaft by means of a spring. The spring exerts a radially inward force, which is assumed constant. The mass of the shoes when revolving causes a radially inward force (centrifugal force). The magnitude of the centrifugal force depends on the speed. When this force exceeds the spring force, the shoe moves outwards and comes into contact with the driven member and thus transmits power.

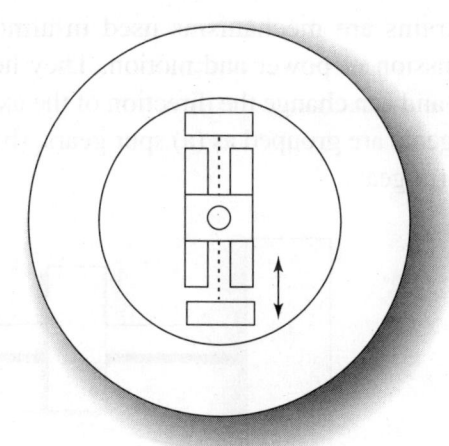

Fig. 3.19 A centrifugal clutch

A simple wedge mechanism works under the principle of friction. Wedge mechanisms are commonly used for stops, locks, and latches.

Special mechanisms commonly used are ratchet mechanisms, integrators, and Geneva mechanisms. Figure 3.20 illustrates the working of these mechanisms. The Geneva mechanism is used for a sequence of operations. The ratchet is an asymmetric mechanism that allows something to turn or move in one direction only. The ratchet works like a locking system when used in conjunction with a Powel.

A simple cam follower system is a wheel with an axle not in the centre. The follower rests on the edge of the cam. As the cam rotates, the follower moves up and down. The contour of the cam is broadly divided into three regions, rise, fall, and dwell. The rise portion of the cam makes it possible to raise the followers; during the fall portion, the follower returns to its original position or reference position. In the dwell portion of a typical cam, the followers do not move relative to the cam axis. Each portion can be described in terms of profile equations—the displacement, velocity, and acceleration profiles.

The Geneva wheel-based mechanism is used for achieving intermittent motion. The mechanism has two wheels. There is a projection called the drive pin mounted on one of the wheels. The other wheel, called the Geneva wheel, has four slots. However, more slots can be designed depending upon the need. When the drive wheel rotates, the projection pin is inserted into the slot, making the Geneva wheel rotate in the appropriate direction. Afterwards, the pin disengages from the Geneva wheel. During one rotation of the drive wheel, the drive pin engages with the Geneva wheel and rotates the Geneva wheel by one step. The Geneva wheel rotates stepwise.

Gear trains are mechanisms used in almost all machines to facilitate the transmission of power and motion. They help increase or decrease torque or speed, and can change the direction of the axis of rotation. Depending upon the teeth, gears are grouped as (a) spur gears, (b) bevel gears, (c) helical gears, and (d) worm gears.

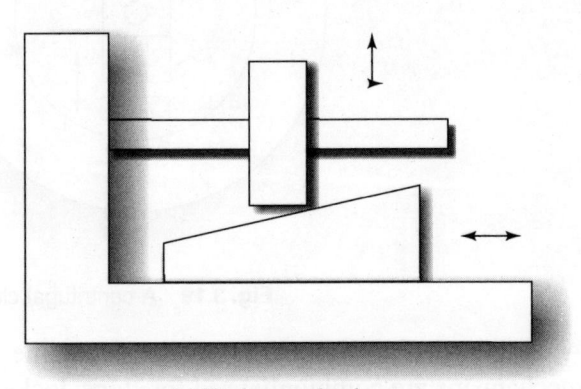

(a) Wedge mechanism

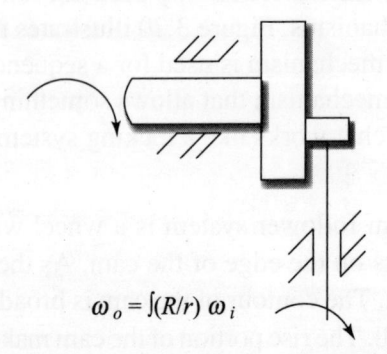

$\omega_o = \int (R/r)\, \omega_i$

where R = radius of drive disc, r = radius of driven disc, ω_i = input angular velocity, and ω_o = output angular velocity.

(b) Mechanical integrator

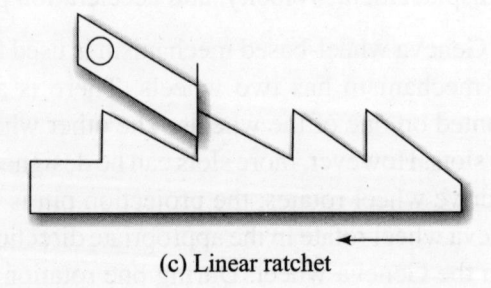

(c) Linear ratchet

Powel

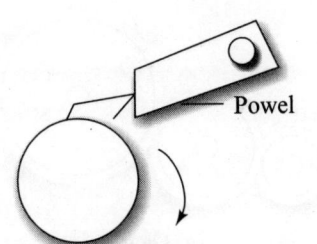

(d) Rotary ratchet

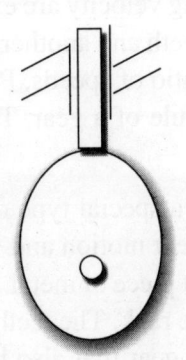

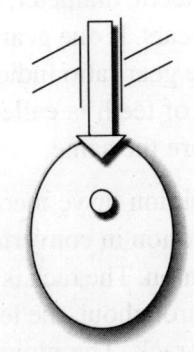

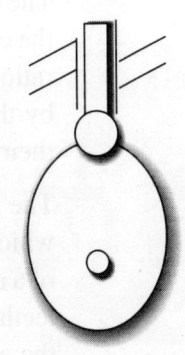

(e) Cam and follower

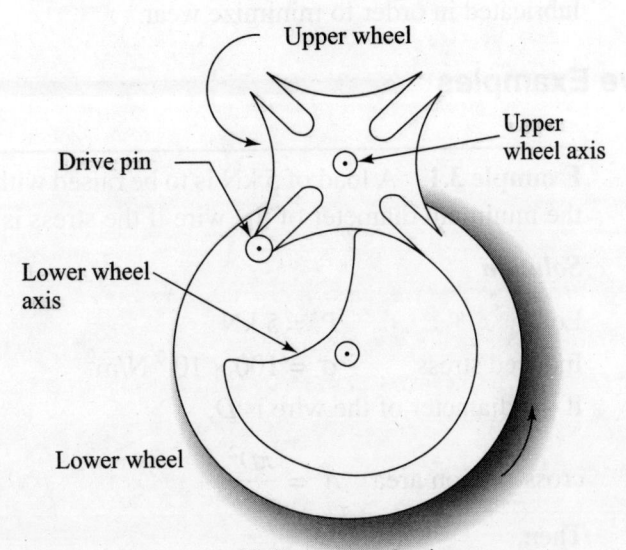

Upper wheel

Upper wheel axis

Drive pin

Lower wheel axis

Lower wheel

(f) Geneva mechanism

(g) Gear train (h) Rack and pinion

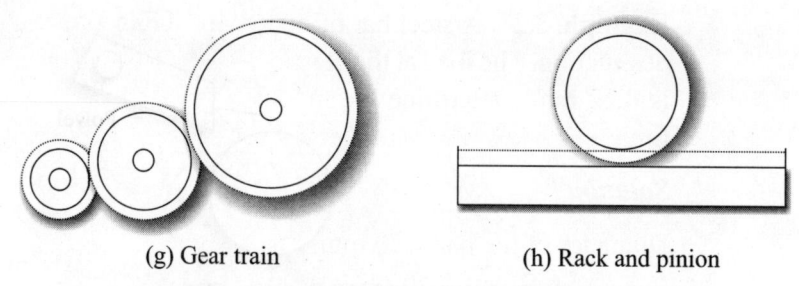

Fig. 3.20 Special mechanisms used in mechatronics

The number of teeth, diameter, and rotating velocity are extremely important in the context of gears. If one gear has 120 teeth and another has 20 teeth, the gear ratio is 6:1. The gear ratio indicates the ratio of speeds. Pitch diameter divided by the number of teeth is called the module of a gear. Two gears can mate if their modules are the same.

The rack-and-pinion drive mechanism is a special type of gear mechanism in which rotary motion in converted into linear motion and vice versa. It consists of a rack and pinion. The rack is a long, flat piece of metal. The flat side contains teeth running throughout the length of the rack. The teeth are perpendicular to the edge of the rack. The pinion is like a gear that also has teeth on it. These teeth run parallel to the length of the shaft. The pinion has rotary motion, whereas the rack gives linear motion just like any gear train. A rack and pinion is lubricated in order to minimize wear.

Illustrative Examples

Example 3.1 A load of 5 kN is to be raised with the help of a steel wire. Find the minimum diameter of the wire if the stress is not to exceed 100 mN/m^2.

Solution

Load $\quad P = 5$ kN

Induced stress $\quad \sigma = 100 \times 10^6$ N/m^2

If the diameter of the wire is D,

cross-section area $\quad A = \dfrac{\pi D^2}{4}$

Then,

$$P/A = 100 \times 10^6$$

Solving,

$$D = 0.0079 \text{ m} = 7.97 \text{ mm}$$

Example 3.2 A steel bar of diameter 20 mm and length 20 cm was tested to destruction. The load at the elastic limit is 65 N and the corresponding extension is 0.22 mm. Determine the modulus of elasticity and the strain at the elastic limit.

Solution

Diameter of the bar = 20 mm

Length of the bar = 20 cm

Elongation at elastic limit = 0.22 mm

Load at elastic limit = 65 kN

Stress = load / area = P/A = 0.2069 kN/m^2

Strain at elastic limit = extension at elastic limit/original length
$$= 0.22/200 = 0.0011$$

Modulus of elasticity = stress/strain
$$= 188.09 \text{ kN/mm}^2$$

Example 3.3 A steel plate 12 mm thick is bent around a drum. If the maximum stress induced in the plate due to bending is 400 N/mm^2, find the radius of the drum. Take $E = 2 \times 10^5$ N/mm^2.

Solution

Thickness of the steel plate = 12 mm

Maximum stress $\quad \sigma = 400$ N/mm^2

Modulus of elasticity $\quad E = 2 \times 10^5$ N/mm^2

If R_1 is the radius of the drum and r is the radius of curvature of the neutral axis of the plate, then
$$R = R_1 + (12/2)$$
$$= R_1 + 6$$

Using the equation for a beam,
$$\sigma / y = E/R$$
$$400/6 = 2 \times 10^5 / R$$

Solving,
$$R = 3000 \text{ mm}$$
$$R_1 = 2994 \text{ mm}$$

Example 3.4 A simply supported beam 4 m long carries a point load of 50 kN at a distance of 1.5 m from each end. If the allowable bending stress is 15.34 kN/cm^2, find the section modulus.

Solution

The beam is symmetrically loaded. Hence, the reaction force on each support, $R = 50$ kN. Bending moment at midspan where the maximum bending moment occurs,

$$M_{max} = R \times 2 - 50 \times 0.5 = 75 \text{ kN m}$$

Using the beam equation,

$$M/I = \sigma/Y$$

$$Z = I/Y = 488.92 \text{ cm}^2$$

Example 3.5 Find the maximum shearing stress developed in a shaft of 40 mm diameter when a twisting moment of 100 kN cm is applied. Also find the cycle of twist in a length of 2 m of shaft. Take $G = 8 \times 10^6$ N/cm^2.

Solution

Diameter of the shaft $d = 40$ mm
Torque applied $T = 100$ kN cm
Length of the shaft $\ell = 200$ cm
Modulus of rigidity $G = 8 \times 10^6$ N/cm^2

Using the torque equation $T/J = G\theta/\ell = 2\sigma/d$,

$$T = \frac{\pi}{32} d^4 \frac{\sigma}{d}$$

Solving,

$$\sigma = 7957.7 \text{ N/cm}^2$$

Angle of twist

$$\theta = \frac{33 \times 10^5 \times 200}{8\pi \times 10^6 \times 10^4} = 0.099 \text{ rad} = 5.699°$$

Example 3.6 A conical friction clutch used for a mechatronic system transmits 90 kW at 1500 RPM. The semi-cone angle is 20° and the coefficient of friction is 0.2. If the mean diameter of the beam surface is 375 mm and the intensity of normal pressure is not to exceed 0.25 N/mm^2, find the dimensions of the conical clutch.

Solution

Power transmitted $P = 90$ kW $= 90 \times 10^3$ W
Speed of the drive shaft $N = 1500$ rpm
Cone angle $\alpha = 20°$

Friction coefficient $\mu = 0.2$
Mean diameter $D = 375$ mm
Normal pressure $p = 0.25$ N/mm^2

Angular velocity $\omega = \dfrac{2\pi N}{60} = 156$ rad

Mean radius $R = \dfrac{D}{2} = 187.5$ mm

Power transmitted = torque multiplied by angular velocity
Torque = $5.77 \; 10^3$ N mm

When the clutch is engaged,
Torque transmitted to driven pulley = $2\pi Rb = \mu pR$

Hence,
Width of the clutch, $b = 52.2$ mm

Let the inside radius of the cone be r_1 and the outside radius r_2. Then
$$r_1 + r_2 = 2R = 375$$
and $r_1 - r_2 = b \sin \alpha = 18$

Solving,
$$r_1 = 196.5 \text{ mm}$$
and $r_2 = 178.5$ mm

Example 3.7 A centrifugal clutch has four shoes which slide in a spider keyway. Each shoe has clearance of 5 mm between shoe and rim. The pull exerted by the spring is 500 N. The mass centre of the shoe is 160 mm from the axis of the clutch. If the internal diameter of the rim is 400 mm, the mass of each shoe is 8 kg, the stiffness of the each spring is 50 N mm, and the coefficient of friction between shoe and rim is 0.3, find the power transmitted by the clutch at 500 rpm.

Solution

Number of shoes	$n = 4$
Clearance between shoe and rim	$c = 5$ mm
Spring pull when the clutch is at rest	$S = 500$ N
Mass centre of the shoe	$r = 160$ mm
Spring stiffness	$s = 50$ N/mm

Friction coefficient $\quad \mu = 0.3$

Speed of the drive shaft $\quad N = 500$ rpm

Angular velocity $\quad \omega = \dfrac{2\pi N}{60} = 52.37$ rad/sec

Operating radius $\quad r_1 = r + c = 165$ mm

Centrifugal force $\quad P_f = m\omega^2 r_1 = 3620$ N

Upward force exerted by the spring,

$\quad P_s$ = initial pulling force + clearance multiplied by stiffness

$\quad\quad = S + cs = 750$ N

Friction force acting tangential on each shoe,

$$F_f = \mu(P_f - P_s) = 861 \text{ N}$$

Torque transmitted $\quad T = n F_f R = 688.8$ N

Power transmitted $\quad = Tw = 36.1$ kW

Exercises

3.1 Define the terms stiffness, compliance, and cross compliance.

3.2 Discuss the different types of lubricating systems.

3.3 Explain the functions of self-lubricating bearings.

3.4 Define the terms hardenability factor, hardenability index, and r parameter.

3.5 Differentiate between trusses and frames.

3.6 Why is heat treatment required? Discuss double tempering treatment.

3.7 Differentiate between

  (i) unilateral and bidirectional tolerances

  (ii) the shaft-based system and the hole-based system.

3.8 Why do we need to scrape the surface of a machine tool? How is it done?

3.9 Discuss the simple mechanisms used in mechatronic systems.

Chapter 4

Electronics in Mechatronics

4.1 Conductors, Insulators, and Semiconductors

According to the definition given by the Institute of Electronics and Electrical Engineers, USA, electronics is the science and technology of passage of charged particles in gases, vacuum, or in semiconductors. Electricity flows in electric motors, incandescent lamps, electric furnaces, and transformers through copper wires or some metallic parts. The fluorescent lamp can be considered as an electronic device as electrons flow in space from one end of the tube to the other, whereas the common incandescent lamp cannot be considered as an electronic device since current flows through the filament in it. The following are the salient features of an electronic device.

1. It can respond to very small control signal.
2. It can respond at a speed far beyond the speed at which mechanical or electrical devices can.
3. It is photosensitive.
4. It can rectify alternate current into direct current.
5. Some electronic devices can produce x-ray radiations.

To understand why some materials behave as conductors, insulators, or semiconductors, the atomic structure of materials has to be understood. All matter is made up of atoms consisting of a nucleus of positive charge. Negatively charged electrons revolve around the nucleus in specific orbits. The positive and negative charges form a strong bond and atoms become stable. As per the Pauli's exclusion principle, when the number of electrons in each orbit of an atom is $2n^2$, where n is the orbit number, the atom becomes very stable. Hence, in stable atoms such as those of halogens, the first orbit has two electrons and the second has eight. The electrons in the outermost orbit are called valence electrons and the outermost orbit itself is called valence orbit. In some materials, stable atoms are obtained by sharing or donating the electrons. Accordingly, the bond formed may be called an ionic bond or a covalent bond. Free electrons are not available in such materials which act as insulators. Insulators cannot conduct thermal energy or electricity. In metallic bonds, the outer valence electrons form an electron cloud or gas and move around the material exhibiting Brownian motion. Hence, metals are good conductors of electricity and thermal energy. In the case of semiconductors the energy requirement is in between that of metals and insulators. Hence, they behave as semiconductors of electricity. Graphite, silicon, and germanium are examples of semiconductors.

Electrons not only revolve around the nucleus but also spin on their own axes. If there are two electrons in an orbit, one electron rotates clockwise and the other rotates anticlockwise. This produces a neutral momentum. In some materials, there is positive momentum, indicating more clockwise spin. Such materials are called paramagnetic materials. If there is a net negative momentum, the material is said to be ferromagnetic.

4.2 Passive Electrical Components

Energy is classified into potential energy and kinetic energy. A body raised above the ground level has a certain amount of potential energy. Similarly, the voltage or electric potential at a point in an electric field can be defined as the work done in bringing one coulomb of positive charge to that point from infinity against the electric field. Voltage is measured in volts. If the work done is 1 J, then the potential at that point is 1 V. An electric current may be defined as the movement of electric charge along a definite path. The current corresponding to the flow of 1 C of charge per second through a conductor is 1 A. One ampere of current is the flow of 6.24×10^{18} electrons per second through a conductor. A current in a conductor produces heating effect, magnetic field, and chemical transformation when passed through a certain solution. The heating effect is made use of in incandescent lamps, electric heaters, and furnaces. The magnetic effect is used in electromagnets and the chemical effect in electroplating.

Direct current (DC) is a unidirectional current which does not change its value appreciably. The DC voltage is a unidirectional voltage like that of a battery. In alternating current or voltage, the circuit direction of current or voltage reverses at regular intervals of time. The sine wave is most frequently used in alternating current or voltage. In this waveform, voltage or current varies from zero to a positive peak, falls back to zero, reaches to a negative peak, and again comes back to zero. This is called one cycle of a sine wave. A cycle is spread over 360° or 2π rad. Domestic electric supply is sinusoidal and has a frequency of 50 cycles/sec, usually expressed as 50 Hz.

Electric energy spent per second is called power. Electric power for a DC supply is given by *VI*, where *V* is the voltage and *I* is the current. Electric power is measured in terms of a unit called watt W. One watt is the power corresponding to the energy spent per second by a current of 1A when the applied voltage is 1 V. For alternating current (AC), power is equal to $VI\cos\phi$, where $\cos\phi$ is called the power factor and ϕ is the phase shift between voltage and current.

4.2.1 Resistor

Resistance is defined as the property of a material due to which it opposes the flow of electricity through it. Metals are good conductors of electricity. They offer very low resistance to the flow of current. The resistance offered by a conductor depends upon the length, cross-sectional area, and nature of materials and the temperature of the conductor. If the effect of temperature can be neglected, then resistance $R = \rho\ell/a$, where ℓ is the length of the conductor, a is the cross-sectional area, and ρ is called the specific resistance, which depends on the nature of the material of the conductor. If the effect of temperature only is considered, the resistance $R_T = R_0(1 + \alpha T + \beta T^2 + \cdots)$, where R_0 is the resistance at room temperature, T is the temperature of the material, and α and β are constants and are material dependent. The symbol for resistance and temperature characteristics are shown in Fig. 4.1. If the material resistance comes to zero for a particular temperature, then it is called a superconducting material. At 4 K, mercury behaves as a superconducting material.

(a) Symbol   (b) Characteristics

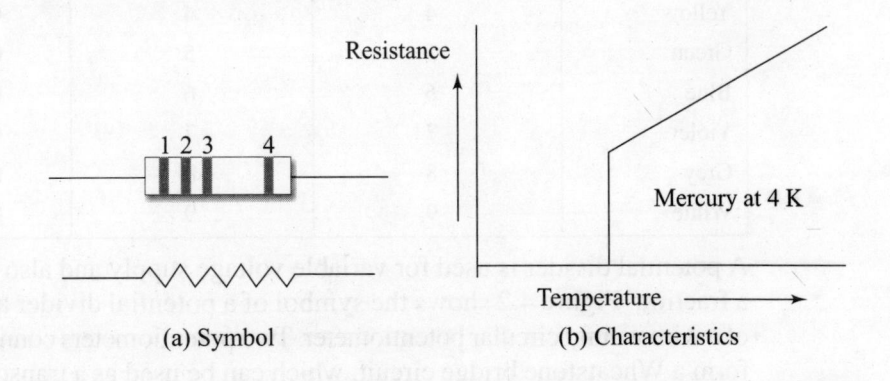

Fig. 4.1 A resistor

The resistances of some resistors are indicated by numbers. This method is used for low value resistors. Most resistors are coded using colour bands. The first band gives the resistance value of the resistor in ohms. The fourth band, which is far away from the usual band, indicates the accuracy of the value. Red in this region indicates 2%, gold indicates 5%, and silver indicates 10% accuracy. Table 4.1 indicates the colour codes, bands, and the corresponding resistance values.

Usually a resistor is made up of a film of carbon or a coil of nichrome wire. Resistors are available in some standard values, power ratings, and tolerance ranges. A nichrome wire has 75% Ni and 25% Cr. Its resistance does not vary more than by 0.005 for a temperature change from 25°C to 125°C. The linearity of the wire wound resistor is within 0.003%. Cermets are used as resistors. Cermets are an abbreviation for ceramic metal. They consist of fine particles of precious metals. They are manufactured by baking in kilns. Ceramic substrates such as aluminium and sterlite are also used. Plastic conductors are made of epoxy, polyester, and other resins blended with carbon powder to make them conductive. Their linearity can be held to 0.025%. When resistors are connected end to end, they are said to be connected in series. The equivalent resistance is equal to the sum of the individual resistances so connected. Resistors are said to be connected in parallel if two or more resistors are connected across each other. The equivalent resistance is equal to multiplication of the individual resistances divided by the sum of individual resistances so connected.

Table 4.1 Colour and resistance value

Colour	Band 1	Band 2	Band 3
Black	0	0	
Brown	1	1	0
Red	2	2	0 0
Orange	3	3	0 0 0
Yellow	4	4	0 0 0 0
Green	5	5	0 0 0 0 0
Blue	6	6	0 0 0 0 0 0
Violet	7	7	0 0 0 0 0 0 0
Grey	8	8	0 0 0 0 0 0 0 0
White	9	9	0 0 0 0 0 0 0 0 0

A potential divider is used for variable voltage supply and also as multiplier of a fraction. Figure 4.2 shows the symbol of a potential divider and the diagram of a wire wound circular potentiometer. Two potentiometers connected in parallel form a Wheatstone bridge circuit, which can be used as a transducer to convert two mechanical inputs into an electrical signal. Refer to Fig. 4.2(c).

E_i = input voltage
E_o = output voltage
$E_o = KE_i$ for $K < 1$

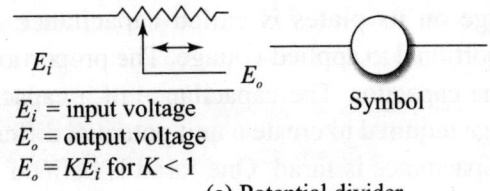

(a) Potential divider

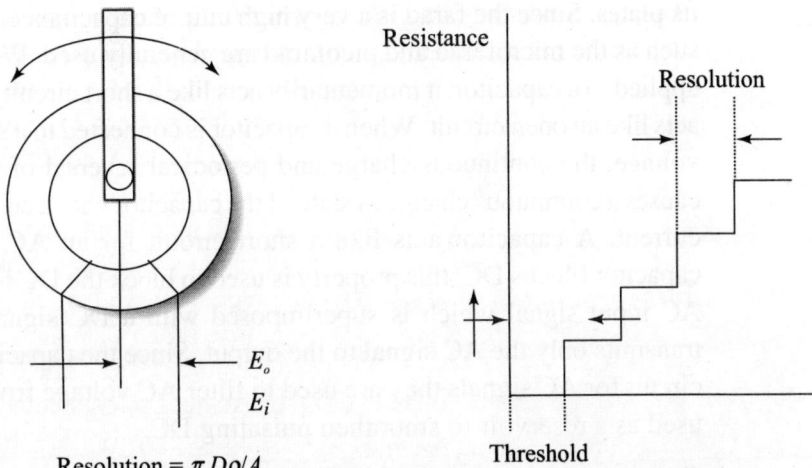

Resolution = $\pi D\rho/A$

where D is the diameter of the core, A is the cross-sectional area of the wire, and ρ is the specific resistance.

(b) Circular potentiometer

When the bridge is balanced,

$$\frac{R_1}{R_2} = \frac{R_3}{R_4} \quad \text{or} \quad \frac{R_1}{R_3} = \frac{R_2}{R_4}$$

θ_i = angular input
θ_o = angular output from a mechanical system

(c) Conversion of two mechanical signals

Fig. 4.2 Potentiometer

4.2.2 Capacitor

A capacitor essentially consists of two conducting surfaces separated by a layer of insulating medium called dielectric. The dielectric is an insulating medium through which an electrostatic field can pass. The main purpose of the capacitor is to store electrical energy. The property of the capacitor by which it stores charge on its plates is called capacitance. The charge of the capacitor is proportional to applied voltage. The proportionality constant is the capacitance of the capacitor. The capacitance of a capacitor is defined as the amount of charge required to create a unit potential difference between its plates. The unit of capacitance is farad. One farad is defined as the capacitance of a capacitor which requires a charge of 1 C to establish a potential difference of 1 V between its plates. Since the farad is a very high unit of capacitance, much smaller units such as the microfarad and picofarad are generally used. When a DC voltage is applied to a capacitor, it momentarily acts like a short circuit and the capacitance acts like an open circuit. When a capacitor is connected to a source of alternating voltage, the continuous charge and periodical reversal of the applied voltage causes a continuous change in state of the capacitor with a continuously changing current. A capacitor acts like a short circuit for an AC voltage. Since the capacitor blocks DC, this property is used to block the DC components from an AC input signal which is superimposed with a DC signal, so the capacitor transmits only the AC signal to the output. Since the capacitor acts like a short circuit for AC signals they are used to filter AC voltage from DC voltage. It is used as a reservoir to smoothen pulsating DC.

Capacitance of a capacitor depends on the dielectric constant k, area of one side of the plate A, number of plates N, and separation of the plate surfaces d:

$$C = \frac{kA(N-1)}{d}$$

Capacitance can be varied by changing any of the variables on the right-hand side of the above equation. Figure 4.3 shows the symbol and construction of capacitor. A capacitor is also called a condenser.

(a) Symbol (b) Construction

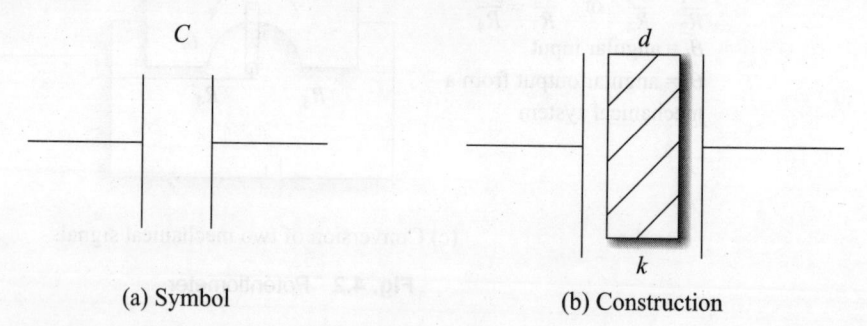

Fig. 4.3 Capacitor

4.2.3 Inductor

A wire wound in the form of a coil makes an inductor. The property of an inductor is that it always tries to maintain a steady flow of current and opposes any fluctuation in it. When a current flows through a conductor, it produces a magnetic field around it in a plane perpendicular to the conductor. When a conductor moves in a magnetic field, an electromagnetic force is induced in the conductor. Imagine a coil of wire in a case connected to a battery. Whenever an effort is made to increase the current through the coil, it is opposed by the instantaneous production of a counter emf. The energy required to overcome this opposition is supplied by the battery. This energy is stored in the inductor in the form of additional magnetic flux produced. At this stage, if an effort is made to decrease the current through the coil, the current decrease is delayed due to the production of self-induced emf. The property of the inductor due to which it opposes any increase or decrease in current by the production of a counter emf is known as self-inductance. It is initially difficult to establish a current through a coil, but once it is established it is equally difficult to withdraw it. Hence inductance is sometimes called electrical inertia. The emf developed is proportional to the rate of current through the inductor. Mathematically, $e = LdI/dt$, where the proportionality constant L is called the self-inductance of the coil, I is the current, and e is the voltage developed.

Mutual inductance arrangements use two coils—one acting as the power coil and the other as the supply output. A mutual inductance profile is used as a sensor for measuring small displacements. A linear variable differential transformer (LVDT) works under the principle of mutual inductance. Figure 4.4 illustrates the working of an inductor.

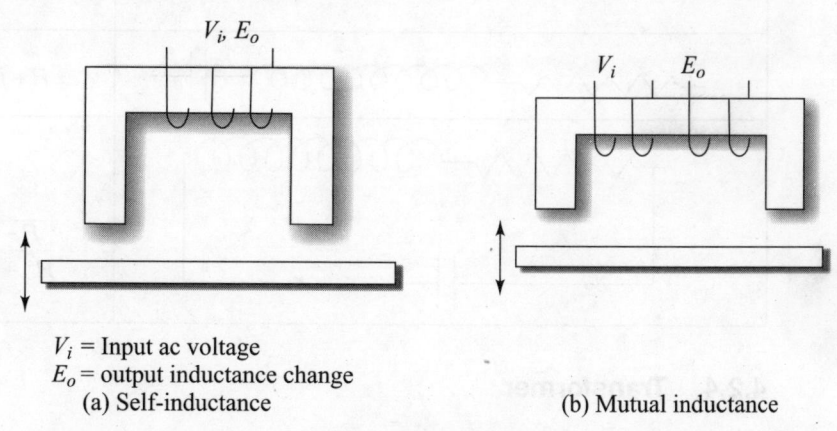

V_i = Input ac voltage
E_o = output inductance change
(a) Self-inductance   (b) Mutual inductance

Fig. 4.4 Inductor

Ohm's law $V = IR$, where V is the supply voltage, I is the current, and R is the resistance, holds good for DC supply. However, it cannot be used for AC supply because of the current and voltage phase relationship. Impedance is defined as AC resistance equal to the ratio of the voltage to the current. When a circuit consists of several elements—resistances, capacitances, and inductances in series or parallel, the resulting impedance can be calculated. Table 4.2 shows the various formulae that can be used to calculate impedance. The impedance of a capacitor is known as capacitive reactance and the impedance due to an inductor is known as inductive reactance. The inductive reactance X_L of an inductance is given by $L\omega$, where $\omega = 2\pi f$ and f is the frequency. The capacitive reactance X_C of a capacitor is given by $X_C = 1/\omega_C$. Inductive reactance is positive, whereas capacitive reactance is negative.

Table 4.2 Impedance formulae

Circuit	Impedance
Resistor	R
Inductor	$j\omega L$
Capacitor	$-j/\omega C$
Resistor–Inductor (series)	$R + j\omega L$
Resistor–Capacitor (series)	$R - j/\omega C$
Resistor–Capacitor (parallel)	$R/(1 + j\omega CR)$
Resistor–Inductor–Capacitor (series)	$R + j(\omega L - 1/\omega C)$
Resistor–Inductor in series, parallel with Capacitor	$\dfrac{1}{\dfrac{R - j\omega L}{R^2 + \omega^2 L^2} + j\omega C}$

4.2.4 Transformer

A transformer is a device by means of which electric power in one circuit is transferred to the another circuit without a change in frequency. It essentially consists of two or more inductive windings wound on the same core. A transformer can raise or lower the voltage in a circuit with corresponding decrease or increase in current. Transformers are effectively used to step up and step down voltages in power stations and substations, respectively, so that power losses can be minimized. Figure 4.5 shows the schematic diagram of a transformer.

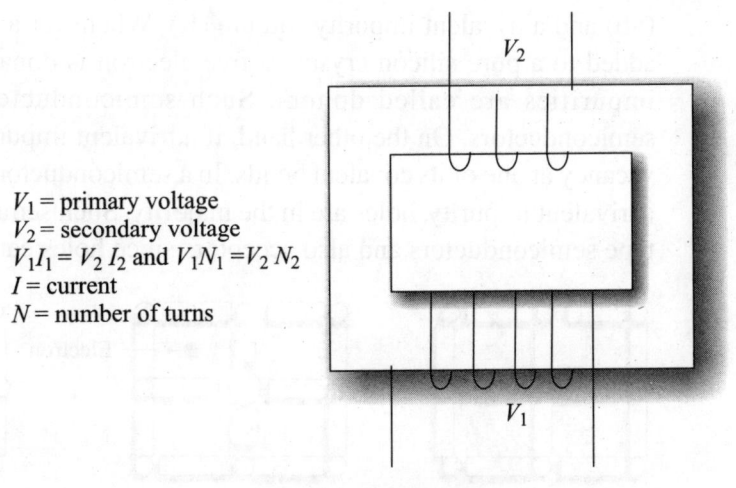

V_1 = primary voltage
V_2 = secondary voltage
$V_1 I_1 = V_2 I_2$ and $V_1 N_1 = V_2 N_2$
I = current
N = number of turns

Fig. 4.5 A transformer

4.3 Active Elements

Active elements have an auxiliary source of power which supplies a major part of the output power while input signals supply only an insignificant portion. There may or may not be a conversion of energy from one form to another. Solenoid drives, electronic amplifiers, servosystems, and digital systems are some examples of active devices.

4.3.1 Semiconductor Devices

Graphite, silicon, and germanium are semiconductor materials. Most commonly Ge and Si are used in semiconductor devices. Silicon and germanium are tetravalent, having four electrons in the outermost orbit. Both the materials are crystalline in nature. At very low temperatures, say 0°C, the crystal behaves like an insulator as there are no free electrons available in it. However, as the temperature increases and reaches room temperature, some of the covalent bonds are broken because of thermal energy, and the conductivity of the crystal improves. The space where the electron can be accepted is called a hole. Even though pure silicon and germanium have free electrons and holes at room temperature, conductivity is poor because the number of free electrons and holes is not sufficient. In order to improve the conductivity of these materials, some impurities are intentionally added. This addition of impurities in a semiconductor is known as doping. Usually either pentavalent impurities such as phosphorous and antimony or trivalent impurities such as iridium, gallium, or boron are added to pure semiconductors. Figure 4.6 shows the crystal structure of pure silicon and the crystal structure after doping with a pentavalent impurity antimony

(Sb) and a trivalent impurity iridium (Ir). Whenever a pentavalent impurity is added to a pure silicon crystal, a free electron is donated. Hence pentavalent impurities are called donors. Such semiconductors are called N-type semiconductors. On the other hand, if a trivalent impurity is added it creates a vacancy at one of its covalent bonds. In a semiconductor formed by doping with a trivalent impurity, holes are in the majority. Such semiconductor are called P-type semiconductors and also acceptors since holes can accept electrons.

(a) Pure silicon (b) Silicon with pentavalent impurity (c) Silicon with trivalent impurity

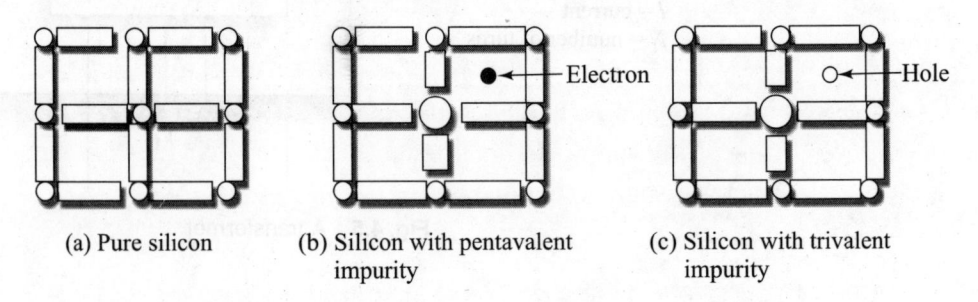

Fig. 4.6 (a) Pure silicon, (b) and (c) silicon with impurities

A point contact diode has a metal base on which a germanium semiconductor wafer is mounted and a tungsten alloy wire is allowed to press against the semiconductor. This device allows current to flow in one direction and is called a diode. The disadvantage of the contact type diode is that the contact area is very small. Junction diodes can be made with a P-type and an N-type semiconductor material. This type of diode is called a P-N junction diode. P-N junction diodes are made by doping a P-type impurity on one side and an N-type impurity on the other side in a single crystal of a semiconductor. The P region is called the anode and the N region the cathode. If a P-N junction diode is connected in a circuit with the anode connected to the positive terminal and the cathode connected to the negative terminal of a battery, it is said to be forward biased. On the other hand, when the polarity of the battery is reversed, the diode is said to be reverse biased. The forward and reverse characteristics of a diode are illustrated in Fig. 4.7. It shows the diode current for various applied voltages plotted against the applied voltage for both forward and reverse bias. For the forward-biased diode, current increases steeply for every increase in applied voltage after an initial small voltage. This small voltage is called the cut-in voltage and depends on the semiconducting materials. For germanium, it is of the order of 0.2 V and for silicon it is around 0.6 V. For reverse bias, the current is very small till the breakdown voltage. After the initial voltage, a large reverse current flows through the diode. It may get damaged if the current is not limited through an external resistor. Beyond the breakdown voltage, the diode is said to be operating in the breakdown region.

The requirement of DC power supply can be met by rectifying AC supply using a rectifier. Since the diode conducts in one direction only, it can be used as a rectifier to convert AC into DC. If AC input is given to a diode during the positive half-cycle, the diode is forward biased. If AC input is given to a diode during the negative half-cycle, it is reverse biased. As the diode conducts when it is forward biased, the positive half-cycle is transmitted to the output and it appears across the load. To improve the average value of DC voltage, centre-tap rectifier with two diodes or bridge rectifiers with four diode circuits are used. The efficiency of a rectifier is measured in terms of the ripple factor. Ripple factor is defined as the ratio of AC voltage to DC voltage. The goal of any power supply design is to reduce this factor as much as possible. The ripple factor is improved in the bridge rectifier configuration.

(a) P-N junction diode

(b) Symbol

(c) Forward-biased diode in circuit

(d) Reverse-biased diode in circuit

(e) Diode characteristics

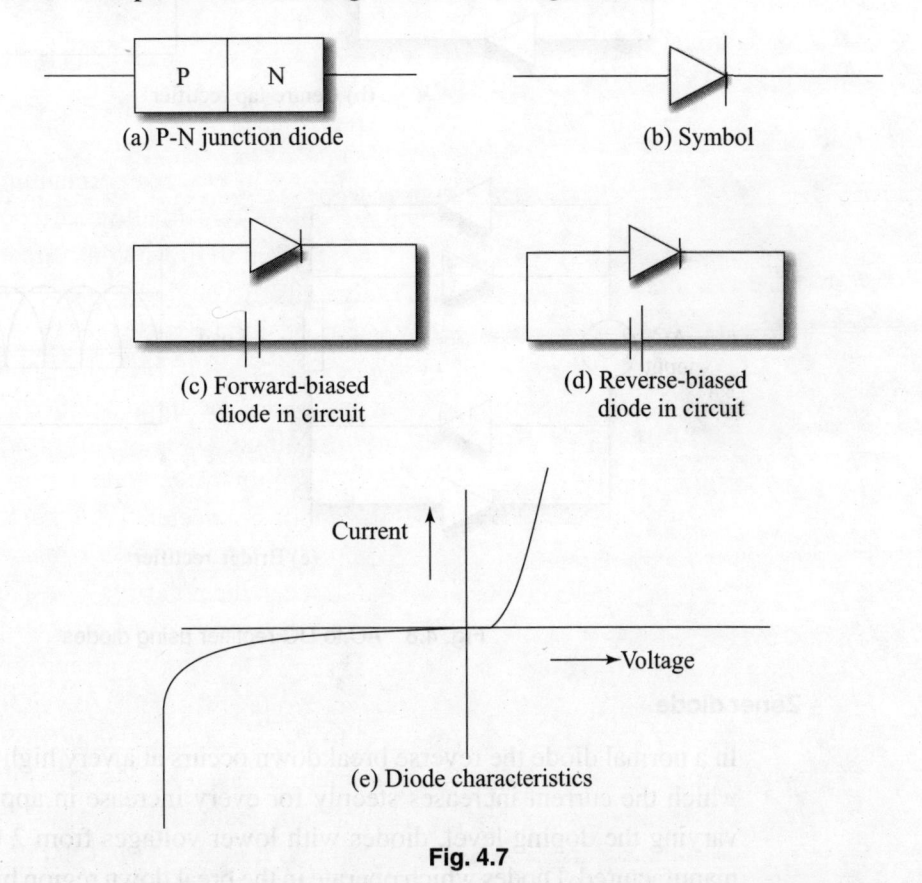

Fig. 4.7

Figure 4.8 illustrates the output waveform for a single diode with load, centre-tap rectifier and bridge rectifier. The bridge rectifier is the most widely used rectifier. Its main advantage is that it eliminates the requirement of a centre-tap transformer. But the rectifier creates problems when the secondary voltage is low.

(a) Half-wave rectifier

(b) Centre-tap rectifier

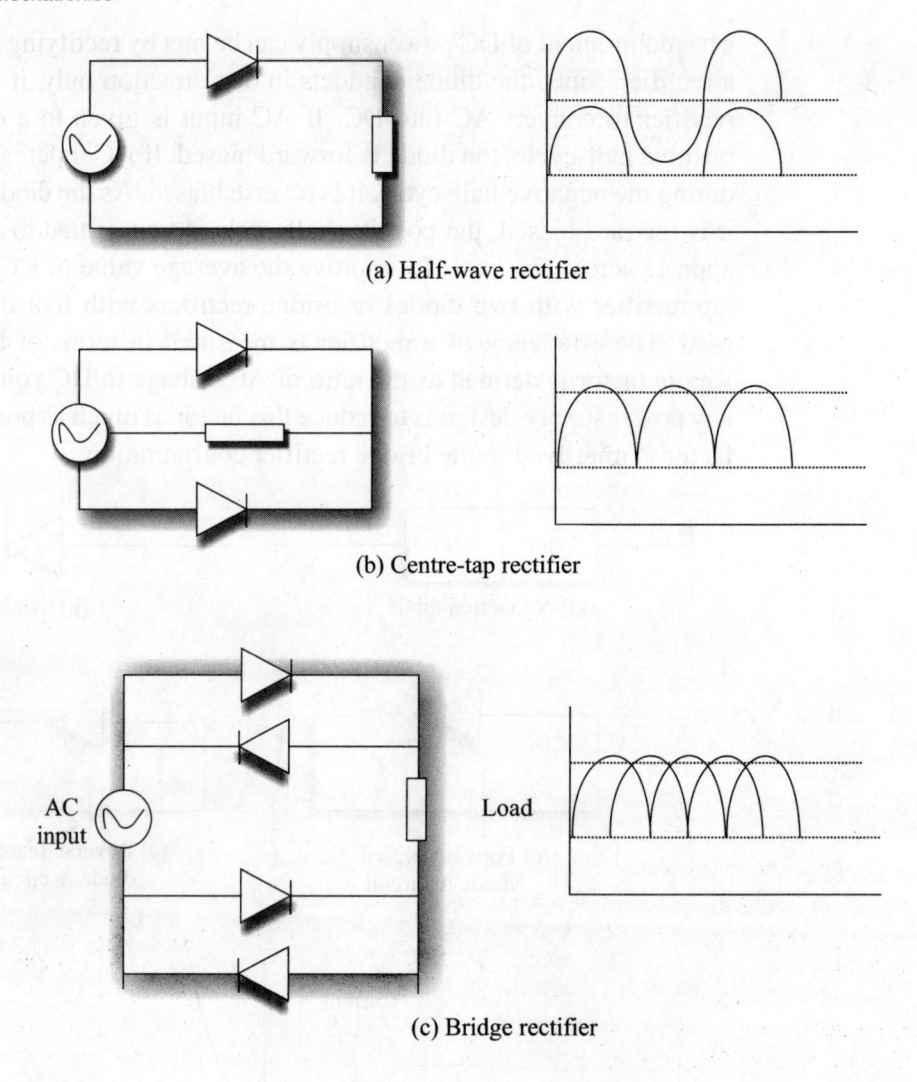

(c) Bridge rectifier

Fig. 4.8 AC to DC rectifier using diodes

Zener diode

In a normal diode the reverse breakdown occurs at a very high voltage, beyond which the current increases steeply for every increase in applied voltage. By varying the doping level, diodes with lower voltages from 2 to 200 V can be manufactured. Diodes which operate in the breakdown region by reverse biasing are called zener diodes. The voltage at which breakdown occurs is known as zener voltage. The zener diode is of great importance for voltage regulators since it can be used as a constant-voltage source by applying the reverse voltage that exceeds the zener breakdown voltage.

Tunnel diode

By increasing the doping level one can obtain breakdown voltage at 0 V. This will happen when the concentration of impurities is increased by more than 1000 times. The diode starts conducting at zero volt in both positive and negative directions. Such a diode is known as tunnel diode or Esaki diode.

Light emitting diode (LED)

In a forward-biased diode, free electrons cross the junction and recombine with holes. Whenever electrons combine with holes, energy is radiated. In a rectifier diode this energy is liberated as heat. But in light-emitting diode, this energy is radiated as light—red, green, yellow, orange, or infrared. LEDs producing the visible radiation are often used in instrument displays, digital readouts, digital clocks, calculators, etc. LEDs producing infrared radiation are used in burglar-alarm systems.

Photodiodes

When a P-N junction of the diode is housed in a glass package, strong light hitting the junction increases the reverse current. A photodiode is a normal diode optimized for its sensitivity to light and is housed in a glass package. The glass window lets the increasing light pass through the housing and hit the P-N junction. This light produces additional holes and electrons, giving rise to higher reverse current. Photodiodes are always used in reverse-biased conductors.

Optoisolator

An optoisolator combines an LED and photodiode in a single package. The LED supply voltage forces a current through the LED. The light from the LED hits the photodiode and sets up a current. If the LED current varies, the output from the photodiode also varies. If the LED current has an AC variation, the photodiode current will have an AC variation of the same frequency. The resistance between the input and output is greater than $10^{12} \, \Omega$.

4.3.2 Transistor

A junction transistor consists of either an N-type semiconductor sandwiched between two P-type semiconductors or a P-type semiconductor sandwiched between two N-type semiconductors. In the former case the transistor is referred to as a PNP transistor and in the latter case as an NPN transistor. The semiconductor junctions thus formed are housed in a hermetically sealed case of either plastic or metal. Three leads are brought out from the three semiconductor regions. The sandwiched semiconductor region is called the base.

The other two regions are called the emitter and the collector. The collector region is large in size and more heavily doped. The transmitter can be considered as two semiconductor diodes connected back to back in a crystal. Negligible current flows through an NPN transistor with both diodes reverse biased. If both diodes are forward biased, a large current will flow in either diode. In an NPN transistor with the emitter diode forward biased and the collector diode reverse biased, there should be a large current in the emitter diode and negligible current in the collector diode. But this does not happen in practice. There is a large current in the emitter diode and an equally large current in the collector diode also. This is referred to as transistor action. The forward-biased emitter diode can be considered as a small resistance and the reverse biased collector diode can be considered as a very high resistance equivalent to the back resistance of a diode. Since collector current is almost equal to emitter current, it is clear that in the low resistance, input is transferred to the high resistance output circuit.

Figure 4.9 explains how a transistor can amplify an input signal. The emitter diode is forward biased and collector diode is reverse biased. Hence the forward-biased emitter can be represented as a resistance R_{eb} having a low resistance value equivalent to the forward resistance of a diode. Similarly, the reverse-biased collector diode can be replaced by another resistance R_{cb} having very high value. The collector current I_c is about 95% of the emitter current I_e and the base current I_b is about 5% of I_c. The input voltage V_i is the voltage across the input resistance R_{eb}. Here $V_i = I_e R_{eb}$, Similarly $V_o = I_c R_{cb} = 0.95\, I_e R_{cb}$. The voltage gain is given by $a = V_o/V_i = 0.95 I_e R_{cb} / I_e R_{eb}$ as $I_c = I_e$. As R_{cb} is very high in comparison its to R_{eb}, the voltage gain is high. Typically R_{cb} is about 500 kΩ and R_{eb} is about 50 Ω. Voltage amplification is given by $a = 0.95 \times 500 \times 10^3 / 50$ = 9500. Calculations show that the transistor can amplify an input voltage 9500 times. In this case, it is assumed that the load resistance is infinite, which is not

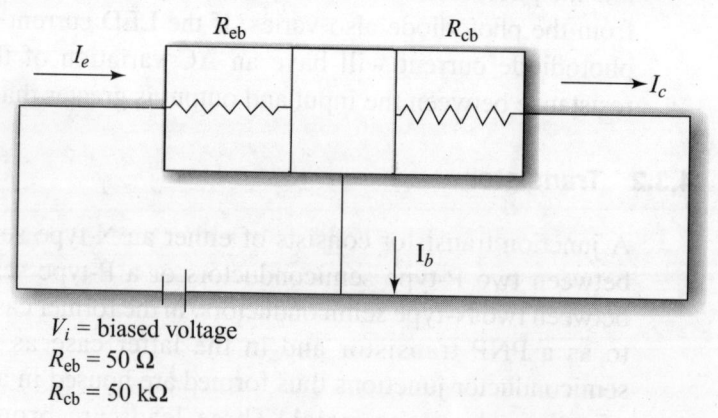

V_i = biased voltage
R_{eb} = 50 Ω
R_{cb} = 50 kΩ

Fig. 4.9 Transistor as amplifier

the case in practice. The load resistance lies between 10000 and 25000 Ω. Hence the voltage amplification lies between 200 and 500 V.

A transistor can be used in three basic configurations, namely, common-base configuration, common-emitter configuration, and common-collector configuration.

When the transistor is connected in the common-emitter configuration, the collector current is much larger than the base current and the typical current amplification is more than 100. As the base current goes on increasing, the collector current and the drop across the collector resistance also increase. When the drop across the collector resistor almost equals the supply voltage, the drop V_{ce} across the transistor is only a fraction of a volt, so small that transistor seems to act like a closed contact. On the other hand, when the base current is very small, the collector current is also small resulting in the whole supply voltage being dropped across the transistor. V_{ce} in this case equals the supply voltage and the transistor acts like an open circuit. Here the transistor works like a switch depending upon the base current. Figure 4.10 illustrates the common-emitter configuration.

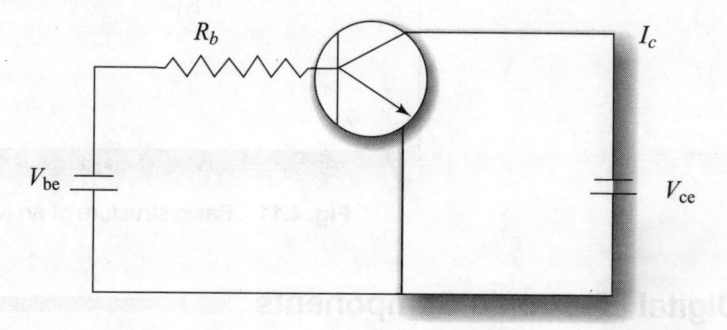

Fig. 4.10 Transistor as switch

4.3.3 Integrated Circuits

Circuits which use separate circuit elements—resistors, capacitors, transistors, diodes, etc.—connected together are called discrete circuits. Printed circuit (PC) boards provide ways to interconnect hundreds of discrete devices in a single board. Integrated circuits (IC) permit hundreds of transistors, diodes, and resistors to be formed and connected together to realize a complex circuit within the size of a small pill. All these circuit elements are made on tiny wafers of silicon. The main advantages of ICs are that they (a) save space, (b) provide improved reliability, (c) improve performance, (d) match devices, and (e) have a low cost.

Depending on the scale of integration, integrated circuits are classified into medium-scale integrators (MSIs), large-scale integrators (LSIs), and very large-scale integrators (VLSIs). The basic structure of an IC consists of four distinct layers. The bottom layer is of P-type silicon, called substrate, over which the IC is built. The second layer is a thin N-type layer grown on the substrate. The third layer is a thin layer of silicon oxide. It acts as a barrier to protect portions of the wafer against impurity penetration. The fourth layer is an aluminium layer which provides necessary interconnection between components. All the components are built in the second layer by a series of diffusion steps. The most complicated fabricated component is a transistor and all other elements are constructed with one or more similar processes. Figure 4.11 shows the basic structure of an IC (metal oxide semiconductor).

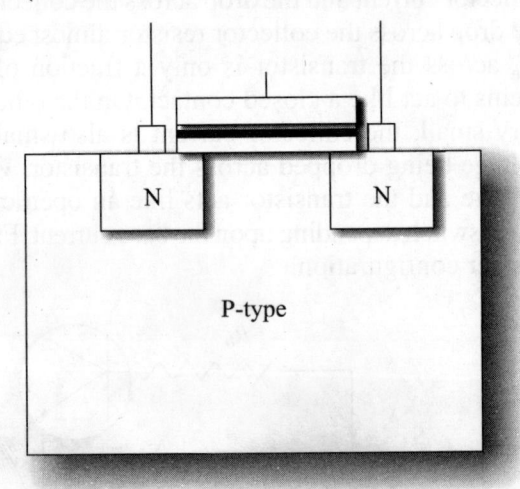

Fig. 4.11 Basic structure of an IC

4.4 Digital Electronic Components

The devices used in digital systems function in a binary manner. These devices exist in two possible states 'on' and 'off' or '1' and '0'. A transistor used as a digital device is used either in the cut-off or ion-saturation region and not in the inactive region. A high voltage, 4 ± 1 V, corresponds to state '1' and a low voltage, 0.2 ± 0.2 V, corresponds to state 0.

4.4.1 Logic Gates

The building blocks of a digital system that control its output based on the conditions of the input are called logic gates. A gate is a logic circuit with one or more input signals but has only one output signal. The input and output signals can consist of either high or low voltages corresponding to '0' or '1', respectively. There are basically three types of gates—OR gate, AND gate, and NOR gate. All other gates are combinations of these three.

The OR gate operates on both commutative and associative laws of addition. The OR operator can be accomplished with switches in parallel. Figure 4.12 shows the symbols of the OR and NOR gates, as well as the corresponding truth tables.

A	0	0	0	0	1	1	1	1
B	0	0	1	1	0	0	1	1
C	0	1	0	1	1	1	0	1
Z	0	1	1	1	1	1	1	1

(a)

A	0	0	0	0	1	1	1	1
B	0	0	1	1	0	0	1	1
C	0	1	0	1	0	1	0	1
Z	1	0	0	0	0	0	1	1

(b)

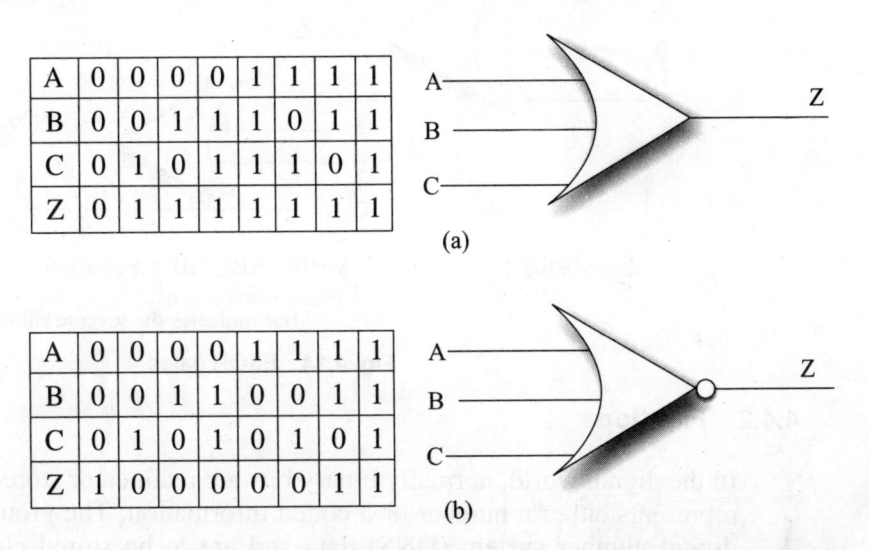

Fig. 4.12 (a) OR gate and its truth table (b) NOR gate and its truth table

The AND operation is a multiplication of two or more signals. The AND gate can be accomplished with switches in series. The associative law holds good in the AND gate. Figure 4.13 shows the symbols of the AND and NAND gates, as well as the corresponding truth tables.

A	0	0	0	0	1	1	1	1
B	0	0	1	1	0	0	1	1
C	0	1	0	1	0	1	0	1
Z	0	0	0	0	0	0	0	1

(a)

A	0	0	0	0	1	1	1	1
B	0	0	1	1	0	0	1	1
C	0	1	0	1	0	1	0	1
Z	1	1	1	1	1	1	1	0

(b)

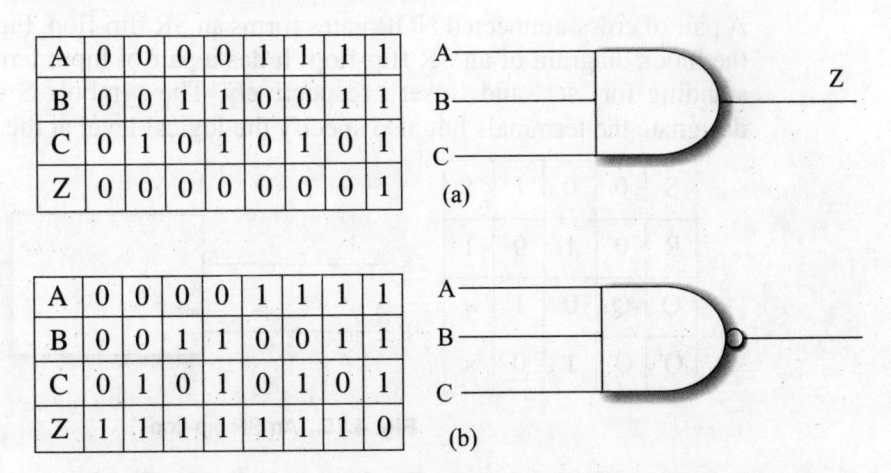

Fig. 4.13 (a) AND gate and its truth table (b) NAND gate and its truth table

The exclusive OR gate is known as EXOR gate and its operation is both commutative and associative. The symbol of an exclusive OR gate is shown in Fig. 4.14.

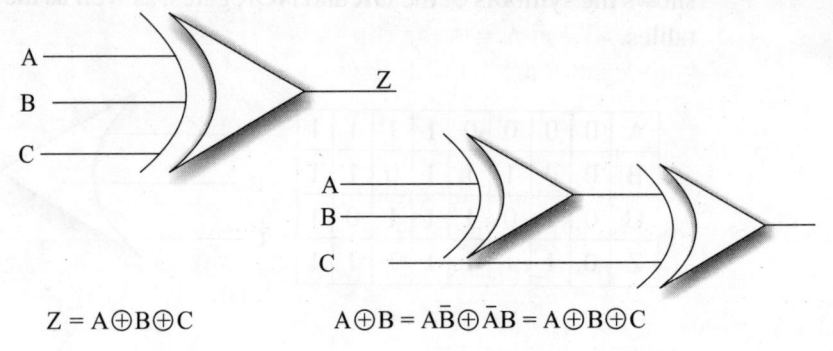

$$Z = A \oplus B \oplus C$$

$$A \oplus B = A\bar{B} \oplus \bar{A}B = A \oplus B \oplus C$$

Bar indicates the reverse value.

Fig. 4.14 EXOR gates

4.4.2 Flip-flops

In the digital world, normally binary bits are retained or stored in a group that represents either a number or a coded information. The group bits are called digital number system (DNS) data and are to be stored electronically. An electronic circuit that retains a single bit of DNS data is called flip-flop. Generally, flip-flops are used as memory devices for storing previous input values. For a given logical level from the output, these elements are used for digital system devices for reset, binary conversion systems, etc. There are three types of flip-flops, namely, SR flip-flops, JK flip-flops, and D flip-flops. Flip-flops are operated using clock cycles.

A pair of cross-connected NOR gates forms an SR flip-flop. Figure 4.15 shows the block diagram of an SR flip-flop. It has a pair of input terminals, S and R standing for 'set' and 'reset', respectively. The symbols S and R not only designate the terminals but also specify the logical level at the terminals.

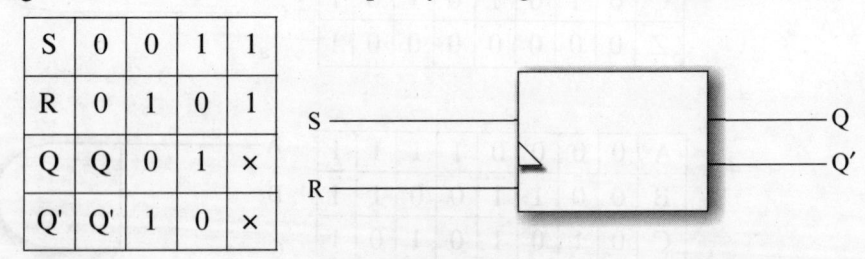

S	0	0	1	1
R	0	1	0	1
Q	Q	0	1	×
Q'	Q'	1	0	×

Fig. 4.15 An SR flip-flop

Assuming $Q' = 0$, when $S = R = 0$, then $Q = 1$. The flip-flop may persist in either of situations, one with $Q = 0$ and $Q' = 1$, the other with $Q = 1$ and $Q' = 0$. The first state refers to the reset state and the second state to the set state. The reset state is called the clear state and is defined as the state in which $Q = 0$. In the desk calculator, on pressing the clear button, all the registers turn to zero, and the machine is ready for another calculation. If $S = 0$ and $R = 1$, the flip-flop can establish itself in only one possible state, the set state with $Q = 0$. Similarly, with $S = 1$ and $R = 0$, the flip-flop will find itself in the set state with $Q = 1$. If both S and R are set to 1, the output will be at logic 0. This state is not used.

A JK flip flop is a clocked flip-flop with modifications using the NAND gate. The other types of gate also serve well. The modification consists of providing an additional input terminal to each gate and there is a connection from the flip-flop output to the input. The data terminals previously identified as J and R are now called K. Figure 4.16 shows the truth table and symbol of a JK flip-flop.

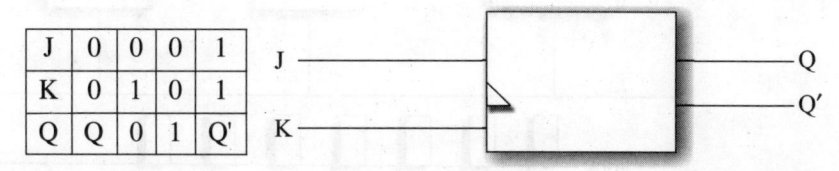

J	0	0	0	1
K	0	1	0	1
Q	Q	0	1	Q'

Fig. 4.16 JK flip-flop

Applications often require to delay a digital signal in time by the duration of one clock cycle. This operation can be performed by using the SR or the JK flip-flop. When clock C makes a positive transition from $C = 0$ to $C = 1$, $Q = 0$ if $D = 0$ and Q becomes 1 if $D = 1$. Q can be changed in no other way. Changes in D when $C = 1$ do not affect Q. Hence the triggering takes place when the clock transition is positive-going. Figure 4.17 shows the D flip-flop.

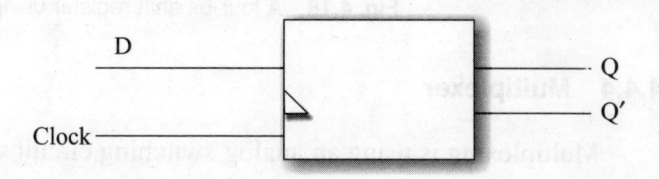

Fig. 4.17 D flip-flop

4.4.3 Shift Register

A logical 1 or 0 is called a bit. A flip-flop can store, remember, or register a single bit. A flip-flop is therefore referred to as a one-bit register. If N bits are remembered, N registered flip-flops are required. When an array of flip-flops has a number of bits in storage, it becomes necessary on occasions to shift bits from one flip-flop to another. An array of flip-flops that permits such shifting is called a shift register. A four-bit shift register constructed with type D flip-flops is shown in Fig. 4.18. More or fewer bits can be accommodated by adding or deleting flip-flops. Interconnections between flip-flops are such that the logic level at a data input terminal is determined by the state of the preceding flip-flop. Digital data may be presented in a serial or parallel form. In the serial form, the data appear in a single line, one bit at a time, often synchronous with a clock. In the parallel form, data appear at one instant of time on as many lines as there are bits. A shift register may serve as a convenient means of transforming serial data to parallel data and vice versa.

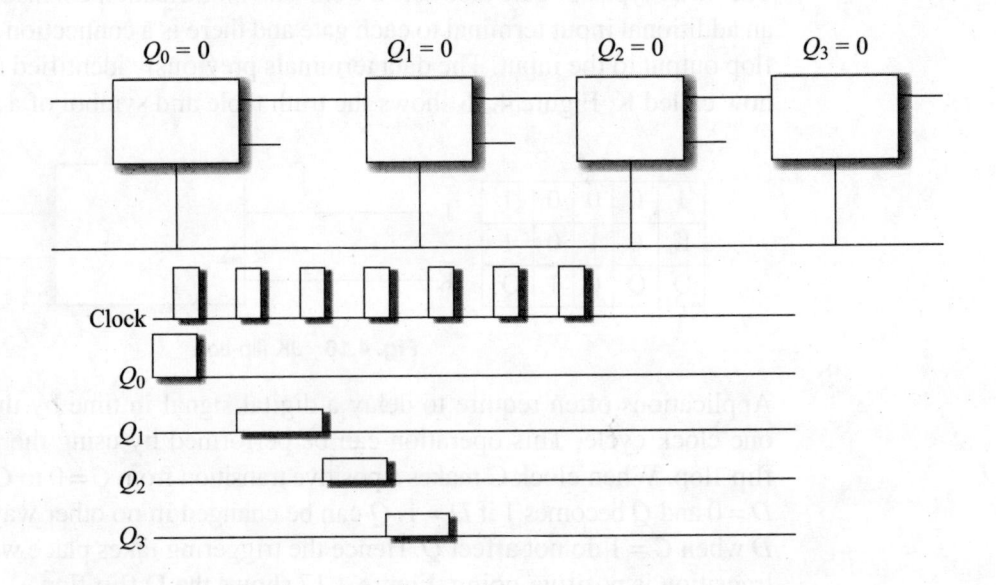

Fig. 4.18 A four-bit shift register using D flip-flops

4.4.4 Multiplexer

Multiplexing is using an analog switching circuit which allows the connection of a number of analog signals at a time to a common load. A multiplexer is shown in Fig. 4.19, where four separate signals are indicated, but the number of signals that may be involved is arbitrary. An approximate set of control logic signals is easily generated in a four-stage ring connector. The usefulness of the multiplex arrangement is a number of information-bearing signals (speech, music, date, etc.) which are to be transmitted over a communication channel through a pair of wires or a radio link to a distinct point. Multiplexing is performed sequentially and each choice is time shared. The timing of various input channels is controlled by a control unit. Commonly, 2×1, 4×2, 8×4, and 16×8 multiplexers are available.

V = input signal
C = control signal
S = switch

(a) Working of a multiplexer

(b) Application of a multiplexer

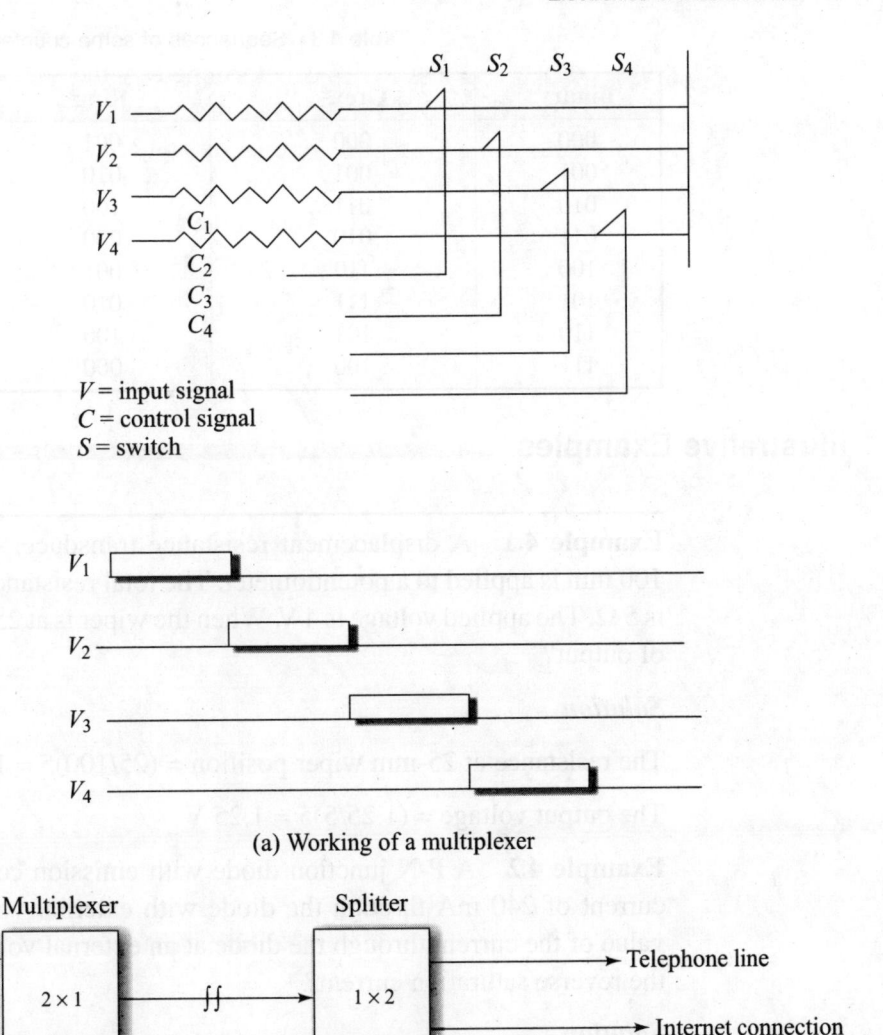

Fig. 4.19 Multiplexer

4.4.5 Counter

A counter is an electronic device that counts. It generates a sequence of count values determined by selected encoding and the status input. The counter can be an up-counter or down-counter. The up-counter starts from a value to the next larger values in the sequence and down counters do the opposite. Counters are constructed using flip-flops. Counters can be (a) binary counters, (b) decode counters, (c) grey counters, and (d) ring counters. Table 4.3 shows different counter sequences for three-bit signals.

Table 4.3 Sequences of some counters

Binary	Grey	Ring	Decode
000	000	001	000
001	001	010	001
010	011	100	010
011	010	000	011
100	110	001	000
101	111	010	001
110	101	100	010
111	100	000	0111

Illustrative Examples

Example 4.1 A displacement resistance transducer with a shaft stroke of 100 mm is applied to a potentiometer. The total resistance of the potentiometer is 5 Ω. The applied voltage is 4 V. When the wiper is at 25 mm, what is the value of output?

Solution

The resistance at 25-mm wiper position = (25/100)5 = 1.25 kΩ

The output voltage = (1.25/5)5 = 1.25 V

Example 4.2 A P-N junction diode with emission coefficient 2 conducts a current of 240 mA through the diode with external voltage 0.8 V. Find the value of the current through the diode at an external voltage of 0.7 V. What is the reverse saturation current?

Solution

Current through the diode

$$I_A = I_R \left[\exp \frac{V_{AK}}{\eta VT} - 1 \right]$$

$$I_A = 240 \times 10^{-3} A$$

External supply voltage, $V_{AK} = 0.8$ V

Emission efficiency, $\eta = 2$

Thermal voltage, $V_T = 26$ mV

Diode current at 0.8 V, $I_{A1} = I_R \left[\exp \dfrac{V_{AK_1}}{\eta VT} - 1 \right]$

Diode voltage at 0.7 V, $\quad I_{A2} = I_R\left[\exp\dfrac{V_{AK_2}}{\eta VT} - 1\right]$

Hence, $\quad I_{A2} = 55$ mA

$I_R = 50$ mA

Example 4.3 A logical function of four variables is $f(A, B, C, D) = (\overline{A} + BC)(B + CD)$. Express the function as a sum of products.

Solution

$$\begin{aligned} f(ABCD) &= (\overline{A} + BC)(B + CD) \\ &= (\overline{A} + BC)B + (\overline{A} + BC)CD \\ &= \overline{A}B + BBC + \overline{A}CD + BCCD \\ &= \overline{A}B + BC + \overline{A}CD + BCD \end{aligned}$$

Example 4.4 Determine the value of resistance for the colour bands brown, black, and red.

Solution

Band one, brown colour = 1

Band two, black colour = 0

Band three, red colour = 00

Hence, the resistance is 1000 Ω.

Example 4.5 Design a personal safe with a combination lock.

Solution

One solution could include the use of a rotary switch connected in series to form a circuit. A three-switch diagram is shown in Fig. 4.20, which illustrates such an arrangement. When the combination of switches is correct, a solenoid activates the bolt of the safe to open.

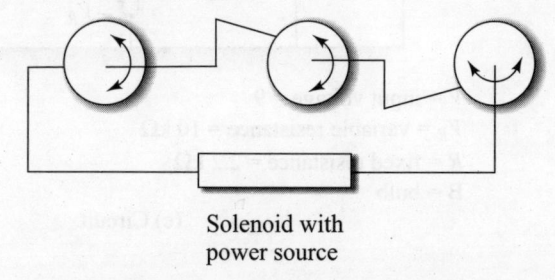

Solenoid with power source

Fig. 4.20 A combination switch

Example 4.6 Design a mechatronic system for indicating water pollution in rivers and canals.

Solution

When the LDR (light-dependent resistor) is dark in polluted water, its resistance is high. Sufficient base current flows to turn on the transistor and the indicator bulb turns off. In less polluted water, the resistance of LDR falls. This allows sufficient base current to flow to turn the transistor on. The transistor collector current passes through the indicator bulb, making it to glow. A simple circuit diagram for the set-up is shown in Fig. 4.21.

Light sensor → Transistor switch → Bulb

(a) Block diagram for design

Bulb   LDR (light-dependent resistor)

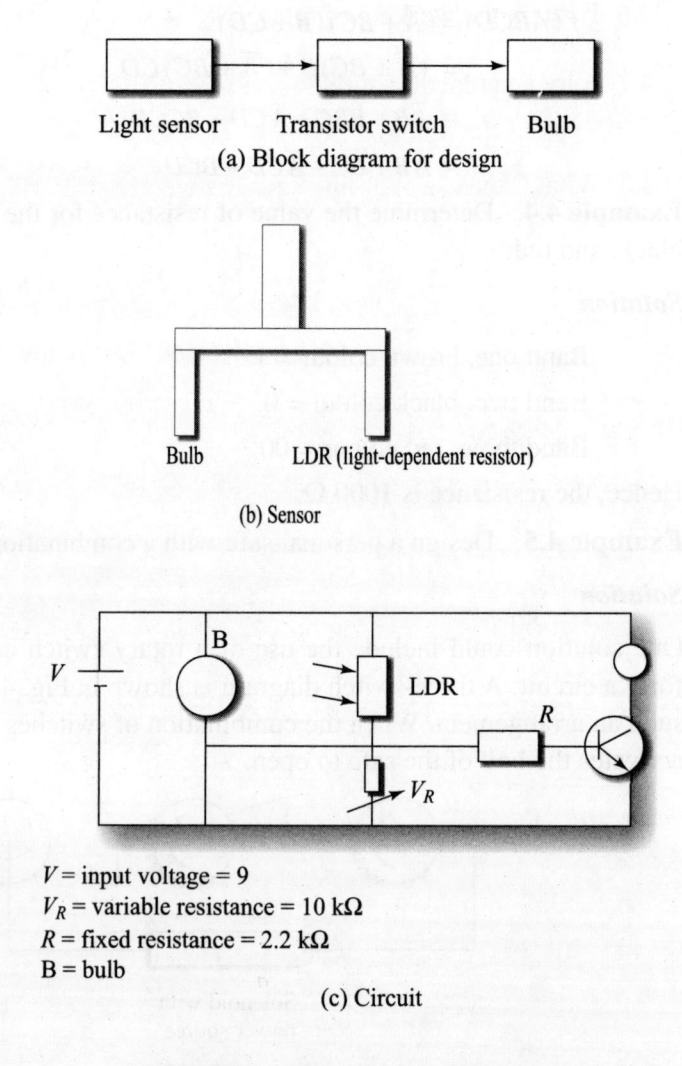

(b) Sensor

V = input voltage = 9
V_R = variable resistance = 10 kΩ
R = fixed resistance = 2.2 kΩ
B = bulb

(c) Circuit

Fig. 4.21 Water pollution indicator

Exercises

4.1 Differentiate between an insulator, a conductor, and a semiconductor.

4.2 Write notes on passive devices used in electronic circuits.

4.3 Distinguish between the zener diode, tunnel diode, LED, and photodiodes.

4.4 Explain the use of transistors as amplifiers.

4.5 Explain the use of transistor as a switching device.

4.6 Explain the working of OR, NOR, AND, NAND, and EXOR gates.

4.7 A logical function of five variables is $f(A, B, C, D, E) = (A + \overline{BC})(\overline{D + BE})$. Express the function as a sum of products.

4.8 What is the use of a flip-flop? Explain the working of a shift register using flip-flops.

Chapter 5

Computing Elements in Mechatronics

Mathematical instruments are used to generate or to solve mathematical functions such as trigonometric functions, hyperbolic functions, exponential functions, quadratic equations, equations of high order, differential equations, integration, etc. In these instruments, signals are continuous, analog, and without time delay. The calculating mechanism involved is similar to the one involved in calculating and adding machines, where all the operations are performed with a discontinuous or a digital signal. The computer is a device that involves the generation and solution of mathematical functions, and involves calculating mechanisms. The computer that operates for the mathematical function and can accept any analog signal is called the analog computer. The computer that can make calculations or digital computations is called the digital computer. The first calculating machine was developed by Pascal in 1642 in order to help his father, who collected taxes in France.

The first step-up drums principle with digital transfer was developed by the mathematician and philosopher Lecibniz in 1672, but he did not develop a machine. Ohhner in 1878 in Sweden fabricated a calculating machine and industries started using such machines. Figure 5.1 illustrates the difference between an analog signal and a digital signal.

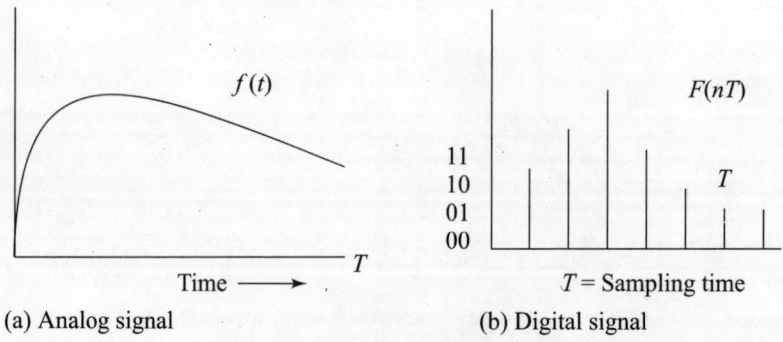

Fig. 5.1 Analog and digital signals

In the last 50 years, semiconductor technology has undergone many changes. Integrated circuits (ICs) appeared on the scene at the end of the 1950s, followed by the invention of the transistor. In the early 1960s, logic gates belonging to what is known as the 7400 series were commonly available as ICs and the technology of integrating the circuits of a logic gate on a single chip became known as small-scale integration (SSI). As semiconductor technology advanced, more than hundred gates were fabricated on one chip, which was called medium-scale integration (MSI). A typical example of MSI is the decoding counter 7490. At the end of the twentieth century, it was possible to fabricate more than 1000 gates on a single chip, which was known as large-scale integration (LSI). That was an era of very large-scale integration (VLSI) and super large-scale integration (SLSI). The lines of demarcation between these different scales of integration are ill-defined and arbitrary. With the advancement of computer technology, the digital computer as well as the analog computer used semiconductor devices. Digital computers are more popular than analog computers, since analog signal solutions can also be easily obtained by calculation through the difference differential equation of a function.

5.1 Analog Computer

The heart of an analog computer is the operational amplifier. Operational amplifiers (op-amps) are commercially available in the form of ICs. These amplifiers are generally intended to operate linearly rather than in a digital switching mode. However, op-amps do serve as components in sample and hold circuits in multiplexers and in circuits used for analog to digital conversion and vice versa.

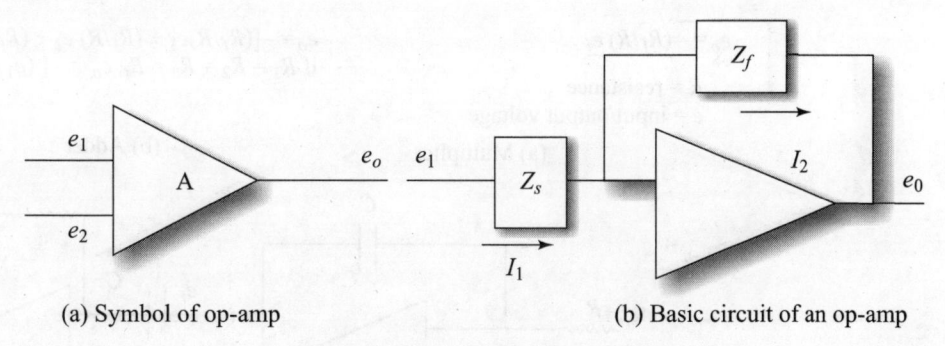

(a) Symbol of op-amp   (b) Basic circuit of an op-amp

Fig. 5.2 The operational amplifier

As shown in Fig. 5.2 (a), an op-amp has two inputs, e_1 and e_2, to accommodate two input signals and a single output signal e_o. The output voltage of an op-amp is equal to the difference between the two input voltages multiplied by a high voltage gain. The voltage gain at low frequency is a large real positive number, and can be as large as 5×10^4.

Referring to Fig. 5.2(b), Z_s is the series impedance and Z_f is the feedback impedance. Using Kirchhoff's law,

Current $\quad I_1 = \dfrac{e_1 - e_2}{Z_s} \quad$ and $\quad I_2 = \dfrac{e_o - e_g}{Z_f}$

Since an op-amp is a high-gain element, there is virtually no current flow. Hence $I_1 = I_2$. Equating the current and rearranging, we get

$$e_o = \frac{Z_f}{Z_s} e_i + \frac{Z_f}{Z_s} e_g + \frac{Z_f}{Z_s} e_g$$

Since $e_g = e_o/A$ and A has very large value,

$$e_o = -\frac{Z_f}{Z_s} e_i$$

This is the basic principle of an op-amp. With appropriate selection of feedback impedance and series impedance, it can be used as a multiplier, adder, integrator, or differentiator. Figure 5.3 illustrates the applications of the op-amp for various operations.

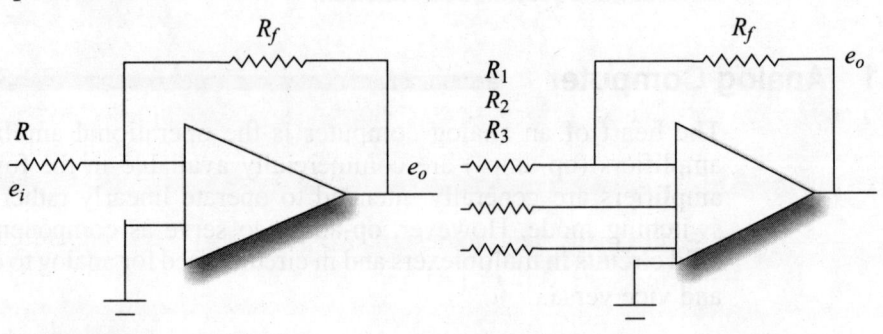

$e_o = -(R_f/R) e_i$

R = resistance
e = input/output voltage

(a) Multiplier

$e_o = -[(R_f/R)e_1 + (R_f/R) e_2 + (R_f/R) e_3)]$
if $R_1 = R_2 = R_3 = R_f$, $e_o = -[(e_1 + e_2 + e_3]$

(b) Adder

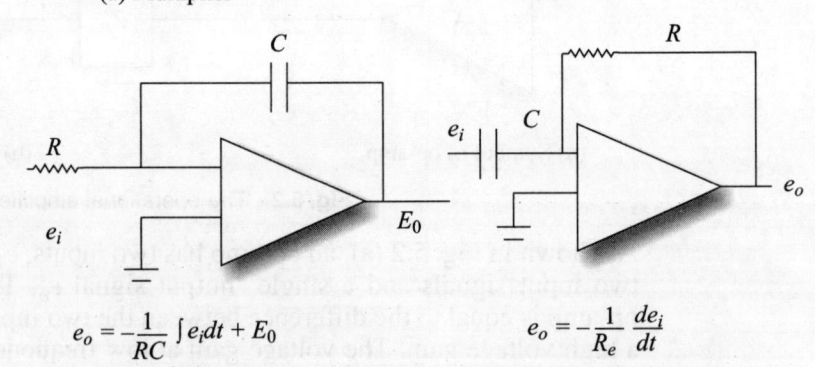

$e_o = -\dfrac{1}{RC} \int e_i dt + E_0$

E_0 – The constant of integration, initial charge of the capacitor.

(c) Integrator

$e_o = -\dfrac{1}{R_e} \dfrac{de_i}{dt}$

(d) Differentiator

Fig. 5.3 Applications of the op-amp

To generate a function $y = t^2$, the op-amp can be used to rig up an analog computer. As a first step, a differential equation has to be obtained for a function $y = t^2$. The differential equation is $d^2y/dt^2 = 2$.

The next step is to decide the number of op-amps required. Here, the order of the differential equation is 2. The analog output t^2 can be obtained with two integrators. Figure 5.4 shows the analog computer diagram to generate t^2.

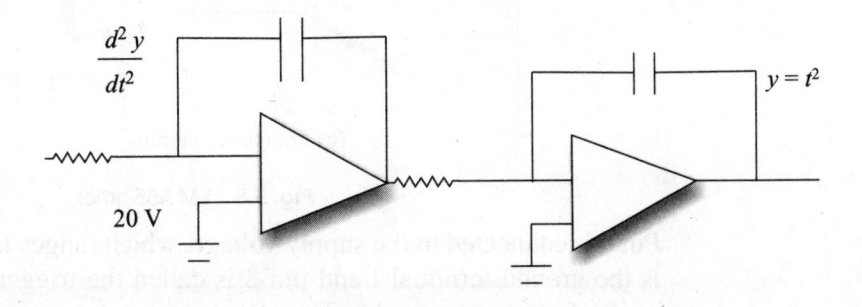

Fig. 5.4 Generation of function t^2 (The resistance and capacitance are 1 Ω and 1pF, respectively.)

5.2 Timer 555

There are many ways to design an oscillator. An oscillator circuit is available in IC form as NE 555 or LM 555, which is called a timer. These can be used for precision time keeping, pulse generation, sequential timing, time delay generation, pulse width modulation, etc.

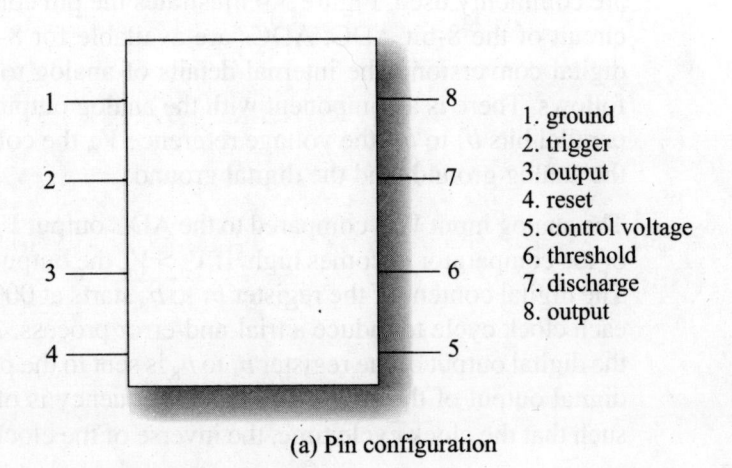

1. ground
2. trigger
3. output
4. reset
5. control voltage
6. threshold
7. discharge
8. output

(a) Pin configuration

(b) Equivalent circuit

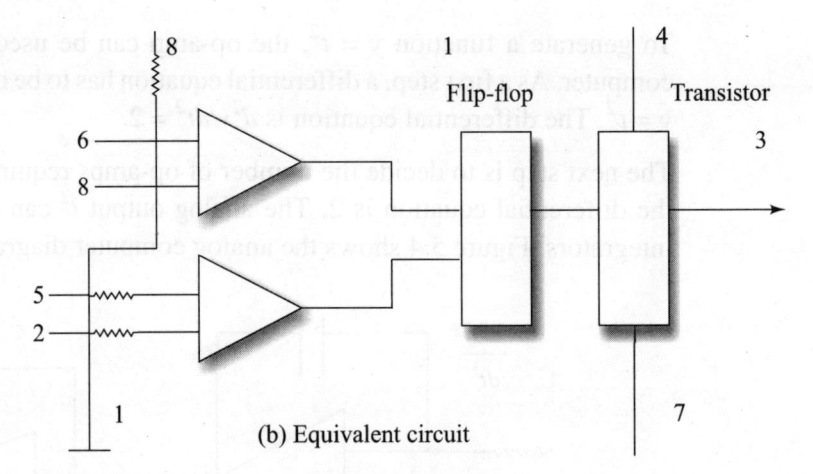

Fig. 5.5 LM 555 timer

Pin 8 is connected to the supply voltage, which ranges from + 5 V to 8 V. Pin 1 is the ground terminal 1 and pin 2 is called the trigger terminal. It is an ON switch for the output pulse. The output is taken from pin 3. Figure 5.5 illustrates the pin configuration and equivalent circuit of an LM 555 timer.

5.3 Analog to Digital Conversion

Analog to digital converters (ADCs) are required to convert analog signals from sensors into a digital form, compatible with a personal computer's digital input. Generally, analog to digital converters are complex and sophisticated systems. There are many possible schemes for A/D conversion and many variations are possible within each scheme. Comparators with register ADCs in chip forms are commonly used. Figure 5.6 illustrates the pin configuration and equivalent circuit of the 8-bit ADC. ADCs are available for 8-, 10-, 16-, 22- and 24-bit digital conversion. The internal details of analog to digital conversion are as follows. There is a component with the analog output V_o, outputs as the digital parallel bits b_1 to b_8, the voltage reference V_r, the collector–ground voltage V_s, the analog ground, and the digital ground.

The analog input V_i is compared to the ADC output V_o and if $V_o < V_i$, the output of the comparator becomes high. If $V_o > V_i$, the output of the comparator is low. The digital content of the register b_1 to b_8 starts at 00000000 and is modified in each clock cycle to induce a trial-and-error process. At the end of this process, the digital output of the register b_1 to b_8 is sent to the pins 13 to 20 as the parallel digital output of the ADC. The clock frequency is often from 0.25 to 80 MHz such that the clock cycle time, the inverse of the clock frequency, is 0.0125 ms.

The conversion of an analog signal into an n-bit digital number leads to a representative that has 2^n distinct digital values. The quantization error or resolution corresponds to the smallest change in the digital conversion. In the case of input voltages V_i larger than V_N, the conversion is saturated by V_N and these input voltages are represented by the largest digital n-bit number. This error due to saturation can be reduced by signal conditioning to match the range of input voltages with one of the selectable scales such that the maximum $V_i < V_N$.

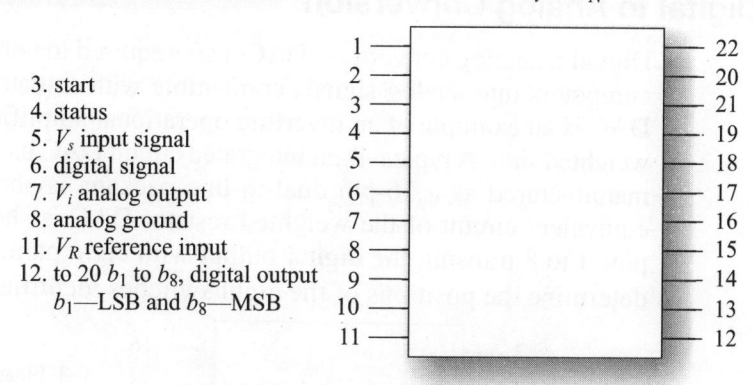

3. start
4. status
5. V_s input signal
6. digital signal
7. V_i analog output
8. analog ground
11. V_R reference input
12. to 20 b_1 to b_8, digital output
   b_1—LSB and b_8—MSB

(a) Pin assignment for 8-bit ADC

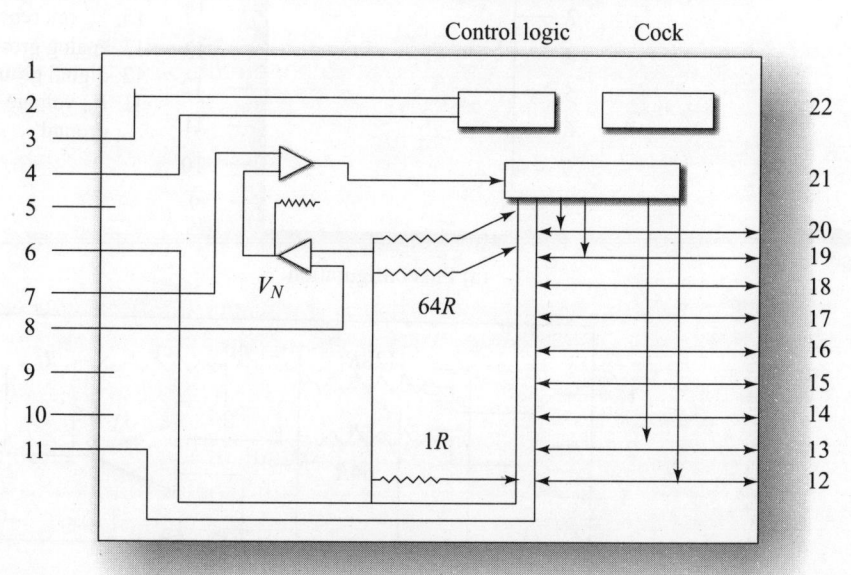

(b) Equivalent circuit

Fig. 5.6 The 8-bit ADC

In the case of input voltages V_i with maximum $V_i < V_N$, the part of the scale between maximum V_i and V_N is not used and the resolution of the digital conversion is reduced. Consequently the analog input voltage V_i has to be normalized on a $0-V_N$ scale, where V_N is the upper limit of the scale. The ideal selection is maximum $V_i \approx V_N$. Often, ADC boards have selectable scales with V_N of 0.1, 1, or 10 V. V_N is determined by the full-scale range voltage given by $V_{\text{fsr}} = 2R_f V_R/R$, where R_f is the feedback resistance of the summing op-amps. Given that the resolution r for a nominal range V_N is the same as the analog output voltage for the LSB, it follows that $r = 2R_f V_R/2^n R$. For V_N chosen equal to $V_{fsr} = 2R_f V_R/R$, r is equal to $V_N/2^n$.

5.4 Digital to Analog Conversion

Digital to analog converters (DACs) are required to convert digital output from computers into analog signals compatible with actuators. A weighted resistor DAC is an example of an inverting operational amplifier used for computing a weighted sum. A typical 8-bit integrated circuit DAC based on weighted resistors manufactured as a 16-pin dual-in-line package is shown in Fig. 5.7(a). The equivalent circuit of the weighted resistor DAC is shown in Fig. 5.7(b). The pins 1 to 8 transmit the digital output with eight parallel bits to the DAC and determine the positions of the eight switches identified by the same symbols.

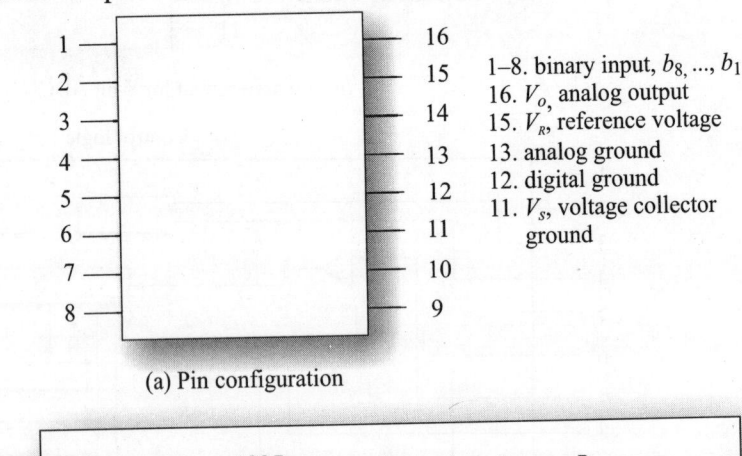

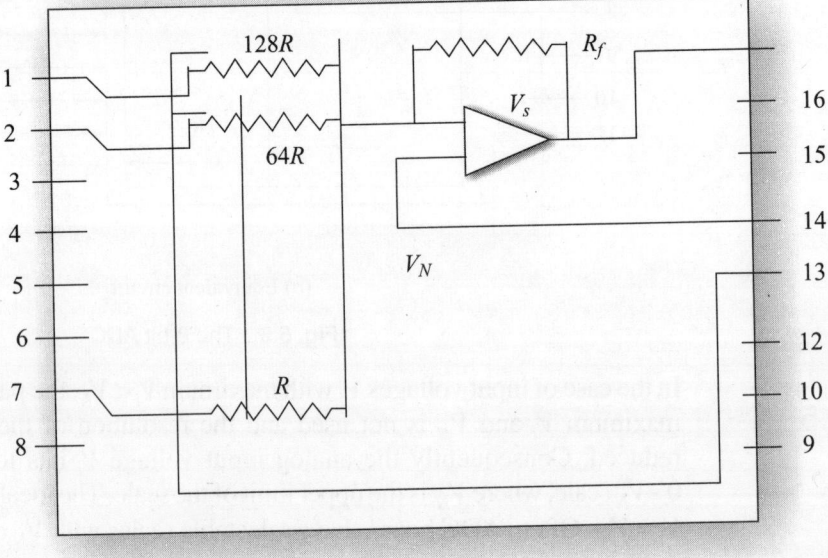

1–8. binary input, $b_8, ..., b_1$
16. V_o, analog output
15. V_R, reference voltage
13. analog ground
12. digital ground
11. V_s, voltage collector ground

(a) Pin configuration

(b) Equivalent circuit

Fig. 5.7 Digital to analog converter

For the value b_i ($i = 1 - 8$), the corresponding switch is connected by a thick line to the common digital ground, pin 12. There is a resistance $2^{i-1}R$. For example, between the reference voltage V_R and the input voltage V_N, for the op-amp, for $b_1=1$, the resistance is $2^0R = R$, while for $b_2 = 0$, the resistance is infinity. For $b_7 = 1$, the resistance is $26R$ and changes to $64R$. The relationship between the input V_R and the output V_o of the inverting amplifier is obtained as

$$V_o = -\frac{R_f}{R} V_R$$

where in this case R_f is the equivalent resistance resulting from all parallel resistances associated with switches in the position corresponding to $b_i = 1$ ($i = 1 - 8$), such that

$$\frac{1}{R_1} = \frac{b_1}{R} + \frac{b_2}{2R} + \cdots + \frac{b_8}{128R}$$

$$\frac{1}{R_1} = \left(\frac{1}{128R}\right)(2^7 b_1 + 2^6 b_2 + \cdots + 2^0 b_8)$$

The foregoing input–output relationship becomes

$$V_o = -\left(\frac{2RV_R}{R}\right)\left(\frac{b_1}{2} + \frac{b_2}{2^2} + \cdots + \frac{b_8}{2^8}\right)$$

A binary number $b_1, b_2, \cdots, b_8$ has the decimal equivalent $(2^7 b_1 + 2^6 b_1 + \cdots + 2^0 b_8)$. The bit b_1 is the most significant bit (MSB) and the bit b_8 is the least significant bit (LSB). The resolution has the same value as the V_o LSB, given the LSB that corresponding to the smallest change in output voltage V_o. For an n-bit DAC,

$$V_o = -\left(2R + \frac{V_R}{R}\right)(2^{n-1} b_1 + 2^{n-2} b_2 + \cdots + 2^0 b_n)\frac{1}{2^n R}$$

$$= -\left(2R\frac{V_R}{R}\right)\left(\frac{b_1}{2} + \frac{b_2}{2^2} + \cdots + \frac{b_n}{2^n}\right)$$

The full-scale range voltage is defined as

$$V_{\text{fsr}} = 2R_f V_R/R$$

5.5 Digital Computer

The abacus is a primitive calculator made up of beads. It was first made in 5000 BC in China. It was the first calculating machine. In about 500 BC, the Greeks began to use the abacus. From the fifteenth century to the beginning of the twentieth century, various calculating machines which used mechanical elements were developed. The first solid state computer was made in 1960. In 1960, computers were accessible and affordable only to large corporations, big universities, and government agencies. Because of advancement in semiconductor technology, the million dollar computing capacity of the 1960s is now available for less than ten dollars in an integrated circuit called the microprocessor. A computer working on a microprocessor is called a microcomputer. Microcomputers are nowadays making an impact on every aspect of our lives and in industrialized societies. The basic structure of a microcomputer is not different from any other computer. The computers communicate and operate using digital signals. Each computer can accept a set of instructions in the form of a binary pattern called a machine language. Since it is difficult for people to communicate with the computer in the language of 0's and 1's, binary instructions are given abbreviated names, called mnemonics, which form the assembly language for a given computer.

A digital computer is a multipurpose, programmable machine that reads binary instruction from its memory, accepts binary data as input, processes data according to those instructions, and provides results as output. The physical components of a computer are called hardware. A set of instructions written for the computer to perform a task is called a program, and a group of programs is called software.

The Motorola 6800 microprocessor has a set of binary codes and mnemonics entirely different from those of 8085/8080A. An assembly language program written for a microprocessor is not transferable to a computer with another microprocessor unless the two microprocessors are compatible in their machine codes. Machine language and assembly language are considered as low-level languages. In the case of assembly language, mnemonics get translated into a binary code using an assembler. The programming languages that are intended to be machine-independent are called high-level languages. These include such languages as FORTRAN, BASIC, PASCAL, and COBOL. The English words are translated into binary language through another program called the compiler or interpreter. The compiler or interpreter requires a large memory space. Different types of computers are designed to serve different purposes. The microcomputer is defined as a computer with microprocessors as the central processing unit. Large computers have various types of microprocessors performing different functions within a large CPU. In 1994, the fifth generation computer was established. Such a computer was called a supercomputer. It is smart, fast, and performs large computations. Typically a supercomputer accomplishes 64,000 processes per second. A microcomputer system consists primarily of four components, the microprocessor, memory, input, and output.

5.6 Architecture of A Microprocessor

The microprocessor manipulates data, controls the timing of various operations, and communicates with input/output devices and memory. The internal logic design of a microprocessor is called its architecture. The microprocessor consists of three basic units. The basic units are the central processing unit (CPU) to recognize and carry out program instructions, the input and output circuit interface, and the memory to retain the program instructions and data. The input/output arrangement is used to send out or feed data. The data bus is used to transfer a word to and from the CPU. The address bus bar carries signals which indicate what data is to be found and selected. The control bus is the means by which signals are sent to synchronize the separate elements. In microprocessors, the memory and various input and output arrangements are all arranged in one chip called the microcontroller. Different types of microprocessors are available. Commonly, a single-chip microprocessor is used. The different types of microprocessors are 8085, 8086, z80, and 68000. Figure 5.8 shows the basic structure of a microprocessor and pin configuration and internal structure of the 8085 chip.

The internal architecture of the 8085 microprocessor determines how an operation can be performed with the given data and also which operations can be performed. These operations are (a) storing 8-bit data, (b) performing arithmetic and logical operations, (c) testing for conditions, (d) sequencing the execution of instructions, and (e) storing data temporarily during execution in the defined read/write memory location.

To perform these operations, the microprocessor requires registers, an arithmetic logic unit, a control logic unit, and internal buses. The 8085 is an 8-bit general-purpose microprocessor capable of addressing 64K of memory. The device has 40 pins, requires a +5 V single power supply, and can operate with a 3-MHz single phase clock.

In the signal line b_0 to b_7, all the b_i's are bidirectional. They are used for a dual purpose, as the low-order address bus as well as the data bus. This is also known as multiplexing the bus. The control groups of signals include two central signals RD and WR, and three status signals IO/M, S_1, and S_0, which identify the nature of the operation. One special signal, ALE, indicates the beginning of the operation. RD reads the control signal. This signal indicates the selected input/output or memory devices to read and data availability in the address bus bar. WR writes data on the data bus in a selected memory or input/output (I/O) location. IO/M is a status signal used to differentiate between I/O and memory operations. When it is high, it indicates input/output operation and when it is low, it indicates memory operation. S_1 and S_2 are status signals, similar to IO/

92 Introduction to Mechatronics

M, which can identify various operations; but they are rarely used in small systems.

(a) Structure of a microprocessor

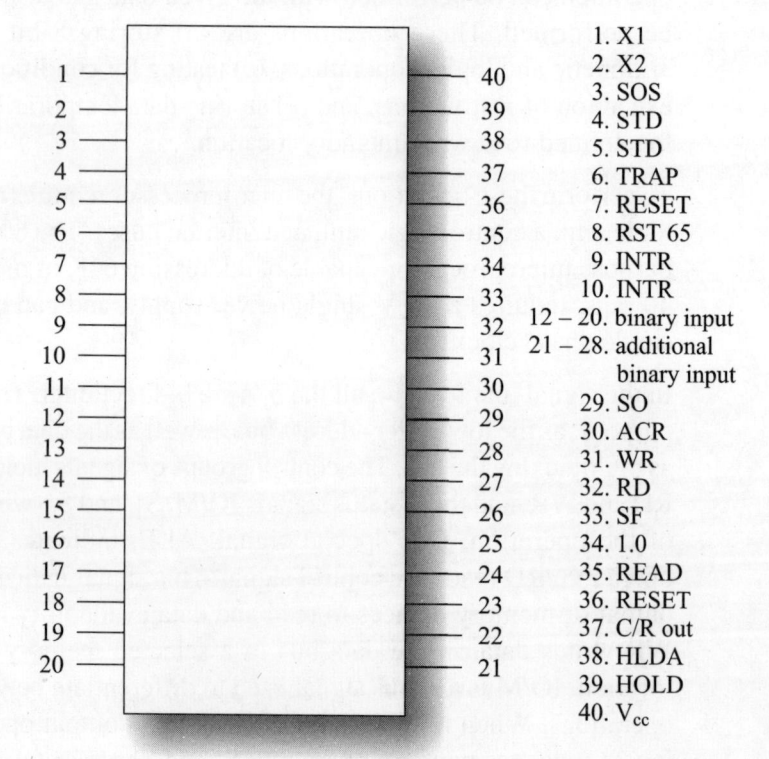

(b) Pin configuration of 8085

1. X1
2. X2
3. SOS
4. STD
5. STD
6. TRAP
7. RESET
8. RST 65
9. INTR
10. INTR
12 – 20. binary input
21 – 28. additional binary input
29. SO
30. ACR
31. WR
32. RD
33. SF
34. 1.0
35. READ
36. RESET
37. C/R out
38. HLDA
39. HOLD
40. V_{cc}

Registers   Decoder

ALU

I/O bus

Address bus

Control bus

Timer

(c) Structure of 8085

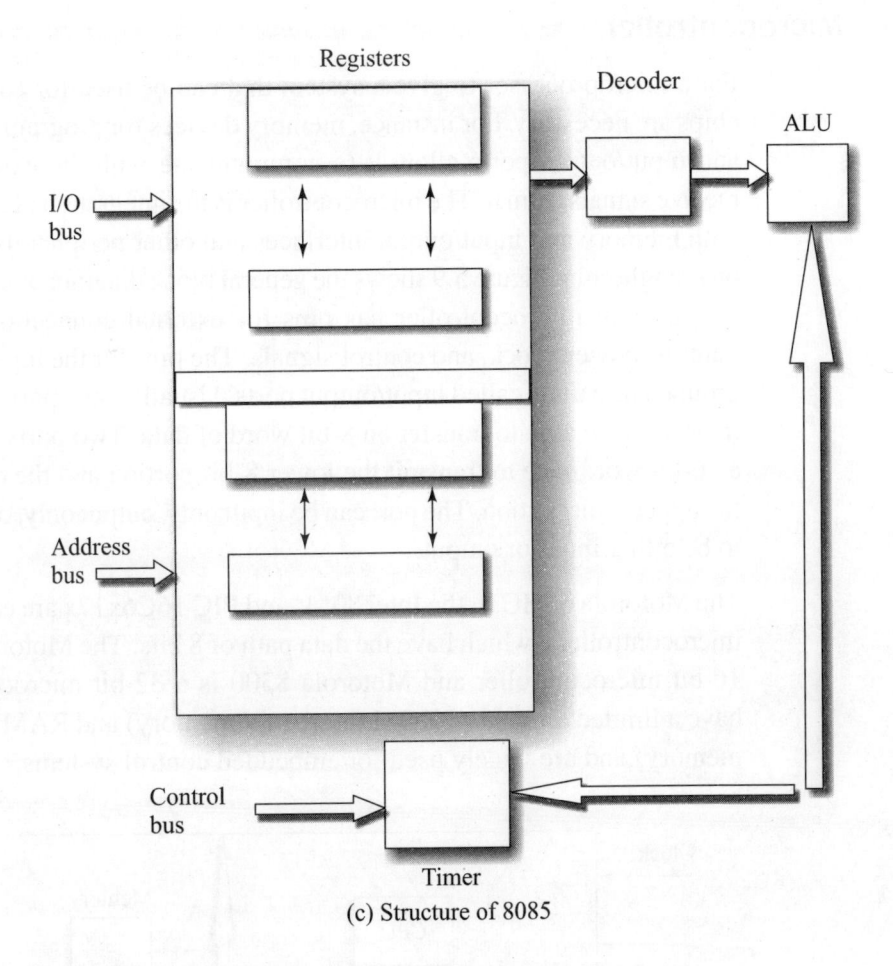

Fig. 5.8 The 8085 microprocessor

V_{cc} is + 5 V and V_{ss} is the ground. X_1 and X_2 are crystals connected to the pin. The frequency is internally divided into two. To operate at 3 MHz, the crystal should have 6 MHz. CLK(out) can be used as the system clock for other devices.

The 8085 microprocessor has five interrupt signals that can interrupt a program execution. INTRA is an interrupt request. It acknowledges the interrupt signal. In addition to the interrupt, three pins—RESET, HOLD, and READY—accept externally initiated signals. INTRA responds to HOLD request. INTRA acknowledges the same kind of signal from HLDA. When the RESET in the signal becomes low, the program counter is set to zero. RESET OUT indicates that the MPU is reset. The signal can be used to reset other devices. For serial transmission, SID (serial input data) and SOD (serial output data) are used.

5.7 Microcontroller

For a microprocessor to give a system that can be used for control, additional chips are necessary. For instance, memory devices for program and data storage and input/output ports allow it to communicate with the external world and receive signals from it. The microcontroller is the integration of a microprocessor with memory and input/output interfaces and other peripherals such as a timer on a single chip. Figure 5.9 shows the general block diagram of a microcontroller. The general microcontroller has pins for external connection of inputs and outputs, power, clock, and control signals. The pins for the input and output are grouped into units called input/output ports. Usually, such ports have eight lines in order to be able to transfer an 8-bit word of data. Two ports may be used for a 16-bit word, one to transmit the lower 8-bit portion and the other to transmit the upper 8-bit portion. The port can be input only, output only, or programmable to be either input or output.

The Motorola 68HC II, the Intel 8051, and PIC 16C6x17x are examples of 8-bit microcontrollers, which have the data path of 8 bits. The Motorola 68HC16 is a 16-bit microcontroller and Motorola 8300 is a 32-bit microcontroller. These have a limited amount of ROM (read only memory) and RAM (random access memory) and are widely used for embedded control systems.

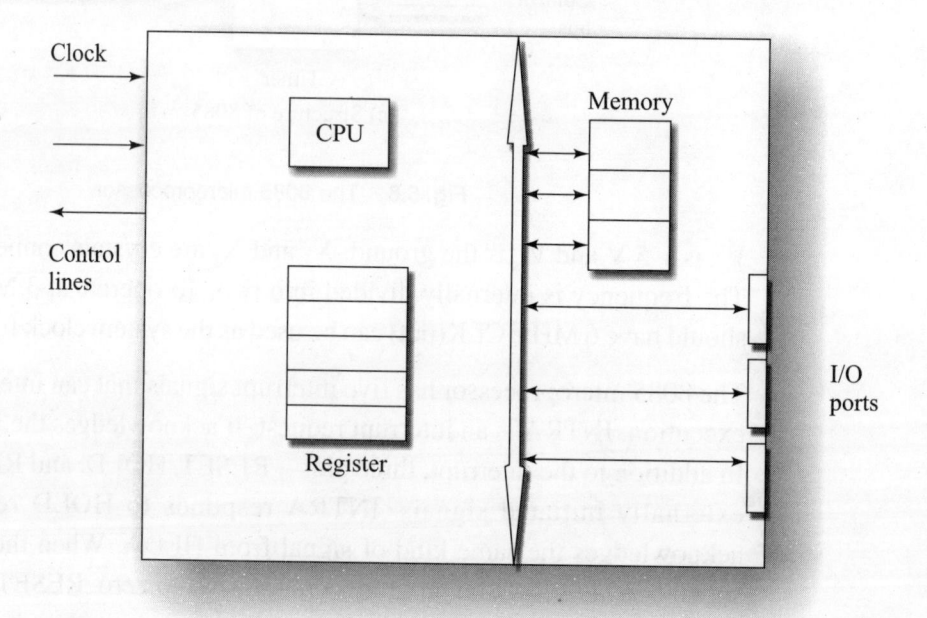

Fig. 5.9 Block diagram of a microcontroller

A microprocessor system with separate memory and input/output chips is more suited for processing information in a computer system. While selecting a microcontroller, the following factors are considered.

1. Number of input/output pins
2. Interfaces required
3. Memory requirement
4. Number of interrupts required
5. Processing speed required

Microcontrollers are used in temperature-measuring systems, domestic washing machines, and many industrial applications.

5.8 Programmable Logic Controller

A programmable logic controller (PLC) is a digital electronic device that uses a programmable memory to store instructions and to implement functions such as logic, sequence, timing, counting, and arithmetic in order to control machines and processes. Such controllers have been specifically designed to make programming easy. The term 'logic' is used because the programming is primarily concerned with the implementation of logic and switch operations. Input devices such as switches and output devices such as motors being controlled are connected to the PLC and then the controller monitors the input and output according to the program stored in the PLC by the operator and so controls the machine or process. The PLC has the great advantage that it is possible to modify a control system without having to remove the connections to the input and output devices, the only requirement being that an operator has to key in a different set of instructions. PLC-operated systems are much faster than relay-operated systems. A PLC is smaller than a computer, but has certain features given below that are specific to their use as controllers.

1. They are rigid and designed to withstand vibration, temperature, humidity, and noise.
2. The interfacing circuit for input and output is inside the controller.
3. They are easily programmed and have an easy programming language. Programming is primarily concerned with logic and switching operations.

PLCs were first conceived in 1968. They are now widely used and extend from small self-contained units for use with perhaps 20 digital input/output to modular systems that can be used for a large number of input/output units. They also handle digital and analog input/output.

The input/output channel provides signal conditioning and isolation functions so that sensors and actuators can generally be directly connected without need for other circuitry. The common output voltages are 24 V and 240 V. The outputs are specified as relay type, transistor type, and triac type. In the relay- type output, the signal from the PLC output is used to operate a relay and so is able to switch currents of the order of few amperes in an external circuit. The relay isolates the PLC from the external circuit and can be used for both AC and DC switching. The relays are relatively slow to operate. The transistor-type output uses a transistor to switch current through the external circuit. This gives a faster switching action. Optical isolators are used with transistor switches to provide isolation between the external circuit and the PLC. The transistor output is only for DC switching. A triac can be used to control external circuits, which are connected to the AC power supply.

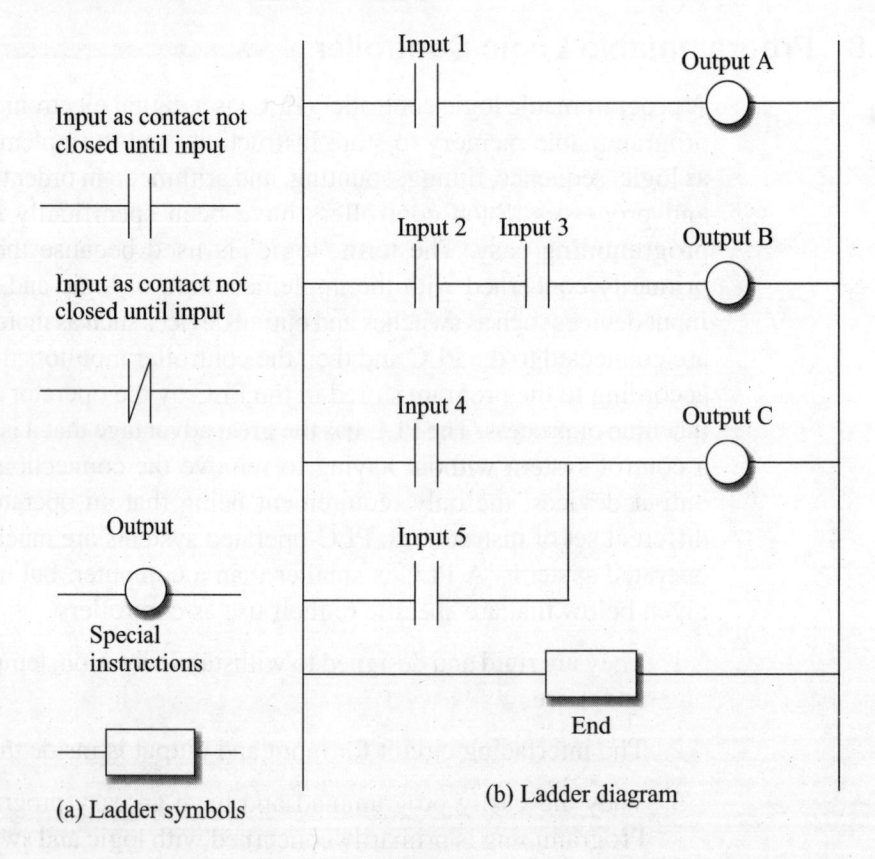

Fig. 5.10

(a) Ladder symbols   (b) Ladder diagram

PLC programming based on the use of the ladder diagram involves writing a program in a manner similar to drawing a switching circuit. The ladder diagram consists of two partial lines representing the power rails. The circuits are connected as horizontal lines. The rugs of the ladder are between the vertical lines. Figure 5.10 shows the basic standard ladder symbols as well as examples of rugs in a ladder diagram. In drawing the circuit line for rugs, inputs must always precede outputs and there must be at least one output on each line. Each rug must start with an input or a series of inputs and end with outputs. The inputs and outputs are members of the ladder diagram. The notations are used depending on the PLC manufacturer. The Mitsubishi F series of PLCs precedes the input element by an X and the output element by a Y as shown in the following example:

Inputs   X400-407, 410-413

          X500-507, 510-513   (24 possible inputs)

Outputs Y430-437

          Y530-537   (16 possible inputs)

5.9 Computer Peripherals

The purpose of the input/output section of a computer is to provide communication with the variety of peripheral devices used with the computer system. Input/output devices have two inverse functions. First, it must interpret the incoming signals and hold them temporarily until they are placed in the main memory of the CPU. Second, the result of the calculation and data processing operation must be transmitted to the appropriate peripheral equipment.

5.9.1 Hardware Devices Used for Computer Storage

The programs and data files are not generally kept in primary storage, but are stored on large-capacity auxiliary devices and loaded in to the main memory as required. The auxiliary devices constitute the secondary storage and are physically external to the computer, which means that the programs and data files are not directly available to the CPU. There are two basic types of secondary storage: (a) sequential access storage and (b) direct access storage.

The sequential storage method has a substantially lower access rate than the direct access storage method. The magnetic tape storage is a prime example of the sequential access storage technology. Data is stored on magnetically coated Mylar tape, similar to the magnetic tape used in audio systems. The recording and reading data are nondestructive, which means that the tape can be erased and reused. Since data are stored sequentially, the access time is relatively slow. However, the low cost per bit and the high capacity of magnetic tapes make it ideal for system backup.

The magnetic drum is a random access storage device with high capacity and high access rates. It consists of a magnetically coated cylinder. During operation, the drum rotates at a constant speed and data are recorded in the form of magnetized spots. The drum can be read repeatedly without causing data loss. Read/write heads are used to read data to and from the drum as it rotates. The drum surface is divided into tracks, each with its own read/write head.

The magnetic disk storage is also a direct access storage device. The storage medium is a magnetic disk. There are several types and sizes of disks, each best suited to a particular set of applications. The flexible floppy diskette comes in several standards—3½ inch, 5½ inch, 8 inch, etc.—and is packaged in a square plastic envelope to protect the magnetic surfaces. Reading and writing are accomplished through access holes in the envelope. Floppy disks are available with either one or both sides used for storing data. The hard disk is a thin metal disk which is coated on both sides with magnetic ferric disks. The data are recorded in the form of magnetized spots on tracks on the disk surface. Several disks are combined into a single disk pack and these are separated by a fixed distance and joined by a vertical shaft. The disk pack is rotated at several thousand revolutions per minute by a disk drive unit. The data are transferred by many sets of read/write heads (one per recording surface) to the appropriate track. The rate of operation is usually several thousand bytes per second.

Bubble memory consists of microscopic magnetic bubbles on a thin crystalline magnetic film. It has high storage density and random address capabilities. However, because of its high cost, the technology is not yet competitive with other memory technologies.

Optical data storage using laser beams is widely used in the form of compact disks (CDs). The data are stored as microscopic bits on a metallic surface. The data are read by directing a laser beam across the surface and measuring the wavelength of light reflected. This reflected light can be converted into electrical pulses. The storage medium is metal film strips, arranged into tracks.

R/WM (read/write memory)

Read/write memory is popularly known as random access memory (RAM). This memory is volatile, meaning that when power is turned off, all the contents are destroyed. Two types of R/W memories are available–static and dynamic. Static memory is made up of flip-flops and stores a bit as voltage. Dynamic memory is made of metal oxide silicon (MOS) transistor gates and stores a bit as a charge.

ROM (read only memory)

The ROM is nonvolatile, meaning that it retains the stored information even if the power is turned off. ROM uses diodes arranged in a matrix format. Four types of ROM are presently available, MASKED ROM, PROM, EPROM, and EEPROM.

Masked ROM

In this ROM, a bit pattern is permanently recorded by the masking and metallization process, which is an expensive and specialized process. The manufacturer produces quantities in thousands, considering the economy.

PROM (programmable read only memory)

This memory has nichrome or polysilicon wires arranged in a machine. The wires can be functionally varied as diode or fuse. This memory can be programmed by the user with a special PROM program that selectively burns the fuse according to the bit pattern required to be stored.

EPROM (erasable programmable read only memory)

The information stored in this memory is semi-permanent. All the information can be erased by exposing the memory to ultraviolet light through a quartz window installed on the chip. The memory chip can be reprogrammed again and again.

EEPROM (electrically erasable PROM)

This memory is functionally similar to EPROM, except that the information can be altered by using electrical signals at the register level rather than erasing all the information. At present, it is expensive and not commonly used. The interface of 1024(1K) bytes read/write memory to an 8085 system, with the many maps from 3000H to 33FFH, uses 2114(1024×4) memory chips and the 74LS318, 32.3 to 8 decoder. The memory chip 2114 is a static memory organized in 1024×4 formats, which means that it has 1024 registers with four input–output lines. Two chips must be connected in parallel to form an 8-bit memory.

5.9.2 Computer Input Devices

The operator input devices are provided with the computer to facilitate convenient communication between the user and the system. Punched cards have been used more during its long existence than any other data recording medium. There are two types of punched cards—the 80-column Hollerith card and the 96 column IBM card.

There are two types of card readers currently in use—the brush reader and the photoelastic reader. In the first case, the punched cards are moved past a cell of conductive brushes. In the latter case, the devices use a cell of photocells and a light source which shines through the hole in the card to produce electrical pulses. The magnetic tape units were introduced because of the demand for fast input/output devices. They can be used for program and data storage and can be interfaced to the computer as both input and output units.

Basically, input devices are classified as (a) cursor control devices, (b) digitalizers, and (c) alphanumeric and other key-based terminals.

The cursor normally takes the form of a bright spot or an arrow on the cathode ray tube screen. The computer is capable of reading the current position of the cursor. A variety of cursor control devices have been employed in computer systems. These include thumbwheels, direction keys on a keyboard terminal, the joystick, trackball, cursor, light pen, and electronic tablets. The first five items in the list provide control over the cursor without direct physical contact on the screen by the user. The last two devices in the list require the user to control the cursor by touching on the screen.

The thumbwheel devices use two thumbwheels, one to control the horizontal position of the cursor, and the other to control the vertical position. The cursor in this arrangement is often represented by the intersection of a vertical and a horizontal line displayed on the screen. The cursor can also be controlled by using the direction keys on the key branch. Four keys are used for each of the four directions in which the cursor can be moved—right or left and up or down. The joystick apparatus consists of a box with vertical toggle state that can be pushed in any direction to cause the cursor to be moved in that direction. The joystick gets its name from the control stick that was used in aeroplanes. The track ball and mouse operation is similar to that of the joystick except that, in this case, an operator ball is rotated to move the cursor in the desired direction on the screen. In the mouse, the body is moved to allow for rotation of the ball due to surface friction, which in turn allows the cursor to move.

The light pen is a pointing device in which the computer seeks to identify the position where the light pen is in contact with the screen. The light pen is a detector of the light on the screen and uses a photodiode, phototransistor, or some other form of light sensor. The tablet and pen in a computer describe electronically sensitive tablets which work in conjunction with an electronic stylus. The light pen and tablet/pen are used for other inputs such as selecting a function from the menu and selecting an operation on the screen for enlargement as well as cursor control.

The digitizer is an operator input device consisting of a large, smooth band and an electronic tracking device which can be moved over the surface to follow the existing lines. It is a common technique for tracking X and Y coordinates from a paper drawing. Such devices contain a switch for the user to record the desired X and Y coordinate positions. The digitizer can be used to digitize line drawings. The user can input data from a rough schematic or long drawing and edit the drawing to the desired level of accuracy. For computer aided drawing, a three-dimensional digitizer can be made use of to draw a wire frame model three-dimensional objects.

Several forms of keyboard terminals are available as input devices for the computer. The most familiar type is the alphanumeric terminal. Such terminals are used to enter commands, functions, and supplemental data to the computer. Some computers make use of special function keyboards to eliminate extensive typing commands. The number of function keys varies from 8 to 80. The transmission speeds are usually selectable from 110 to 960 band. A band is a unit representing the number of discrete signal changes per second. For a binary system, it is equal to number of bits per second.

5.9.3 Computer Output Devices

All the computer terminals available today use the cathode ray tube (CRT) as the display. The operation of the CRT is illustrated in Fig. 5.11. A cathode emits a high-speed electron beam onto a phosphor-coated glass screen. The electrons

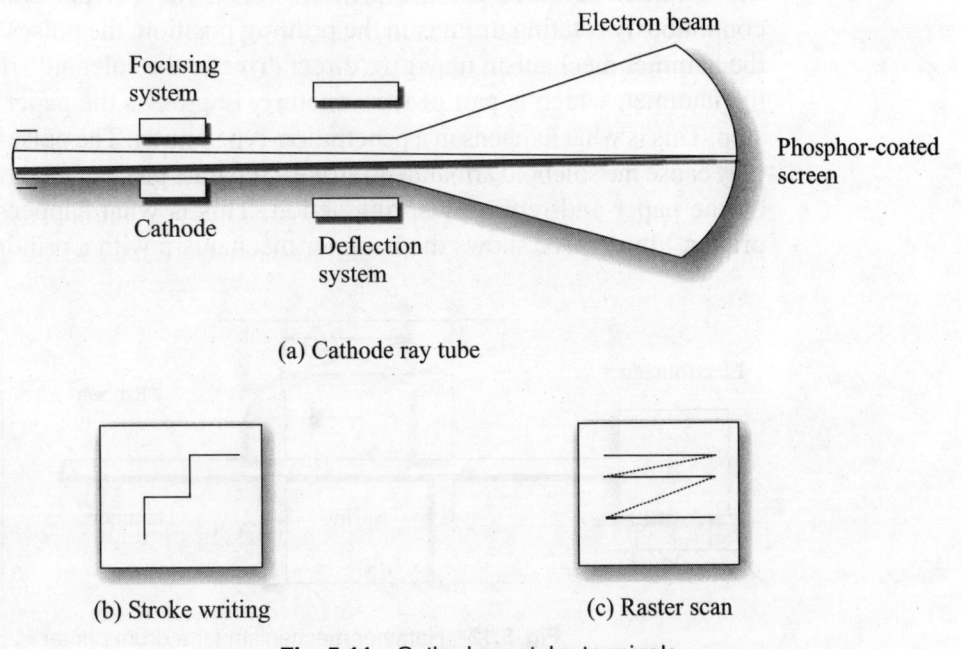

(a) Cathode ray tube

(b) Stroke writing

(c) Raster scan

Fig. 5.11 Cathode ray tube terminals

energize the phosphor coating, causing it to glow at the points where the beam makes contact. By focusing the electron beam, changing its intensity, and controlling its point of contact against the phosphor coating through the use of a deflection system, the beam can be made to generate a picture on the CRT screen. There are two basic techniques used in current computer terminals and CRT screens. They are (a) stroke writing and (b) raster scan.

The stroke writing technique includes line drawing, random positioning, vector writing, stroke writing, and directed beam. The stroke writing system uses an electron beam, which operates like a pencil to create a line image on the CRT screen. The image is constructed out of a sequence of straight line segments.

There are various types of output devices used for permanent record in conjunction with the cathode ray tube terminal. The line printer is the most commonly used terminal device used in computers. There are three types of line printers—the drum printer, the chain printer, and the optical printer.

The basic component of a drum printer is the drum with the complete character set embossed on its periphery. Opposite to each character position on the drum, there is a print hammer. The hammer is fixed and the required character is aligned with the hammer as the required output. The speed of the modern drum printer is in the range of 600 to 1300 lines/min. There are various types of print hammer mechanisms, and those with the least number of moving parts are the most reliable. Normally, a voltage is sent to operate a solenoid associated with the hammer selected for the printer. When the correct character on the continuously rotating drum is in the printing position, the pulses which operate the hammer mechanism may give direct drive to the solenoid armature, so that the hammer, which is part of the armature, impresses the paper up to a preset stop. This is what happens in a penetration-type printer. The pulse to the solenoid may cause the solenoid armature to strike a separate piece of metal which projects on the paper and return by spring action. This is what happens in a ballistic printer. Figure 5.12 shows the hammer mechanism with a printing hammer.

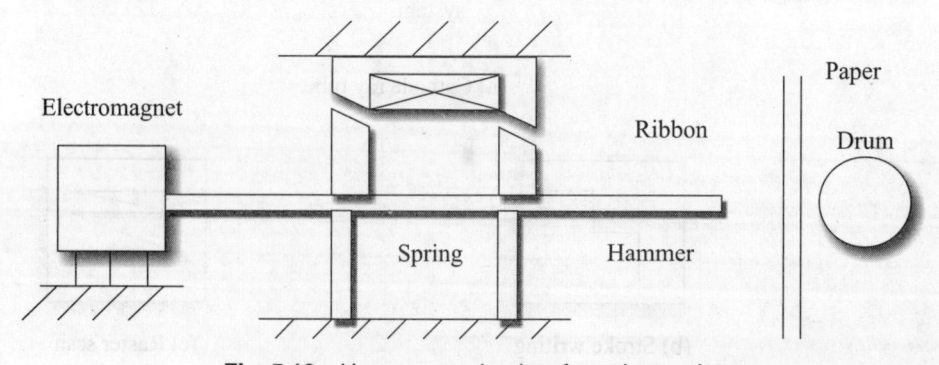

Fig. 5.12 Hammer mechanism for a drum printer

The chain printer has a revolving horizontal chain to carry the tape slugs along the line to be printed. As each character reaches its position, it is printed on the paper by an electromagnetically driven hammer striking the back of the paper in the position corresponding to the required character. The maximum operating speed of the chain printer in practice is somewhat lower than that of the drum printer. The time required between character alignments in a line is the same. The standard model IBM 1403 printer has 48 characters repeated five times around the chain. Each set has 16 slugs with 3 characters on each slug. The printer speed is 1100 lines/min.

In a typical data processing shop, the volume of output is sufficient to justify a high-speed line printer. Such a unit prints entire lines at one cycle, 80 characters or 132 characters per line, at the rate that may exceed 1000 lines/min. These units are expensive and the cost is generally proportional to the printing rates. The mosaic print of characters attached with line printers gives high-speed printing. The dot matrix printer is an example of a line printer. The characters are formed with a large number of dots. The dots form the pixels. The term pixel is used for the smallest addressable dot on any display device. 7×5 matrix dots build a character. The input data to the printer is usually in digital ASCII (American Standard Code for Information Interchange) format. It enables all the keyboard characters as well as some control functions. A dot matrix printer has a print head which consists of either 9 or 24 pins in a vertical line. Each pin is controlled by an electromagnet, which when turned on propels the pin on to the striking ribbon. This transfers a small blot of ink on to the paper behind the ribbon. A character is formed by moving the print head in horizontal lines back and forth across the paper and firing the appropriate pins. Figure 5.13 shows the working principle of a dot matrix printer.

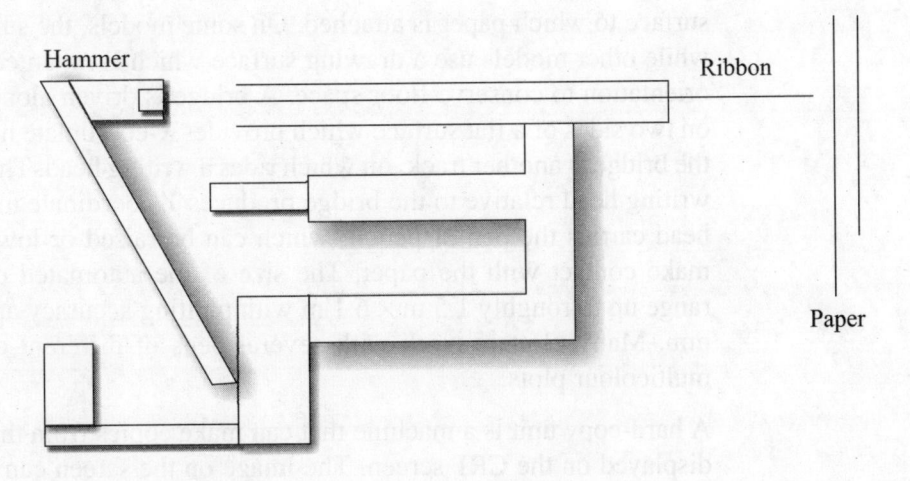

Fig. 5.13 Dot matrix printer head mechanism

Optical printers usually work on the following principle. A character to be printed is displayed on a cathode ray tube screen and is projected onto the surface of the drum, which is continuously rotating. The surface of the drum is light-sensitive and when a suitable powder is sprinkled on to the surface, it adheres to the pattern of the projected image of the character. A continuous sheet of paper makes contact with the drum at one point. The pattern of the character traced by the powder on the drum surface is transferred to the paper, giving a permanent image on the paper. The drum is wiped and the next character may be projected and printed.

For the graphic output from the computer, various types of output devices are used. The output devices are the (a) pen plotter, (b) hard-copy unit, and (c) electrostatic plotter and computer output to microfilm.

The accuracy and quality of a hard-copy plot produced by a pen plotter is better than that produced by a dot matrix printer. The pen plotter uses a mechanical ink pen using either wet ink or a ball point to write on paper through relative movement of the pen and paper. There are two basic types of pen plotters currently in use—(a) drum plotters and (b) flat-bed plotters.

The drum plotter is generally less expensive than the flat-bed plotters. It uses a round drum usually mounted horizontally, and a slide, which can be mounted along the track axially with respect to the drum. The paper is attached to the drum and the pen is mounted on the slide. The relative action between the pen and paper is achieved by coordinating the rotation of the motion of the slide. The drum plotter is faster than the flat-bed plotter and can make drawings of unlimited length. The width however is limited by the length of the drum. These lengths typically range from 216 mm to 1067 mm.

The flat-bed plotter is more expensive than the drum plotter. It uses a flat drawing surface to which paper is attached. On some models, the surface is horizontal, while other models use a drawing surface which is mounted in nearly vertical orientation to conserve floor space. A bridge is driven along the track located on two sides of a flat surface which provides X-coordinate motion. Attached to the bridge is another track, on which rides a writing head. The movement of the writing head relative to the bridge produces Y-coordinate motion. The writing head carries the pen or pencil, which can be raised or lowered as desired to make contact with the paper. The size of the automated drafting tables can range up to roughly 1.5 m × 6.1 m with plotting accuracy approaching ±0.025 mm. Many plotters work with several pens of different colours to achieve multicolour plots.

A hard-copy unit is a machine that can make copies from the same image data displayed on the CRT screen. The image on the screen can be duplicated in a matter of seconds. The hard copies produced from these units are not suitable as final drawings, because the accuracy and quality of reproduction is not as good as the output of a pen plotter. Most hard-copy units are dry silver copies that use light-sensitive paper exposed through a narrow CRT window inside the copier. The window is typically 216 mm × 12 mm.

The electrostatic copier offers a compromise between the hard copy and pen plotter in terms of speed and accuracy. It consists of series wire styli mounted on a bar which spans the width of charge sensitive paper. The styli have a density of up to 200 per linear 25 mm. The paper is gradually moved past the bar and the styli are activated to place dots on the paper. The dots overlap each other to achieve continuity. A limitation of the electrostatic plotter is that the data must be in the raster format. If the data is not in the raster format, some type of conversion is required to change them into the required format. The conversion mechanism is usually based on a combination of software and hardware.

The computer output to microfilm units (COM) reproduces the drawing on a microfilm. It is an expensive device but the advantage is its storage capability. The reduction in the size of each drawing to microfilm achieves a significant storage benefit. The microfilm can be easily retrieved and photographically enlarged to full size. The disadvantage of the COM process is that the user cannot write notes on microfilm as possible with a paper copy. The enlargement of the microfilm is not of as a high quality as that of the output from a pen plotter.

Illustrative Examples

Example 5.1 Convert the binary number 11100011 into its decimal equivalent.

Solution

$$1 \times 2^7 + 1 \times 2^6 + 1 \times 2^5 + 0 \times 2^4 + 0 \times 2^3 + 0 \times 2^2 + 1 \times 2^1 + 1 \times 2^0 = 227$$

Example 5.2 Convert the fractional binary number 0.1101 into its decimal equivalent.

Solution

$$1 \times (1/2^1) + 1 \times (1/2^2) + 0 \times (1/2^3) = 1 \times (1/2^4)$$
$$= 1 \times 0.5 + 1 \times 0.25 + 0 \times 0.125 + 1 \times 0.0625 = 0.8125$$

Example 5.3 Convert the binary number 10001101 into its hexadecimal equivalent.

Solution

Table 5.2 shows a four-bit binary equivalent of hexadecimal (HD) digits.

Table 5.2 Hexadecimal digits and equivalent binary digits

HD	0	1	2	3	4	5	6	7	8
Byte	0000	0001	0100	0011	0100	0101	0111	1000	1001

HD	A	B	C	D	E	F
Byte	1010	1011	1100	1101	1110	1111

From the above table, 1000 in the binary system is the equivalent of 7 in the hexadecimal system and 1101 corresponds to D. Hence, the required hexadecimal number is 7D.

Example 5.4 Convert hexadecimal F10A into a binary number.

Solution

Referring to Table 5.2, F is the equivalent of 1111, 10 = 00010000, A = 1010. Hence, the binary number is 1111000100001010.

Example 5.5 A-12 bit, 10-V range digital to analog converter is to be designed to operate over a temperature range of 25–75°C. What is the value of one least signal bit (LSB). What is the initial accuracy? Find the maximum offset drift allowance.

Solution

One least significant bit (LSB) = $10\text{ V}/2^{12}$

$$= 2.44\text{ mV}$$

Initial accuracy = LSB/2

$$= 1.22\text{ mV}$$

The maximum offset drift allowed = 1.22/(75 − 25)

$$= 24.3\ \mu\text{V}/°\text{C}$$

Example 5.6 A 10-bit counter comparator analog to digital converter uses a clock cycle of 1 MHz and has a resolution of 10 mV/step. Determine the maximum possible conversion time and speed.

Solution

Maximum number of steps = $2^n - 1 = 2^{10} - 1$

$$= 1023$$

Maximum conversion time = $1023/10^6$

$$= 1023\ \mu\text{s}$$

Maximum speed of conversion = 1/1023

= 997 bits/sec

Example 5.7 For an ADC with $n = 3$, $R_f = R = 6\,k\Omega$, the input voltage scale is 0–10 V and $V_r = 5$ V. (a) Calculate the full-scale range and nominal scale. (b) Calculate resolution. (c) Calculate the amplification gain using an inverting amplifier for signal conditioning, assuming that voltage output from a piezoceramic transducer can be up to 500 V.

Solution

We have $\qquad V_{fsr} = 2R_f V_R/R = 10\text{ V} = V_N$

Resolution, $r = V_N/2^n = 10/8 = 1.25$ V

A gain of $10/500 = 0.02$ reduces the maximum voltage 500 V to 10 V, achieving an ideal matching of the domain of the input voltage from 0 to 500 V, with the 0 to 10 V ADC scale

Example 5.8 A 12-bit weighted resistor DAC has $R_f = 5\,k\Omega$, $V_R = 5$ V, and $R = 5\,k\Omega$. Calculate the value of V_o for (a) the least significant bit, (b) the most significant bit, (c) the maximum value of the digital input, (d) the output for binary input 100010010001, and (d) the full-scale range voltage V_{fsr}.

Solution

For LSB, the binary input is 000000000001 such that $V_{oLSB} = -2(5)(1/2^{12})$ $= -0.00244$ V $= 2.44$ mV.

The resolution is also 2.44 mV.

For MSB, the binary input is 100000000000, $V_{oMSB} = -2(5)(1/2^1) = -5$ V.

The maximum value of the digital input is 111111111111 such that

$$V_o = -2(5)\left(\frac{1}{2^1} + \cdots + \frac{2}{2^{12}}\right) = -9.999756\text{ V}$$

For the binary input 100010010001,

$$V_o = -2(5)\left(\frac{1}{2^1} + \frac{1}{2^5} + \frac{1}{2^8} + \frac{1}{2^{12}}\right) = -5.35\text{ V}$$

$$V_{fsr} = 2\,R_f\,V_R/R = 10\text{ V}$$

Example 5.9 Design a circuit using an op-amp for weighing soap cakes in a soap manufacturing company. The machine is to be designed to sense 2N weights, 1 N weights, and zero weight. LED2 lights when 2N weight is present, and LED1 lights when 1N weight is present and at zero weight.

Solution

A simple circuit indicator for weight differences is shown in Fig. 5.14. R_1, R_2, R_3, R_4, R_5, R_6, and R_7 are potentiometer dividers. V_R is a variable resistance which varies as the weight on the weight pan changes. When a force of 1N acts on the pan, it causes voltage accuracy of V_R to 4.0 V, but less than 4.8 V. The output of the op-amp becomes positive and LED1 lights. When the weight is 2N, voltage accuracy of V_R is 4.8 V and output of the op-amp becomes positive. Hence, LED2 lights.

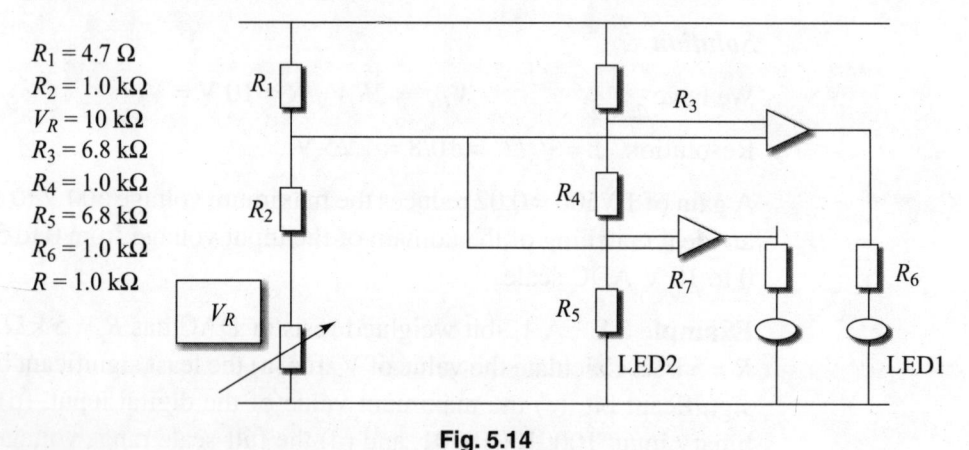

$R_1 = 4.7\ \Omega$
$R_2 = 1.0\ k\Omega$
$V_R = 10\ k\Omega$
$R_3 = 6.8\ k\Omega$
$R_4 = 1.0\ k\Omega$
$R_5 = 6.8\ k\Omega$
$R_6 = 1.0\ k\Omega$
$R = 1.0\ k\Omega$

Fig. 5.14

Example 5.10 Design a tone generator using a 555 timer.

Solution

As the pin 3 of the timer has one stable state, a 0 V output can be converted into one of 9 V. However, it goes back to 0 V after a predetermined period of time. The tone generator has application in alarms, musical instruments, beepers, etc. The node produced by the speaker can be changed by adjusting the resistance V_R. The circuit diagram is shown in Fig. 5.15.

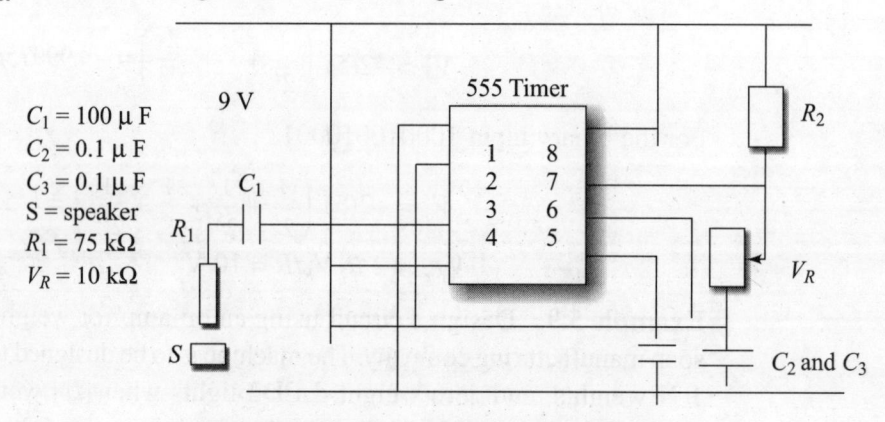

$C_1 = 100\ \mu F$
$C_2 = 0.1\ \mu F$
$C_3 = 0.1\ \mu F$
S = speaker
$R_1 = 75\ k\Omega$
$V_R = 10\ k\Omega$

Fig. 5.15 Circuit diagram of a tone generator

Example 5.11 Sketch a block diagram for the temperature control of a heater using a PLC.

Solution

The input corresponding to temperature could be a thermocouple output, which after amplification can be fed to the analog to digital converter into the PLC. The PLC is programmed to give an output proportional to the error from the reference temperature and measured temperature. The output is then fed through a digital to analog converter to actuate the heater in order to reduce the error.

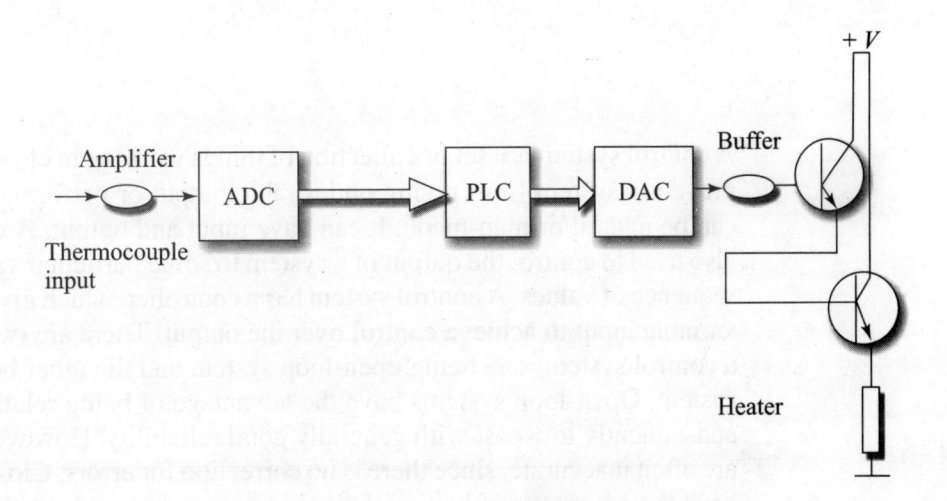

Fig. 5.16 Temperature control of heater with PLC

Exercises

5.1 Differentiate between analog and digital computers.

5.2 Generate an analog computer solution for the differential equation

$$\frac{d^2x}{dy^2} + 5\frac{dx}{dt} + 8x = 10 \sin t$$

5.3 Explain the working of A/D and D/A converters.

5.4 Explain the structure of an 8085 microprocessor.

5.5 Explain the function of a microprocessor.

5.6 What are the advantages of the PLC over the microprocessor?

5.7 Explain a ladder diagram with an example.

5.8 Explain different storage devices used in computers.

5.9 Explain the working of a three-dimensional digitizer.

5.10 Explain the working of a dot matrix printer.

Chapter 6

Systems Modelling and Analysis

A control system is a set or collection of things working in close coordination to prevent a system from giving undesirable output or performance. The system can be natural or man-made. It can have input and output. A control system is also used to control the output of a system to some particular value or particular sequence of values. A control system has a controller, which gives a manipulated variable input to achieve control over the output. There are two basic forms of a control system: one being open-loop system and the other being closed-loop system. Open-loop systems have the advantage of being relatively simple and consequently low cost with generally good reliability. However, these systems are often inaccurate, since there is no correction for errors. Closed-loop systems have the advantage of being relatively accurate in matching the actual input to the required values. However, they are complex and hence costly. Closed-loop control systems are sensitive to disturbance, instability, response speed, and parameter variation. Any fluctuations of these values should be suitably responded to by the system to ensure that the output quality is maintained without any error.

An unstable system can be made stable and its response speed brought to satisfactory levels with the help of a feedback control system. Similarly, any variation in the parameters of system elements should not affect the accuracy of the output. Figure 6.1(a) graphically illustrates a system and Figs 6.1(b) and (c) illustrate control systems of open-loop and closed-loop type, respectively.

A fan with regulator can be considered as an open-loop system. The control knob positions of the regulator give the various desired speeds of the fan and the regulator acts as a controller that manipulates voltage as a variable parameter to exercise control over the speed of the fan. Depending on the voltage input to the fan motor, the fan rotates at a different speed. The occupants of the room can adjust the regulator knob to make the fan run at a desired speed. This system acts as manual, closed-loop system.

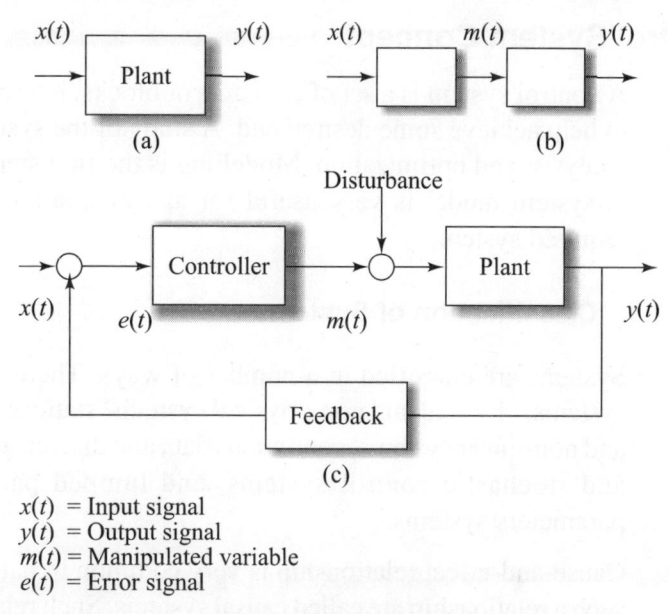

Fig. 6.1 Control systems

$x(t)$ = Input signal
$y(t)$ = Output signal
$m(t)$ = Manipulated variable
$e(t)$ = Error signal

Consider a water-heating system where an electric heating element connected to a voltage supply generates heat and raises the temperature of the water. If the desired temperature is set on a calibrated dial of the regulator system, it acts as an open-loop system. The dial is calibrated for a particular ambient temperature. When the ambient temperature changes significantly, the controlled temperature deviates from the desired value. In this situation, the desired objective of control system is not realized. This open-loop water heating system can be converted into a closed-loop system by adding the function of the measurement of the controlled temperature to the system. This function will measure the water temperature and compare it with the desired value of the controlled temperature. Any mismatch will produce an error signal to suitably control the current flow through the heating element to give a constant temperature output. Any sudden removal of water from the water-heating system can be considered as disturbance to the system. The closed-loop system will accordingly adjust the voltage input to the heater so that the water-heating system is prevented from burning. Otherwise the system would become unstable. If the user wants to reset the output temperature from 50°C to 60°C, the system should react to the change within 4–5 min. The response speed of any system should be within a specified time. If the temperature of the heater increases, its resistance also varies; this in turn increases or decreases its resistance. This is known as parameter variation of a control systems. Parameter variation of the system does not affect the desired output value if one uses a closed-loop system.

6.1 Control System Concept

A control system is a set of elements or blocks, interconnected in such a way as to help achieve some desired end. A study of the system consists of modelling, analysis, and optimization. Modelling is the first step in the study of a system. A system model is very useful for analysis, simulation, and designing of a required system.

6.1.1 Classification of Systems

Systems are classified in a number of ways. There are causal and non-causal systems, physical and non-physical systems, static and dynamic systems, linear and non-linear systems, continuous data and discrete data systems, deterministic and stochastic control systems, and lumped parameters and distributed parameters systems.

Cause-and-effect relationship is very common in nature. Systems governed by such a relationship are called causal systems. Such relationships are alternatively called input–output or stimulus–response relations. Examples of this type of relationship are as follows:

1. Newton's second law of motion, which gives acceleration equal to force divided by mass.
2. Hook's law, which gives displacement equal to force divided by stiffness.
3. Ohm's law, which gives current equal to voltage divided by resistance.
4. Bernoulli's law, which gives outflow equal to the area of the flow multiplied by velocity.

All the causal systems can be represented as cause-and-effect relationships. On the other hand, systems which do not obey the cause-and-effect relationship are said to be non-causal. Typically, such systems are anticipatory in nature. For example, stock market speculation, chess game played between a human and computer, and the sixth sense (of human in danger) are all examples of non-causal response.

A physical system comprises physical elements such as springs, valves, gear trains, capacitors, batteries, etc. On the other hand, there are systems which have no physical existence. Even the components of such systems do not have any existence in this material and physical world. Such systems are non-physical systems. Systems of philosophy, mathematical logic, etc. are typical examples of non-physical systems.

A static system is one that does not change with time. Such systems are governed by algebraic equations. Systems following Ohm's law or Hook's law are static systems. Contrary to static systems, dynamic systems change with time and, therefore, require differential or difference equations to characterize them. An automobile system and a system for cooling of hot milk or discharge of water from tank are examples of dynamic systems.

In a linear system the input–output relation is linear, if $y(t)$ is a function of $x(t)$, where $y(t)$ is the output and $x(t)$ is the input, the system is said to be linear, if the commutative law and additive law are satisfied. Mathematically,

$$\text{if } y(t) = f(x), \text{ then } \alpha y(t) = f(\alpha, x) \text{ (commutative law)}$$

and

$$\text{if } y_1(t) = f(\alpha, x) \text{ and } y_2(t) = f(\beta, x), \text{ then } y_1(t) + y_2(t) = f(\alpha x + \beta x)$$
(additive law)

where $f(\cdot)$ indicates a function. If the output–input relation does not satisfy the commutative and additive laws, the system is said to be non-linear.

Systems that have analog signal input and output are said to be a continuous data systems. On the other hands, systems operating on binary signal, 1 or 0, are said to be discrete data systems. Discrete data systems make it possible to transfer information over large distances without loss of information. These systems achieve this by decoding and reconstructing the signal. In physical systems every signal input is external. Distribution and measured outputs have association with certain amount of randomness, called noise. Such uncertainties are consequential in many practical cases since all signals are a function of time. This assumption gives a deterministic model of the control system, wherein the signals must be treated as random functions of time. Stochastic models are used to characterize the system behaviour.

The significant variables in a system are distributed in space, and they vary with the spatial coordinate and time. The resulting dynamic model is called a distributed parametric model and consists of partial differential equations with time and space coordinates as independent variables. In the lumped parametric model, matter is assumed to be lumped at some discrete part of the space, or space is subdivided into cells and matter is assumed to be lumped at the cells. The resulting dynamic models are ordinary differential equations with time as the only independent variable. For example, take the assumption that solid is a rigid body. There is parametric lumping on all its masses and its mass centres result in a lumped parameter model of dynamics. Considerable simplification is achieved in a theoretical model, if lumped parameter models are used.

If the parameters of the system are varying with time, the system is called a time varying system, for example, guidance and control of rockets. The mass of a rocket changes with time due to the continuous depletion of fuel and reduction in gravitational force as the rocket moves away from the earth. The complexity of the control system design increases considerably if the parameters are time varying.

6.1.2 Modelling of Systems

Modelling is the first step in the study of a system. It enables one to understand the construction and operation of a system. It helps to analyse the behaviour of the system under various operating conditions. A model is defined as a representation of some or all characteristics or aspects of the behaviour of a system or a device or an object.

There are basically three types of models: mathematical, physical, and logical. Mathematical models may be linear or non-linear, static or dynamic, continuous or discrete data, deterministic or stochastic. Physical systems can be represented in electrical analogs or scale model. Scale models can be made for network analysis, aircraft and vessels, dams and valleys, buildings and townships, etc. Logical system models may be heuristic or computational. A computer can be used for modelling logical systems. Mathematical models consist of equations or mathematical expressions describing the principle of operation and characteristics of the components, elements, or the entire systems. Such mathematical expressions are obtained by applying the basic laws of physics such as Newton's laws of motion, the principle of conservation of energy, momentum and mass, the laws of thermodynamics, Faraday's laws of electromagnetic induction, Ohm's and Coulomb's laws, etc.

Dynamic systems can be modelled by differential equations. The governing differential equations can be solved by classical solutions by determining the complimentary function and particular integral, or by an analog computer, or by the Laplace transform technique. The first two methods are time consuming, whereas in the Laplace domain the differential equation can be simplified very easily to solve the problem. In the classical control approach the transfer function concept is used to obtain the output response. Transfer function is defined as the ratio of the Laplace transform of the output to the Laplace transform of the input, assuming all the initial conditions are zero. The transfer function concept of design can be applied only for the linear, time-invariant and single-input–single-output (SISO) system. In the modern control theory, a system is represented through a state-space representation. The state-space representation of a system permits to consider initial values. It can be applied to multiple-input–multiple-output (MIMO) systems. Moreover, the state-space representation gives the internal detail of the system. It is very easy to get a transform function model, by obtaining governing differential equation of the system, taking Laplace transform, and writing the ratio of output to input (for details of Laplace transform refer to Appendix A).

6.1.3 Transfer Function of Physical Systems

Physical systems that we frequently come across can be modelled as zeroth, first, second, or higher order transfer functions. In some special cases the integrator transfer function is also used.

6.1.3.1 Zeroth order model

In the zeroth order model the transfer function is a constant K. Mathematically

$$\frac{Y(s)}{X(s)} = K$$

Figure 6.2 gives examples of the zeroth order system.

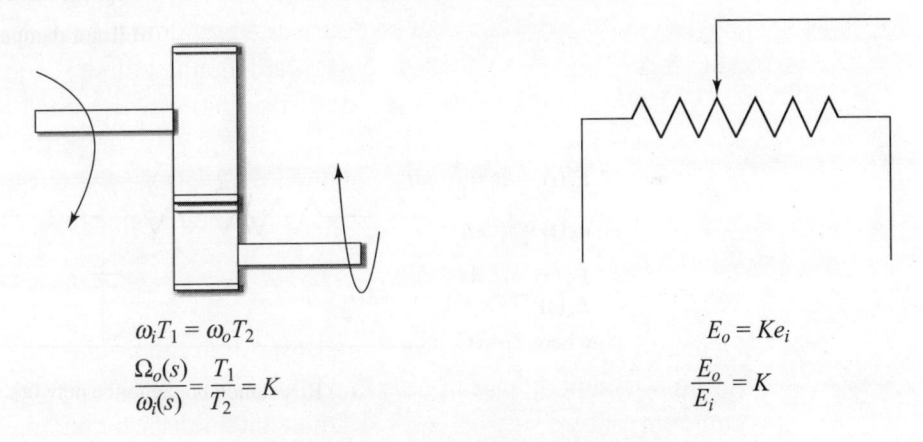

$$\omega_i T_1 = \omega_o T_2$$
$$\frac{\Omega_o(s)}{\omega_i(s)} = \frac{T_1}{T_2} = K$$

$$E_o = K e_i$$
$$\frac{E_o}{E_i} = K$$

Fig. 6.2 Zeroth order systems

6.1.3.2 First order system

The transfer function of the first order system is of the form

$$\frac{Y(s)}{X(s)} = \frac{1}{T_1 s + 1}$$

where s is the Laplace operator. Many of the systems in practice are first order systems. The spring damper system, inertia damper system, resistance, capacitance network, thermal systems, and liquid level systems are all examples of first order systems. Figure. 6.3 illustrates the schematic diagram of the first order system and its transfer function.

$$c\frac{dy}{dt} + k(y-x) = 0$$
$$\frac{Y(s)}{X(s)} = \frac{1}{T_1+1}$$

where $T_1 = c/k$

(a) Spring damper system

$$J\frac{d\omega}{dt} + c\omega(t) = T(t)$$
$$\frac{\omega(t)}{T(t)} = \frac{1/c}{T_1 s + 1}$$

where $T_1 = J/c$

J = moment of inertia
t = torque
c = damping coefficient
k = spring stiffness

(b) Rotor damper system

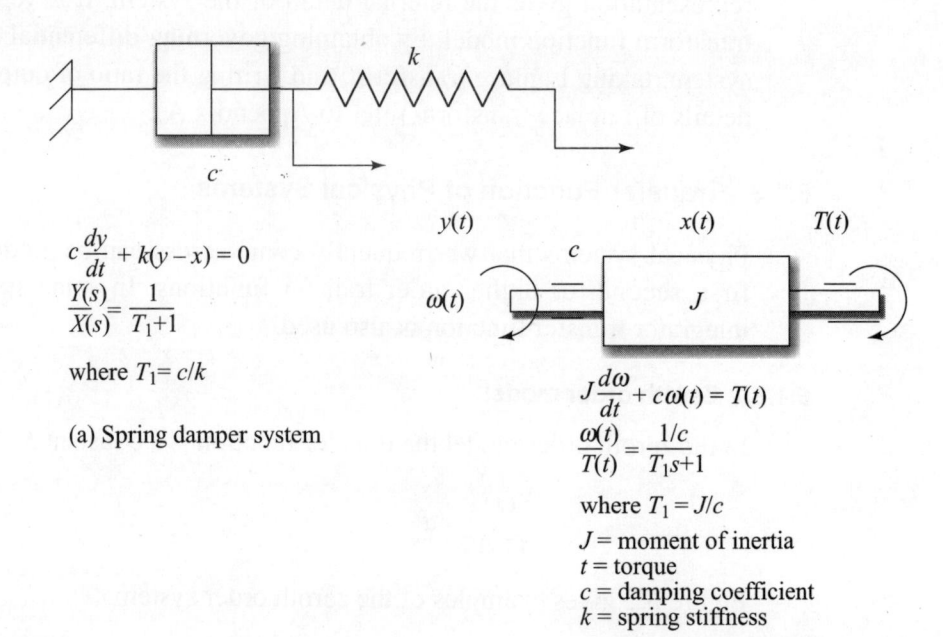

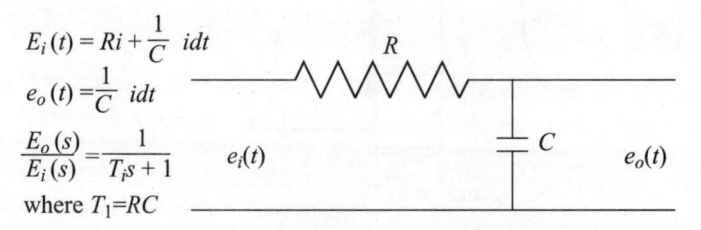

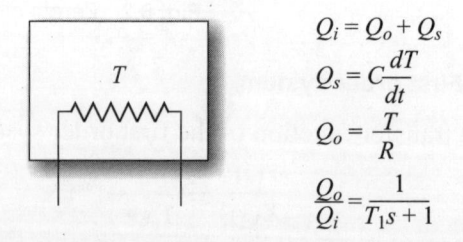

$$E_i(t) = Ri + \frac{1}{C}\int i\,dt$$
$$e_o(t) = \frac{1}{C}\int i\,dt$$
$$\frac{E_o(s)}{E_i(s)} = \frac{1}{T_1 s + 1}$$

where $T_1 = RC$

(c) Resistance-capacitance network

$$Q_i = Q_o + Q_s$$
$$Q_s = C\frac{dT}{dt}$$
$$Q_o = \frac{T}{R}$$
$$\frac{Q_o}{Q_i} = \frac{1}{T_1 s + 1}$$

Q_o—heat output, Q_s—heat accumulated Q_i—heat input, $T_1 = RC$, T—temperature of the plant, R—resistance offered by the body, and C—capacitance to store energy in the body. Q represent incremental change in heat flow from equilibrium.

(d) Electrical heater

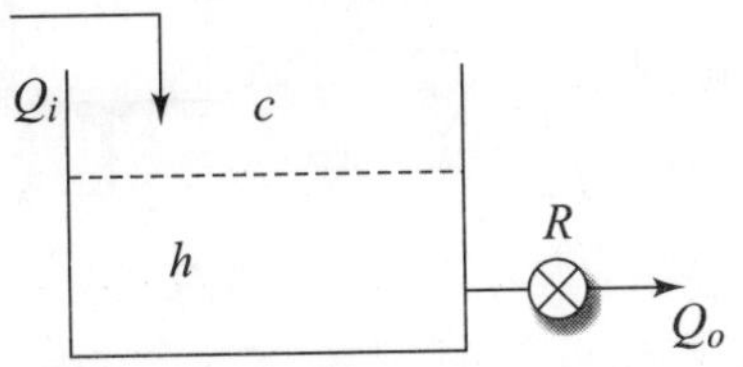

$$Q_i = Q_o + Q_s$$
$$Q_s = c\frac{dh}{dt}$$
$$Q_o = \frac{h}{R}$$
$$\frac{Q_o}{Q_i} = \frac{1}{T_1 s + 1}$$

Q_o—output flow, Q_s—liquid accumulated, Q_i—input flow $T_1 = Rc$, h—incremental head from the available head at steady state condition, R—resistance offered by the valve, c—capacity of the tank, which is equal to the cross sectional area of the tank.

(e) Liquid level system

Fig. 6.3 First order systems

6.1.3.3 Second order system

The study of second order systems may be generalized by the introduction of a certain parameter to the study of the first order systems. The transfer function of a second order system in the canonical form is

$$\frac{Y(s)}{X(s)} = \frac{1}{(T_1 s + 1)(T_2 s + 1)} = \left[-\varsigma \pm \sqrt{\varsigma^{2-} - 1}\right]$$

$$= \frac{1}{\dfrac{s^2}{\omega_n^2} + \dfrac{2\varsigma}{\omega_n} s + 1}$$

where T_1 and T_2 are time constants, ω_n is the natural frequency of the system, and ς is the damping ratio. It can be easily shown that

$$\frac{1}{\omega_n^2} = T_1 T_2$$

and

$$\frac{2\varsigma}{\omega_n} = T_1 + T_2$$

Using the relationship $(T_1 + T_2)^2 - (T_1 - T_2)^2 = 4T_1 T_2$, the time constants T_1 and T_2 can be determined.

The spring-mass-damper system [block diagram shown in Fig. 6.4(a)] and resistance-inductance-capacitance network [block diagram shown in Fig. 6.4(b)] are examples of second order systems.

$$m\frac{d^2y}{dt^2} + c\frac{dy}{dt} + k(y-x) = 0$$

$$\frac{Y(s)}{X(s)} = \frac{1}{\frac{s^2}{\omega_n^2} + \frac{2\varsigma}{\omega_N}s + 1}$$

$$\omega_n = \sqrt{\frac{k}{m}} \text{ and } \varsigma = 2\sqrt{km}$$

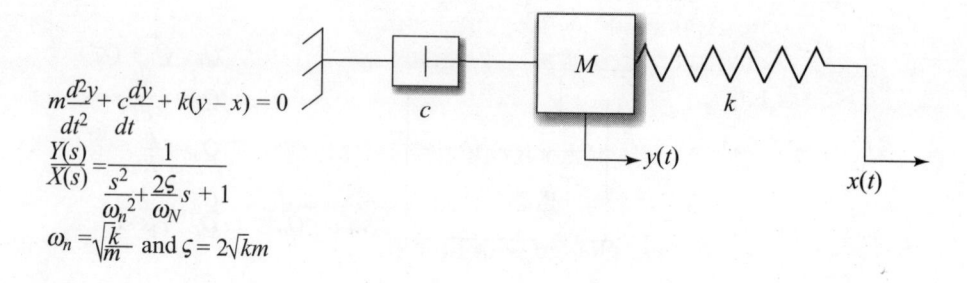

c—damping coefficient
k—spring stiffness
M—mass
$y(t)$—output displacement
$x(t)$—input displacement

(a) Spring-mass-damper system

$$e_i = Ri + L\frac{di}{dt} + \frac{1}{c}\int i\, dt$$

$$e_o = \frac{1}{c}\int i\, dt$$

$$\frac{E_o(s)}{E_i(s)} = \frac{1}{\frac{s^2}{\omega_n^2} + \frac{2\varsigma s}{\omega_n} + 1}$$

where $\omega_n = \sqrt{L/c}$

R—resistance, L—inductance, C—capacitance, e_i—input voltage, e_o—output voltage.

(b) Resistance-inductance-capacitance system

Fig. 6.4 Second order systems

6.1.3.4 Higher order system

In general, the higher order transfer function is of the form

$$\frac{Y(s)}{X(s)} = \frac{1}{(T_1 s + 1)(T_2 s + 1)\cdots(T_n s + 1)}$$

where $T_1, T_2, \ldots, T_n$ are time constants. The response of the particular system for a particular input depends on time constants. It can be easily shown that two time constants near to the origin of the Laplace domain dominate the system response (explained in the Section 6.3), which allows the higher order system to reduce to a second order system for analysis and design.

6.2 Standard Test Signals

The reference signal of some control systems is known to the designer ahead of time. However, command signals are not known fully in advance. It is difficult to express the actual input signal mathematically by simple functions. The characteristics of actual signals that severely strain a control system are listed below:

1. A certain shock (impulse function)
2. A sudden change (step function)
3. Linear change with time (ramp function)
4. Faster change with time (parabolic function)
5. Sudden sinusoidal force input (sinusoidal function)

The dynamic behaviour of a system can be adequately judged and compared under the application of standard test signals. Standard test signals are impulse, step, ramp, parabolic, and sinusoidal inputs. If the input is sinusoidal, the behaviour analysis is known as frequency domain analysis. For all other types of input, the behaviour analysis is known as time domain analysis. Figure 6.5 illustrates different types of standard input signals in time scale and their corresponding Laplace transforms.

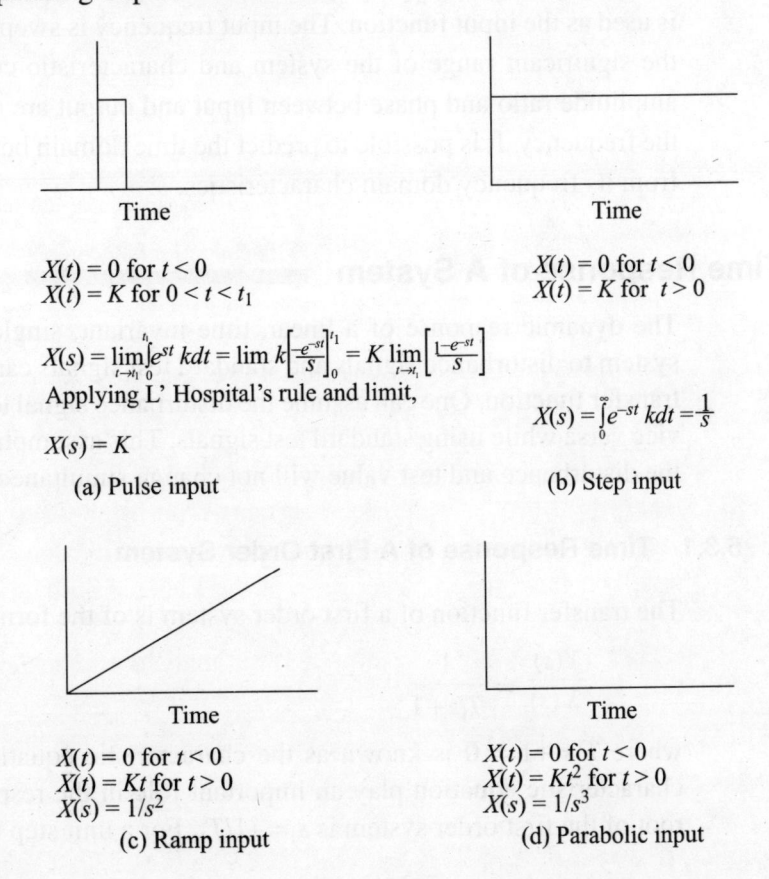

$X(t) = 0$ for $t < 0$
$X(t) = K$ for $0 < t < t_1$

$X(s) = \lim_{t \to t_1} \int_0^{t_1} e^{-st} k \, dt = \lim k \left[\frac{-e^{-st}}{s}\right]_0^{t_1} = K \lim_{t \to t_1} \left[\frac{1 - e^{-st}}{s}\right]$

Applying L' Hospital's rule and limit,

$X(s) = K$

(a) Pulse input

$X(t) = 0$ for $t < 0$
$X(t) = K$ for $t > 0$

$X(s) = \int_0^\infty e^{-st} k \, dt = \frac{1}{s}$

(b) Step input

$X(t) = 0$ for $t < 0$
$X(t) = Kt$ for $t > 0$
$X(s) = 1/s^2$

(c) Ramp input

$X(t) = 0$ for $t < 0$
$X(t) = Kt^2$ for $t > 0$
$X(s) = 1/s^3$

(d) Parabolic input

$X(t) = 0$ for $t < 0$
$X(t) = K \sin \omega t$ for $t > 0$
$X(s) = K/(s^2 + \omega^2)$

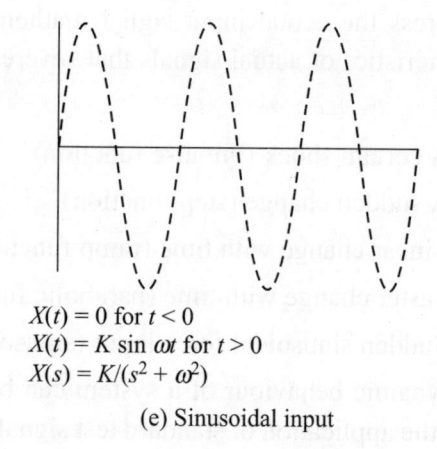

(e) Sinusoidal input

Fig. 6.5 Standard test functions

If $K = 1$, the standard inputs are called unit pulse, unit step, unit ramp, or unit parabolic.

In the frequency domain analysis, a sinusoidal function with variable frequency is used as the input function. The input frequency is swept from zero to beyond the significant range of the system and characteristic curves in terms of the amplitude ratio and phase between input and output are drawn as functions of the frequency. It is possible to predict the time domain behaviour of the system from its frequency domain characteristics.

6.3 Time Response of A System

The dynamic response of a linear, time-invariant, single-input–single-output system to disturbance signals and standard test signals can be obtained from its transfer function. One can assume the disturbance signal to be equal to zero and vice versa while using standard test signals. This assumption is valid since both the disturbance and test value will not change simultaneously.

6.3.1 Time Response of A First Order System

The transfer function of a first order system is of the form

$$\frac{Y(s)}{X(s)} = \frac{1}{T_1 s + 1}$$

where $T_1 s + 1 = 0$ is known as the characteristic equation. The roots of the characteristic equation play an important role in the response behaviour. The root of the first order system is $s_1 = -1/T_1$. For a unit step input, $X(s) = 1/s$,

$$Y(s) = \frac{1}{s}\frac{1}{T_1 s+1}$$

$$y(t) = \left[L^{-1}\frac{1}{s}\frac{1}{T_1 s+1}\right] = L^{-1}\frac{A}{s} + L^{-1}\frac{B}{T_1 s+1}$$

where

$$A = \frac{1}{T_1 s+1} \quad \text{at } s=0 \quad (=1)$$

and

$$B = \frac{1}{s} \quad \text{at } s = -\frac{1}{T_1} \quad (=-T_1)$$

$$y(t) = 1 - e^{-t/T_1}$$

The response of $y(t)$ for unit step input is shown in Fig. 6.6(a). While $y(t)$ increases exponentially as time increases, there is always a steady state error between input and output. If the time constant is negative, the response will be $y(t) = K(-1 + e^{t/T_1})$, since the root of the characteristic equation is positive, $+1/T_1$, which will lead to an increase in the amplitude as the time lapses and the system will become unstable. For the system to be stable, roots of the characteristic equation should be a negative real number. Figure 6.6(b) shows the response for the positive roots of the characteristic equation.

Similarly, the ramp response is obtained by substituting $X(s) = K/s^2$. Then

$$Y(s) = \frac{1}{s^2}\frac{1}{T_1 s+1} = \left[\frac{T_1}{s^2} - \frac{T_1^2}{s} + \frac{T_1^2}{T_1 s+1}\right]$$

Taking the term-by-term Laplace transform, one gets

$$y(t) = T_1 t - T_1^2 + T_1^2 e^{t/T_1}$$

The ramp response is shown in Fig. 6.6(c), wherein it is seen that $y(t)$ increases linearly with time and there is a displacement lag or error of T_1^2.

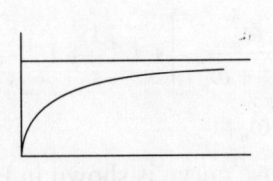

(a) Step response

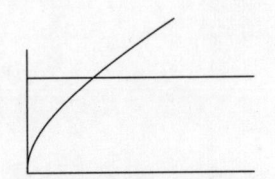

(b) Response for positive root

(c) Response for ramp input

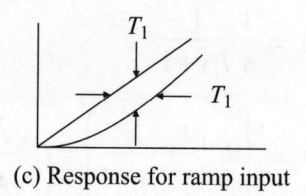

Fig. 6.6 Response of a first order system

6.3.2 Time Response of A Second Order System

The study of the second order system may be generalized since it is also important for higher order systems. A higher order system can also be reduced to a second order system and the response of the reduced second order system is almost the same as the higher order system. The canonical form of any second order system is

$$\frac{Y(s)}{X(s)} = \frac{1}{(T_i s+1)(T_s s+1)} = \frac{\omega_n^2}{s^2 + 2\varsigma\omega_n s + \omega_n^2} = \frac{1}{\dfrac{s^2}{\omega_n^2} + \dfrac{2\varsigma}{\omega_n}s + 1}$$

The characteristic equation of a second order system can be represented as $s^2 2\varsigma\omega_n s + \omega_n^2 = 0$. The roots of the characteristic equation play an important role in the response and depend on the damping ratio ς. The roots of the characteristic equation are as given below:

$$s_{1,2} = \frac{-2\varsigma\omega_n + \sqrt{4\varsigma^2\omega_n^2 - \omega_n^2}}{2} = \omega_n\left(-\varsigma \pm \sqrt{\varsigma^2 - 1}\right)$$

There are five possible responses from the system depending on the value of ς. The five cases are as follows.

CASE 1: $\varsigma = 0$

If $\varsigma = 0$, the roots of the characteristic equation are a complex pair with only an imaginary part. That is, $s_{1,2} = \pm j-\omega$. For the unit step input $X(s) = 1/s$,

$$y(t) = L^{-1}\left[\frac{\omega_n^2}{s(s^2 + \omega_n^2)}\right] + L^{-1}\frac{1}{s} + L^{-1}\frac{\omega_n^2}{s^2 + \omega_n^2}$$

$$= 1 - C\sin\omega_n t$$

where C is a constant. The response curve is shown in Fig. 6.7.

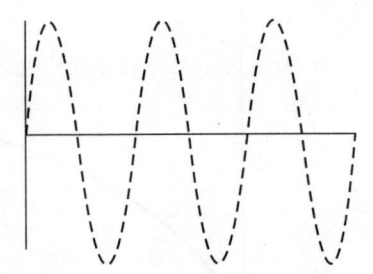

Fig. 6.7 Sinusoidal oscillation

CASE 2: $\varsigma > 1$

When $\varsigma > 1$, the system is said to be over-damped. In this case the roots of the characteristic equation are real and different. Roots are $s_{1,2} = \omega_n \left[-\varsigma \pm \sqrt{\varsigma^2 - 1} \right]$. For the unit step input $X(s) = 1/s$ and taking inverse Laplace transform,

$$y(t) = 1 - C_1 e^{s_1 t} - C_2 e^{s_2 t}$$

where $C_1 = s_2/(s_2 - s_1)$ and $C_2 = s_1/(s_2 - s_1)$ are constants and s_1 and s_2 are the roots of the characteristic equation. Its response curve is shown in Fig. 6.8.

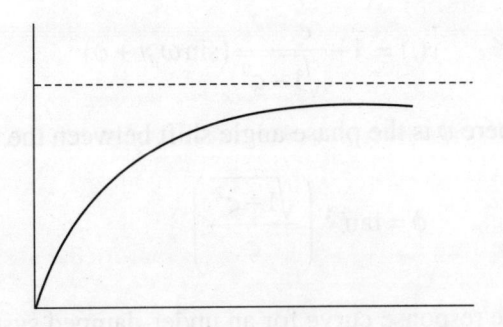

Fig. 6.8 Over-damped response

CASE 3: $\varsigma = 1$

The roots are real and equal in this case. This system is known as a critically damped system. The root is $s_1 = -\varsigma \omega_n$. For unit step input $X(s) = 1/s$ and taking inverse Laplace transform of $Y(s)$,

$$y(t) = 1 - e^{-\omega_n t} - \omega_n t e^{\omega_n t}$$

The response curve for this case is shown in Fig. 6.9.

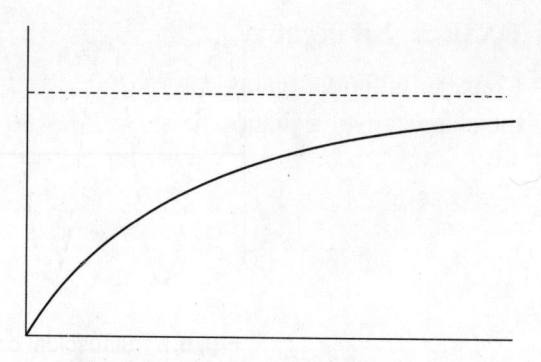

Fig. 6.9 Critically damped response

CASE 4: $\varsigma < 1$

A system in this condition is called an under-damped system. Most of the systems fall in this category since they are oscillatory and steady state errors can be brought to a minimum. Roots of the equation in this case are a complex pair with the real part σ and the imaginary part ω_d; that is,

$$s_{1,2} = -\varsigma\omega_n \pm j\sqrt{1-\varsigma^2} = -\sigma \pm j\omega_d$$

For the unit step input $X(s) = 1/s$ and taking inverse Laplace transform of $Y(s)$,

$$y(t) = 1 - \frac{e^{-\varsigma\omega_n t}}{\sqrt{1-\varsigma^2}}(\sin\omega_d t + \phi)$$

where ϕ is the phase angle shift between the input and output and is given by

$$\phi = \tan^{-1}\left(\frac{\sqrt{1-\varsigma^2}}{\varsigma}\right)$$

The response curve for an under-damped system is shown in Fig. 6.10.

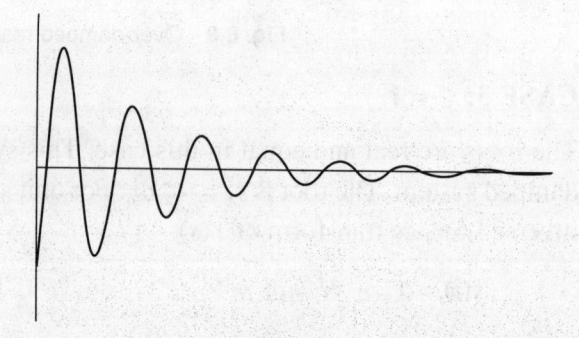

Fig. 6.10 Response curve for an under-damped system

CASE 5: ς is negative

Consider an under-damped system with a negative damping ratio. The roots of the characteristic equation have a positive real part with a complex pair. That is,

$$s_{1,2} = +\varsigma\omega_n \pm j\sqrt{1-\varsigma^2} = -\sigma \pm j\omega_d$$

For step input $X(s) = K/s$,

$$y(t) = 1 - \frac{e^{+\varsigma\omega_n t}}{\sqrt{1-\varsigma^2}}(\sin\omega_d t + \phi)$$

The response grows positively as the time lapses and the system collapses as the amplitude increases beyond a limiting value. Hence, for stability, the roots of the characteristic equation should be a negative real number or complex pair with a negative real part. The response curve for this case is shown in Fig. 6.11.

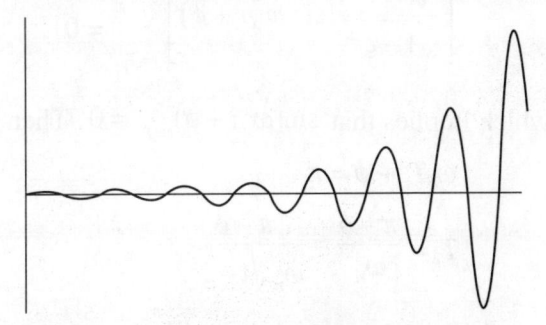

Fig. 6.11 Response curve for positive roots

6.3.3 Characteristics of an Under-damped System

The study of the control system in time domain essentially involves the equation of transient and steady state response of the system. The nature of the transient response of the linear control system is revealed by any of the standard test signals. Step signal is usually used for the analysis of the transient response as well as the steady state response. The steady response depends on both the system and the type of input.

The transient response of an under-damped system exhibits damped oscillation before reaching the steady state. The transient response characteristic of an under-damped system for step input is commonly specified by the rise time T_r, delay time T_d, peak time T_p, peak overshoot M_p, and settling time T_s (see Fig. 6.12). The rise time T_r is the time required to raise the response from 10% to 90% of the input value. The peak time T_p is the time required to raise the response to the overshoot level. The settling time T_s is the time required to settle the response to a specified value of the input. The commonly specified value is ±2% of the input. The delay time T_d is the time required to achieve 50% response of the required value. The overshoot M_p is the difference between the maximum overshoot and the input value. Usually M_p is defined as a percentage of the overshoot over the input value. Mathematically, the transient response performance specification can be determined by the natural frequency and damping ratio. We know that the response of an under-damped second order system for unit step input is given by

$$y(t) = 1 - \frac{e^{-\varsigma\omega_n t}}{\sqrt{1-\varsigma^2}}(\sin\omega_d t + \phi)$$

When $t = T_r$, as per the definition of rise time, $y(T_r) \approx 1$. Hence

$$\left[\frac{e^{-\varsigma\omega_n t}}{\sqrt{1-\varsigma^2}}(\sin\omega_d t + \phi)\right]_{t=T_r} = 0$$

which implies that $\sin(\omega_d t + \phi)_{t=T_r} = 0$. Then

$$\omega_d T_r + \phi = \pi$$

$$T_r = \frac{\pi - \phi}{\omega_d} = \frac{\pi - \phi}{\omega_n\sqrt{1-\varsigma^2}}$$

where ϕ is the phase angle shift between the input and output and is given by

$$\phi = \tan^{-1}\left(\frac{\sqrt{1-\varsigma^2}}{\varsigma}\right)$$

As per the definition of delay time, $T_d = T_r/2$.

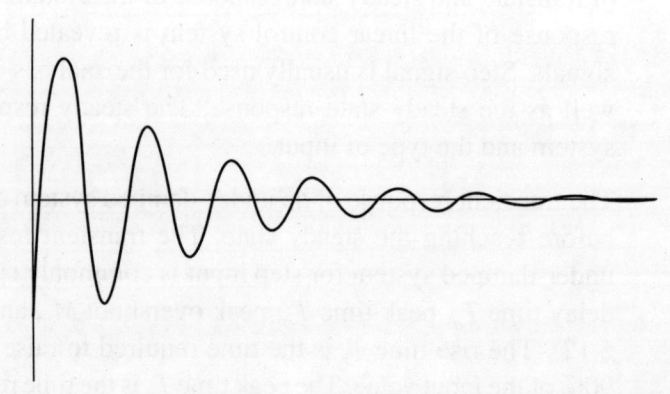

Fig. 6.12 Response of an under-damped second order system

The peak time T_p occurs when response reaches its maximum value. Hence,

$$\frac{dy}{dt} = 0$$

or

$$\left[\varsigma\omega_n \frac{e^{\varsigma\omega_n t}}{\sqrt{1-\varsigma^2}}(\sin\omega_d t + \phi) - \frac{e^{-\varsigma\omega_d t+\phi}}{\sqrt{1-\varsigma^2}}(\cos\omega_d t + \phi)\omega_d\right]_{T=T_p} = 0$$

That is,

$$\left[\varsigma\omega_n \sin(\omega_d + \phi) - \omega_d \cos(\omega_d t + \phi)\right]_{T=T_p} = 0$$

Since $\phi = \tan^{-1}\left(\frac{\sqrt{1-\varsigma^2}}{\varsigma}\right)$,

$$\sin\phi = \sqrt{1-\varsigma^2}$$

and $\cos\phi = \varsigma$

Also $\omega_d = \sqrt{1-\varsigma^2}$

So $\left[\omega_n \cos\phi \sin(\omega_d + \phi) - \sin\phi \cos(\omega_d t + \phi)\right]_{T=T_p} = 0$

$$\sin\omega_d T_p = 0$$

$$T_p = \frac{\pi}{\omega_d} = \frac{\pi}{\omega_n\sqrt{\varsigma^2-1}}$$

Maximum overshoot $M_p = y(T_p) - x(T_p)$.

$$y(T_p) - 1 = \left[-\frac{e^{-\varsigma\omega_n T_p}}{\sqrt{1-\varsigma^2}}(\sin\omega_d T_p + \phi)\right]$$

Substituting

$$T_p = \frac{\pi}{\omega_d} = \frac{\pi}{\omega_n\sqrt{\varsigma^2-1}}$$

we get

$$M_p = \frac{e^{-\varsigma\pi/(1-\varsigma^2)}}{\sqrt{1-\varsigma^2}}\sin(\pi + \phi) = \frac{e^{-\varsigma\pi/(1-\varsigma^2)}}{\sqrt{1-\varsigma^2}}\sin\phi$$

Since $\sin \phi = \sqrt{1-\varsigma^2}$,

$$M_p = e^{\sqrt{-\varsigma\pi/(1-\varsigma^2)}}$$

Percentage overshoot is

$$\frac{y(T_p) - x(T_p)}{x(T_p)} \times 100\% = e^{\sqrt{-\varsigma\pi/(1-\varsigma^2)}} \%$$

$$\varsigma^2 = (\ln M_p^2)/(\ln M_p)^2 + \pi^2$$

Table 6.1 shows the percentage of overshoot for various values of the damping ratio.

Table 6.1 Overshoot versus damping ratio

% Overshoot	Damping ratio
100	0
80	0.1
60	0.2
40	0.3
20	0.5
10	1.0

By the definition of settling time, the amplitude of the output $y(t)$ is equal to 0.95 for a 5% of the tolerance band. That is,

$$\left[1 - \frac{e^{-\varsigma\omega_n t}}{\sqrt{1-\varsigma^2}}\right]_{t=T_p} = 0.95$$

$$T_s = \frac{1}{\varsigma\omega_n} \ln\left(0.05\sqrt{1-\varsigma^2}\right)$$

A common design practice is to approximate the settling time to four times the time constant, which gives a tolerance band of 2%. That is

$$T_s = \frac{4}{\varsigma\omega_n}, \quad 0 \leq \varsigma \leq 0.8$$

6.3.4 Steady State Response

In specifying the steady state response characteristic of the system, it is common to specify the steady state error of the system to one or more of the standard test signals. The steady state error ε_{ss} indicates the error between the actual output and the desired output as time t tends to infinity. Mathematically,

$$\varepsilon_{ss} = \lim_{t \to \infty} x(t) - Y(t)$$

$$\varepsilon_{ss} = \lim_{s \to 0} s[x(s) - Y(s)]$$

The second equation is obtained by using the mean value theorem. When the response is considered for disturbance $x(s) = 0$, $\varepsilon_{ss} = \lim_{s \to 0} sY(s)$. The steady state error for a second order under-damped system for the step input is

$$\varepsilon_{ss} = \lim_{t \to \infty} [1 - y(t)] = \lim_{t \to \infty} \frac{e^{\varsigma \omega_n t}}{\sqrt{1-\varsigma^2}} \sin(\omega_d t - \phi) = 0$$

For unit ramp input $x(s) = 1/s^2$, the response of a second order system is

$$y(t) = t - \frac{2\varsigma \omega_n t}{\omega_n \sqrt{1-\varsigma^2}} \sin(\omega_d t - \phi)$$

Its steady state error is

$$\varepsilon_{ss} = \lim_{t \to \infty} [t - y(t)] = \frac{2\varsigma}{\omega_n}$$

An important step is that the best value of the damping ratio should be chosen for any control system for better performance of a mechatronic system. A smaller damping ratio decreases the rise time, but increases the peak overshoot. Many systems are designed for a damping ratio in the range of 0.4 to 0.7. If rise time considerations allow, ς close to 0.7 is the most obvious choice because it results in a minimum normalized settling time. If the requirement is to design an extremely accurate control system whose steady state errors are extremely small, the damping ratio should be as small as possible. Because the steady state error is proportional to the damping ratio, values close to 0.1 are not reasonable for such applications. Disadvantages of a relatively long settling time have to be tolerated. A good compromise between rise time and overshoot is $\varsigma = 1$.

6.3.5 System Type Number

The difference between the input signal $x(t)$ and output $y(t)$ is the error signal $e(t)$. For a stable system, the transient component of response $e(t)$ will essentially die away, leaving only the steady state component. The steady state error then is given by

$$\varepsilon_{ss} = \lim_{s \to 0} s[x(s) - y(s)] = \lim_{s \to 0} sx(s)\left[1 - \frac{Y(s)}{X(s)}\right]$$

$Y(s)/X(s)$ is the open-loop transfer function of the closed-loop system, which is of the form

$$\frac{Y(s)}{X(s)} = \frac{K \prod_i^m (s + Z_i)}{s^N \prod_i^m (s + P_i)}$$

The term $1/s^N$ corresponds to N number of pure integration. If $N = 0$, the system is called type zero system; if N is equal to 1, the system is called type 1 system; and if N is equal to 2, the system is called type 2 system. When s takes the value of Z_i ($i = 1, 2, 3, \ldots, m$), the transfer function goes to zero. Such values are called zeros of the system. Zeros are indicated as '0'. When s takes the values of $-P_i$ ($i = 1, 2, \ldots, n$), the transfer function goes to infinity. Such values are called poles of the system and are denoted by 'x'. Steady state errors for various system numbers and input are given in Table 6.2. The addition of an integrator to the system tends to decrease the steady state error. However, integrators have destabilizing effects on the system. Thus, control system design is usually a trade-off between the steady state accuracy and acceptable related stability.

Table 6.2 Steady state error for various system numbers

Type of input	Type 0	Steady state error Type 1	Type 2
Unit step	Finite	0	0
Unit ramp	Infinite	Finite	0
Unit parabolic	Infinite	Infinite	0

6.4 Block Diagram Manipulation

Consider the feedback control system shown as a block diagram in Fig. 6.13. The transfer function of the system is indicated by $G_1(s)$, the transfer of the controller is indicated by $G_2(s)$, and the feedback element transfer function is $H(s)$. Equation $G(s) = G_1(s) \times G_2(s)$ is called the forward path transfer function and $H(s)$ is called the feedback path transfer function. $G(s) \times H(s)$ is known as the open-loop transfer function of an open-loop system, whereas $Y(s)/X(s) = G(s)/\{1 + G\,H(s)\}$ is called the open-loop transfer function of a closed-loop system. Referring to Fig. 6.13, $Y(s) = G(s) \times E(s)$, $E(s) = X(s) - H(s) \times Y(s)$, which gives $Y(s)/X(s) = G(s)/\{1 + G\,H(s)\}$. The equation $1 + G(s)\,H(s) = 0$ is known as the characteristic equation.

(a) Closed-loop system

(b) Canonical form of a closed-loop system

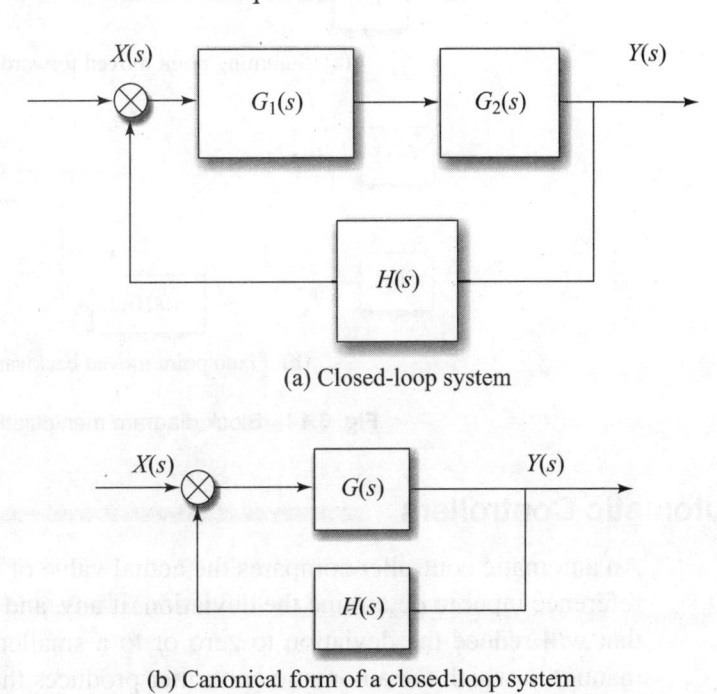

Fig. 6.13 Block diagram of feedback control system

Control system analysis requires one to calculate the closed-loop transfer function of a system through block diagram manipulation. The transfer function can be obtained for each component of the system. It is important to note that a block diagram can be connected in series only when the output of one block is not affected by the next following block. If there is any loading effect between the components, it is necessary to combine these components into a single block. Any number of blocks representing non-loading components can be replaced by a single block. A complicated block diagram involving a feedback loop can be simplified by a step-by-step rearrangement. Simplification of a block diagram by rearrangement reduces the effort needed for subsequent mathematical analysis. The simplification procedure for a block diagram is called *block diagram manipulation*. The basic rules and guidelines for block diagram manipulation are illustrated in Fig. 6.14. Adjusting the summing point or fixed point negative feedback loop containing feedback $H(s)$ will help to manipulate the block diagram.

(a) Summing point moved forward

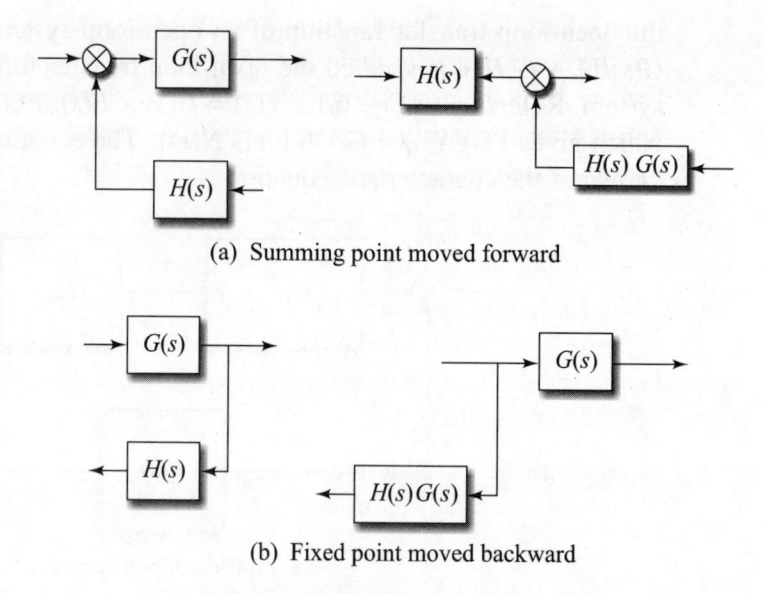

(b) Fixed point moved backward

Fig. 6.14 Block diagram manipulation

6.5 Automatic Controllers

An automatic controller compares the actual value of the plant output with the reference input to determine the deviation, if any, and produce a control signal that will reduce the deviation to zero or to a smaller (negligible) value. The manner in which the automatic controller produces the control signal is called *control action*. Industrial controllers are classified on the basis of their control actions as follows:

1. Two-position or on-off controllers
2. Proportional controller
3. Integral controller
4. Proportional-plus-integral (PI) controller
5. Proportional-plus-derivative controller
6. Proportional-plus-integral-plus-derivative controller

In the two-position control system the actuating element has in many cases simply on and off slots. On-off controllers are relatively simple and inexpensive. Hence, they are widely used in industry and domestic control systems. The output signal from this type of controller remains at a maximum or minimum value, depending on the actuating error signal. Mathematically, $m(t) = m_1$ for $e(t) > 0$ and $m(t) = m_2$ for $e(t) < 0$. The difference between m_1 and m_2 is known as the differential gap.

In a controller with proportional control action the relation between the output of the controller is $m(t) = K_P e(t)$ and the transfer function is $M(s)/E(s) = K_P$, where K_P is the proportional gain which can be adjusted. The advantage of the proportional controller is that it can bring the system to stability. However, a steady state error is always present with the proportional controller.

In a controller with integral control action the value of the controller output $m(t)$ is the integral of error signal. That is, $m(t) = K_I \int e(t)$, where K_I is an adjustable constant called integral gain. The transfer function of the integral controller is $M(s)/E(s) = K_I/s$. The advantage of the integral controller is that it reduces the steady state error equal to zero, but the response is very slow due to the integral action.

In a controller with derivative action the value of the controller output $m(t)$ is changed at a rate proportional to the error signal. Mathematically,

$$-\frac{dm(t)}{e(t)} = K_D s$$

where K_D is the derivative gain.

Proportional-plus-integral controllers bring the system to stability because of their proportional action and the steady study state error for disturbance being equal to zero due to integral action. However, there is some delay in response due to the integral action. The controller action of a proportional-plus-integral controller is defined by $m(t) = K_P e(t) + K_I \int e(t) dt$ and the transfer function of the controller is given by

$$\frac{M(s)}{e(s)} = K_P + \frac{K_I}{s}$$

Proportional-plus-integral-plus-derivative (PID) controller is a combination of proportional, integral, and derivative control actions. This combined action has the advantage of the three individual control actions. The equation of a PID controller with the combined action is given by

$$m(t) = K_P e(t) + K_I \int e(t) dt + K_D \frac{de(t)}{dt}$$

The transfer function of a PID controller is given by

$$\frac{M(s)}{e(s)} = K_P + \frac{K_I}{s} + K_D s$$

where K_P, K_I, and K_D are proportional gain, integral gain, and derivative gain, respectively. The system becomes stable due to the proportional control action, steady state error for disturbance being zero due to integral action, and the correction of the error being faster due to the derivative action. However, the PID controller is expensive.

Derivative controller action as an individual controller is not used since the controller action is rate of error signal and hence the output is not going to respond for error. For some industrial applications where tachogenerator output is used for impact and jerk compensation, proportional-plus-derivative control actions are used.

6.6 Frequency Domain Analysis

The phrase 'frequency response' means the steady state response of a system to a sinusoidal input. In the frequency response method, the frequency of the output signal varies over a certain range, keeping the amplitude constant. One advantage of the frequency response is that it can use the data obtained from measurements on a physical system without deriving its mathematical model. Frequency response method was developed in 1930s and 1940s by Nyqist, Bode, Nicholas, and others. It is the most powerful method in the classical control theory. The Nyqist stability criterion enables us to investigate both absolute and relative stabilities of a linear closed-loop system. In designing a closed-loop system, one can adjust the frequency response characteristic of the open-loop transfer function by using several design criteria in order to obtain an acceptable transient response characteristic for the system. An example of input and output sinusoidal signals is shown in Fig. 6.15. The frequency response of a stable, linear system can help configure the dynamic response to any form of input. Thus the control system's performance can be described in frequency response terms. The correlation between transient response and frequency response through the Fourier integral transform is an important basis of most design procedures and criteria. Usually the Laplace domain behaviour is interpolated in terms of frequency response characteristics, design in frequency domain and the frequency response is then translated into the time domain.

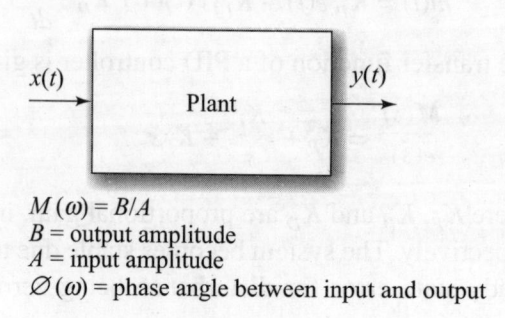

$x(t) \rightarrow$ Plant $\rightarrow y(t)$

$M(\omega) = B/A$
B = output amplitude
A = input amplitude
$\varnothing(\omega)$ = phase angle between input and output

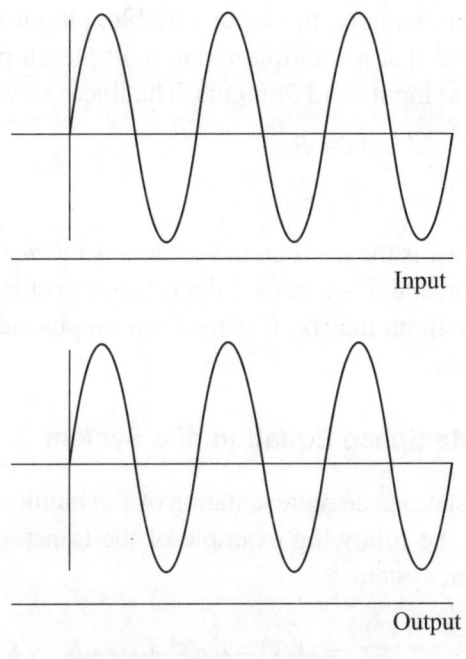

Fig. 6.15 Frequency response

6.7 Modern Control Theory

The current trend in engineering systems is greater design complexity. This has been necessary to facilitate performance of complex tasks and help achieve good accuracy. Such complex systems may have multiple inputs and multiple outputs and may be time varying. Because of the necessity to meet the increasing stringent requirements on the performance of control systems such as turbo-jet engines, nuclear power plants, there has been an increase in the system complexity and the use of computers in the modern control theory. This has led to a new approach to design and analyse control systems. Modern control theory was developed around 1960. This approach is based on the concept of state space and is applicable to linear and non-linear, time-varying systems. The state variable approach is applicable to biological systems, economic systems, and social systems. Often, the state variables of a dynamic system are the variables comprising a smallest set of variables that determine the state of a dynamic system. State variables are represented as $x_1, x_2, ..., x_n$. If n state variables are needed to completely describe the behaviour of the system, then these n states can be considered as n components of a vector x. Such a vector is called state vector. Any state can be represented by a point in the state space, where space is an 'n'-dimensional coordinate system. The state space analysis is concerned with three types of variable that are involved in the modelling of a dynamic system, namely, the input variable, output variable, and state variable. Let us assume that a multiple-input–multiple-output system involves n integrator(s) with m inputs and l outputs. The linear state equation for this is of the form

$$x' = Ax + Bu$$

$$y = Cx + Eu$$

where x is the $n \times 1$ state vector, u is the $m \times 1$ input vector, y is the $l \times 1$ output vector, and d is the $n \times 1$ disturbance vector. A is the $n \times n$ state matrix, B is the $n \times m$ input matrix, C is the $l \times n$ output matrix, and E is the $n \times n$ disturbance matrix.

6.7.1 State Space Equation of a System

The state space representation of a dynamic system can be obtained as illustrated with the following example of the transfer function of a single-input–single-output system.

$$\frac{Y(s)}{X(s)} = (b_0 s^n + b_i s^{n-1} + \cdots + b_{n-1} s + b_n) / (s^n + a_1 s^{n-1} + \cdots + a_n)$$

where $X(s)$ is the single input and $Y(s)$ is the single output and a_i's and b_i's are constants. Partioning the transfer function,

$$\frac{Y(s)}{X(s)} = \frac{Y(s)\xi(s)}{\xi(s)U(s)}$$

The transfer function in the differential equation form is

$$u(t) = \xi^n(t) + a_1 \xi^{n-1}(t) + \cdots + a_n \xi(t)$$

where the superscripts indicate the order of differentiation. Now define the state variables as

$$x_1 = \xi(t)$$
$$x_2 = \dot{x}_1 = \xi^1(t)$$
$$\vdots$$
$$x_n = \dot{x}_{n-1} = \xi^{n-1}(t)$$

From this, we get

$$\dot{x}_n = -a_n x_1 - a_{n-1} x_2 - \cdots - a_1 x_n$$

where $B = \begin{bmatrix} 0 \\ 0 \\ \vdots \\ 1 \end{bmatrix}$

$$A = \begin{bmatrix} 0 & 1 & 0 & . & 0 \\ 0 & 0 & 0 & . & 0 \\ . & . & . & . & . \\ -a_n & -a_{n-1} & . & . & -a_1 \end{bmatrix}$$

since $Y(s)/\xi(s) = b_0 s^n + b_1 s^{n-1} + \cdots + b_n$.

The output equation for the system is

$$y(t) = Cx(t)$$

where the output coefficient matrix is

$$C = [c_0 \quad c_1 \cdots c_n]$$

The block diagram in the state space representation is illustrated in Fig. 6.16.

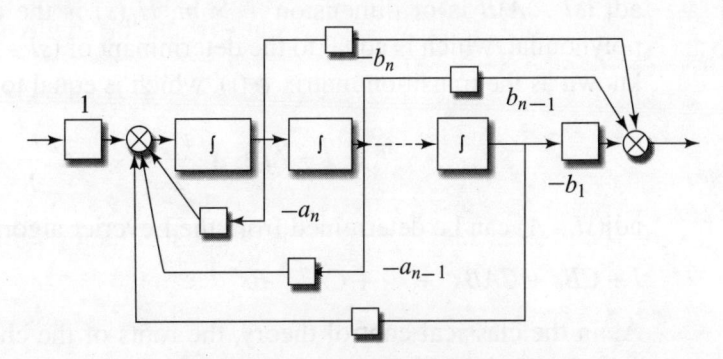

(a) Block diagram showing different states

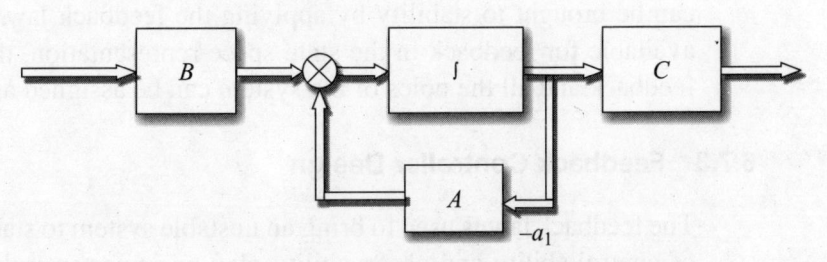

(b) Block diagram of representation

Fig. 6.16 State space representation

6.7.2 Transfer Function from State Space Representation

Correlation between the transfer function and the state space equation is important. To derive the transfer function of a multiple-input–multiple-output system, consider the following state space equation:

$$x' = Ax + Bu$$

$$Y = Cx$$

Usually the state space representation of the system is indicated by the state matrices triplet (A, B, C). For a change in the reset value, the disturbance vector is assumed to be zero. Taking the Laplace transform of the state space equation,

$$sX(s) - AX(s) = BU(s), (sI - A) X(s) = BU(s)$$

Hence $X(s) = (sI - A)^{-1} BU(s)$. Knowing $X(s)$, the output vector $Y(s)$ is $C(sI - A)^{-1} BU(s)$, that is,

$$Y(s) = \frac{W_0(s)}{H_0(s)} U(s)$$

where $W_0(s)$ is known as the numerator of the transfer function matrix C, adj $(sI - A)B$ is of dimension $\ell \times m$, $H_0(s)$ is the open-loop characteristic polynomial, which is equal to the determinant of $(sI - A) = 0$, and $L^{-1}(sI - A)$ is known as the transition matrix $\phi(t)$, which is equal to

$$e^{At} = I + At + A^2 \frac{t^2}{2} + \cdots + A^n \frac{t^n}{n} = L^{-1} \text{adj}(sI - A)$$

adj$(sI - A)$ can be determined from the Leverier algorithm, which is equal to

$$I + CBs + CABs^2 + \cdots + CA^{n-1} Bs^{n-1}$$

As in the classical control theory, the roots of the characteristic equation det $(sI - A) = 0$ should be negative real numbers or complex numbers with negative real parts. If the system is unstable or does not meet the specification, the system can be brought to stability by applying the feedback law. If all the states are available for feedback in the state space representation, then it is called state feedback and all the poles of the system can be assigned arbitrarily.

6.7.3 Feedback Controller Design

The feedback law is used to bring an unstable system to stability,. The concepts of controllability and observability play an important role in the design of a controller in state space. The concepts of controllability and observability were introduced by Kalman. The conditions of controllability and observability may give the existence of a complete solution to the control system design problem.

The solution to this problem may not exist if the system considered is not controllable or observable. Although most of the physical systems are controllable and observable, the mathematical model should be established.

A system is said to be controllable at a time t_0 if it is possible by means of an unconstrained control vector to transfer the system from any initial state $x(t_0)$ operator to any other state in some finite interval of time.

A system is sad to be observable at time t_0 if with the system in state $x(t_0)$ it is possible to determine this state from the observation of the output over a finite time interval.

The solution to the state-space equation $x' = Ax + Bu$ is

$$x(t) = e^{At} x(0) + \int_0^t e^{At(t-\tau)} BU(\tau) d\tau$$

where $x(0)$ is the initial state at $t = t_0$ and t is the finite time interval $t_0 < t < \tau$. When time is required to final the state, applying the definition of complete state of controllability,

$$x(t_1) = -e^{At} x(0) + \int_0^t e^{At(t-\tau)} BU(\tau) d\tau$$

If the system is completely state controllable, then given the initial state $x(0)$ the above equation must be satisfied. This requires the rank of matrix $U = [A\ AB\ \ldots\ A^{n-1}B]$ to have a full rank n. U is called the controllability matrix.

Considering the output equation of the state space representation, the output vector is

$$y(t) = Ce^{At} x(0)$$

If the system is completely observable, the output $y(t)$ over a time interval $0 < t < \tau$, $x(0)$ is uniquely determined from the above equation. It can be shown that it requires the rank of $V = [(C^T\ (CA)^T\ \ldots (CA^{n-1})^T]$ to be a full rank. The matrix V is called the observability matrix. The system should be controllable and observable to design a controller and to bring the system to stability.

For stability, the roots of the characteristic equation $\det(sI - A) = 0$ should be negative real numbers or a complex pair with a negative real part. Using the state feedback law, all the roots of the characteristic equation can be placed arbitrarily at desired locations. The design method to place the roots at desired locations is called the *pole placement* or *pole assignment technique*. Consider a controllable and observable linear system of the state space representation (A, B, C). Applying the feedback law $u(t) = v(t) - K\,x\,(t)$, where $v(t)$ is the reference vector, the state space equation becomes

$$x = Ax + b[v(t) - KCx(t)]$$
$$y = Cx$$

Taking the Laplace transform

$$(sI - A + BKC)x(s) = Bv(s)$$
$$Y(s) = Cx(s)$$

Then

$$X(s) = (sI - A + BKC)^{-1} Bv(s)$$
$$Y(s) = C(sI - A + BKC)^{-1} Bv(s)$$
$$Y(s) = \frac{W_1(s)}{H_1(s)} v(s)$$

where $W_1(s) = C \, \mathrm{adj}(sI - A + BKC)$, C is called the $l \times m$ numerator transfer function matrix of the closed-loop system and $H_1(s) = \det(sI - A + BKC) = 0$ is the characteristic equation of the closed-loop system. By the proper choice of $m \times l$ coefficient of matrix K, the poles of the closed-loop system can be placed arbitrarily. Since the characteristic equation is a non-linear equation, pole assignment becomes difficult. Few iterative algorithms are available for pole assignment. However, these algorithms do not guarantee the exact pole placement. We now explain the linearization of the problem and equations for the design of a controller.

Consider a closed-loop characteristic equation $H_1(s) = \det(sI - A + BKC) = 0$.

Assume the feedback in the dyadic form $K = kf^T$, where k is the $m \times 1$ vector and f^T is the $1 \times l$ vector. The dyadic form is the outer product of two vectors. Hence

$$H_1(s) = |sI - A| \, |I + (sI - A)^{-1} B k f^T C|$$

We use the determinant identity $|I + ab| = 1 + ba$, where a and b are vectors of dimensions $n \times 1$ and $1 \times n$, respectively. This is an inner product of two vectors.

$$H_1(s) = |sI - A| \, [1 + f^T C(sI - A)^{-1} B k]$$
$$= H_0(s) + f^T W_0(s) k$$

Assuming either vector f^T or k maximum of m or l poles can be assigned arbitrarily. If the output coefficient matrix is an identity matrix, all the states of the system are available for measurement. Then, assuming the vector k arbitrarily, by the choice of the vector f^T, all the poles of the system can be assigned arbitrarily. This pole assignment procedure is called state feedback. If all the states are not available for measurement, this pole assignment procedure is called the pole assignment with output feedback.

6.8 Sequential Control System

There are many situations where control is exercised by items being switched on or off at particular preset times or value in order to control the process and give sequence of operations. For example, after step 1 is complete, step 2 starts, and when step 2 is complete, step 3 starts, etc. The term 'sequential control' is used if control is such that control actions are strictly ordered in a time- or event-driven sequence. Such a control could be obtained by an electrical circuit with sets of relays and cam-operated switches, which are wired up in such way as to give the required sequence of control actions. Such hard-cored circuits have now been replaced by microprocessor-controlled systems, with the sequence being controlled by a software program. For an illustration of sequential control, consider the domestic washing machine, in which a number of operations are carried out in a predetermined and prefixed sequence. These operations may involve a pre-wash cycle, wherein the clothes in the drum are given a wash in cold water, followed by the main wash cycle. The main wash cycle involves a washing operation in hot water, a rinse operation (wherein clothes are rinsed in cold water a number of times), and a spin operation (which removes water from the clothes). Each of these operations involves a number of steps. The operating sequence is called a program and the sequence of instructions in each program is defined and built into the controller unit. As an another example, consider an aircraft turbine engine, for which it is necessary to achieve maximum thrust with a minimum engine weight. All components should operate at mechanical or thermal limit for all the engines at critical operating engine conditions. The operating conditions depend on the speed of the aircraft as also the altitude of the flight. Depending on the operating conditions, control parameters have to set sequentially to control each limiting parameter and limit the engine thrust so that none of the limits is exceeded. In the most recent design methods, the military aircraft use multi-variable approach to achieve robustness. Robustness means that all the limiting conditions of a system are met at different operating conditions at one controller setting.

6.9 Digital Control System

The past two decades have seen a rapid increase in the use of digital computers in control systems. The main advantages of using digital control systems are as follows:

1. Digital control systems help achieve the highest levels of accuracy of the control system.
2. The digital signal is less susceptible to noise and drift by power supply.

3. The cost of configuration can be scaled down by adopting the VLSI design methods.
4. Digital control systems provide flexibility to modify the controller by way of changes in program software.
5. Digital control systems can be used for multiple purposes. These not only control the output but also can be used to collect data, carry out complex combinations, and monitor the status of the control system.
6. Digital control systems can handle the non-linear, time-varying and adaptive control schemes.
7. Artificial intelligence can be easily incorporated into the digital control systems.

Digital control systems can be classified into two types:

- Direct digital control systems
- Sampled data control systems (also known as discrete data control systems

In the direct digital control system, a computer is directly used to control the output. A computer algorithm is used for the designed compensator and controller in direct digital control system. Discrete data control systems or sampled data control systems use plant model as digital systems using the Z transform.

6.9.1 Direct Digital Control System

Figure 6.17 depicts a block diagram of a computer-controlled system or direct digital control system showing the configuration of the control scheme. The analog feedback signal coming from the sensor is usually of low frequency. It may often include higher frequency noise. Low-pass filtering is often needed to allow good control performance. The signal coming from the sensor is converted into a digital form using an analog-to-digital (A/D) conversion system. The converter is preceded by a sample-and-hold (S/H) device. The S/H device keeps the input to the A/D converter constant during conversion. A digital-to-analog (D/A) conversion system converts the sequence of numbers in the numerical code into a piecewise continuous time signal. The output of the D/A converter is fed to the plant through the actuator to control its dynamics. The real-time clock in the computer synchronizes all the events of A/D conversion and D/A conversion.

The selection of the best sampling rate for a digital control system is the result of a compromise among many factors. The basic motivation to lower the sampling rate is cost. A decrease in the sampling rate means more time is available for control calculations.

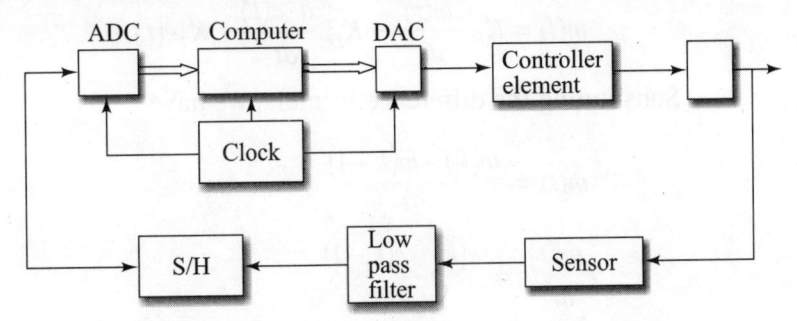

Fig. 6.17 Configuration of direct digital control scheme

6.9.2 Computer Algorithm for Controller or Compensator

Consider the analog compensation with a proper transfer function. The problem at hand is to find a digital algorithm for any error signal $e(t)$ input and the analog output signal $u(t)$ from the compensator or controller. This can be done by transforming a continuous signal with a discrete signal by approximating the derivative in a differential equation representing the transform function by finite differences. From the Taylor expansion, the backward difference approximation is given by

$$\frac{d\,e(t)}{dt} = \frac{e(k) - e(k-1)}{T}$$

$$\frac{d^2 e(t)}{dy} = \frac{e(k) - 2e(k-1) + e(k-2)}{T^2}$$

where k represents the present signal, $(k-1)$ is the previous signal, and $(k-2)$ the previous to previous signal.

The use of backward difference formula for the PID controller computer algorithm implementation is illustrated below. We consider the simplest formula for a PID controller:

$$\frac{M(s)}{e(s)} = K_P + \frac{K_I}{s} + K_D s$$

where K_P, K_D, and K_I are the proportional, derivative, and integral controller gains. In the time domain, the PID controller is

$$m(t) = K_P e(t) + K_D \frac{de(t)}{dt} + K_I \int e(t) dt$$

Differentiating, we get

$$m(t) = K_P \frac{de(t)}{dt} + K_D \frac{d^2e(t)}{dt^2} + K_I e(t)$$

Substituting the difference formula, we have

$$m(t) = \frac{m(k) - m(k-1)}{T}$$

$$\frac{de(t)}{dt^2} = \frac{e(k) - e(k-1)}{T}$$

$$\frac{d^2e(t)}{dt^2} = \frac{e(k) - 2e(k-1) + e(k-2)}{T^2}$$

After simplification,

$$m(k+1) = \left(\frac{K_D}{T} + K_P + K_I T\right)e(k) - \left(\frac{2K_D}{T} + K_P\right)e(t) + \frac{K_D}{T}e(k-2) + m(k)$$

The coefficients K_P, K_D, and K_I can be set easily with the computer since these are numerical values and after each interval of time, a manipulated variable is computed. One of the advantages of the direct digital control system method is that the sampling rate need not be selected until the basic feedback design is completed. One can carry out the analog design and then discretize the analog compensator. If it is not yielding a satisfactory result, one can select a different sampling period without having to redo the design.

6.9.3 Discrete Data Control System

The sampled data control system involves transformation of data from continuous analog signal data into the discrete value form. Referring to Fig. 6.18, momentary open and close of the switch every interval of T seconds produces a pulse train of signals as $f(1T), f(2T), ..., f(nT)$. Then the signal is no more an analog signal, but the function may be represented as a modulated pulse train.

For transfer function analysis of a continuous signal one can choose the Laplace transform defined by $L[f(t)] = F(s) = \int_0^\infty e^{-st} f(t) dt$. Similarly, for a discrete data signal, a transformation procedure is used which is defined as the Z transform of the pulse train, $f(nT)$:

$$Z[f(nT)] = F(z) = \sum_{n=0}^{\infty} f(nT) Z^{-n}$$

where *n* can take any value 0, 1, 2, 3, ... (for details of the Z transform refer to Appendix B).

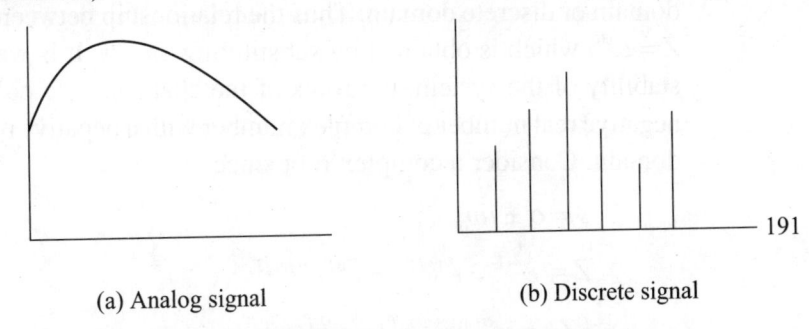

(a) Analog signal (b) Discrete signal

Fig. 6.18 Sampling of continuous signal

The Z transform of $f[(n-1)T]$ is equal to

$$Z[f(n-1)T] = \sum_{n=0}^{\infty} f[(n-T)T]Z^{-n}$$

substituting $n = (n+1)$ since the signal started one sample period earlier. Hence

$$Z[f(n-1)T] = \sum_{n=0}^{\infty} f(nT)Z^{-1}Z^{-n} = Z^{-1}F(Z)$$

Similarly the Z transform of $f(n+1)T$ is $Z[F(Z)]$ and $f(n+2)T = Z^2[F(Z)]$.

The sequence of samples represented by a Z transform may be obtained by polynomial division. Consider the Z transform $F(z) = 4Z / (Z^2 - Z)$. The polynomial division gives

$$F(Z) = Z^0 + 4Z^{-1} + 4Z^{-2} + 2Z^{-3} + \cdots$$

which indicates $f(0t) = 0, f(1t) = 4, f(2t) = 4, f(3T) = 2$, etc. Repeated steps of long division do not give a closed-form expression for the sequence represented by the Z transform. Although, in principle, as many terms in the sequence as desired may be found by the long division.

The relationship between Z transforms is used to understand the stability of the system. If $f(t) = e^{at}$ for $t > 0$, the Laplace transform of $f(t)$ is $F(s) = 1/(s+a)$, the roots of the characteristic equation are at $s = -a$. The Z transform of the pulse train signal $f(nT)$ is

$$F(Z) = \sum_{n=0}^{\infty} e^{-ant} Z^{-4} = \frac{1}{1 - e^{at}Z^{-1}}$$

The root of the characteristic equation in the Z transform is e^{-at}. This means that the roots $s = -a$ in the Laplace domain correspond to a root at $Z = e^{-at}$ in the Z domain or discrete domain. Thus the relationship between the S and Z planes is $Z = e^{-st}$, which is obtained by substituting $-a = s$. It is well known that for the stability of the system, the roots of the characteristic polynomial should be a negative real number or complex number with a negative real part in the Laplace domain. Consider a complex root since

$$s = \sigma \pm j\omega_d$$

$$Z = e^{-sT} = e^{\sigma \pm /\omega_d} = e^{\sigma T} e^{\pm j\omega_d T}$$

$$|Z| = |e^{\sigma T}||e^{+j\omega_d T}| = |e^{\sigma T}|$$

If $\sigma = 0$, the roots lie in the imaginary axis of the s plane. Then $|Z| = 1$ in the Z plane, which means that the imaginary axis in the s plane corresponds to the unit circle in the Z plane. If $\sigma < 0$, the roots are negative real numbers in the s plane and the system is stable, corresponding to the roots inside the unit circle in the Z plane. Hence, the system is said to be stable when the roots of the characteristic equation in the Z plane lie inside the unit circle. If $\sigma > 0$, the roots are positive real numbers in the D plane and the system is unstable. Hence, if the roots of the characteristic equation lie outside the unit circle in the Z plane, the system is unstable.

Digital-to-analog conversion requires continuous time representation of discrete time signals. This process is known as reconstruction. The sampled sequence $f(nT)$ and impulse train $f^*(nT)$ are useful representations of the samples in the analog domain. The objective of reconstruction is to recover the analog signal $f(t)$ from the samples $f(nT)$. A delayed impulse with a different amplitude provides a delayed pulse than that holding the amplitude. A zeroth order hold is commonly used. Figure 6.19 illustrates the reconstruction of the signal from the pulse train data.

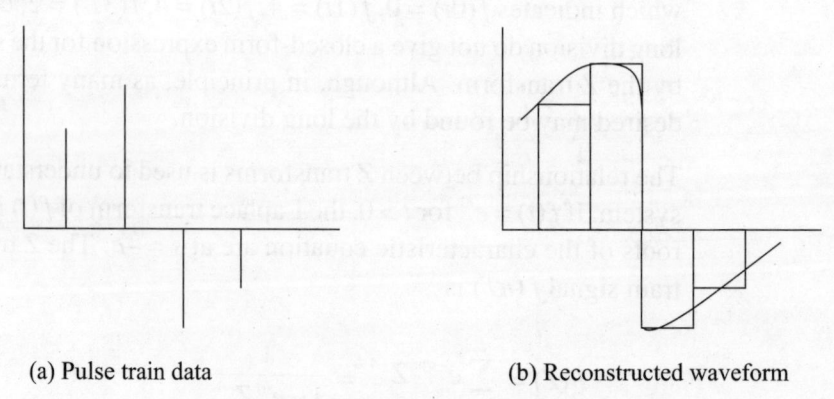

(a) Pulse train data   (b) Reconstructed waveform

Fig. 6.19 Reconstruction of signal

Illustrative Examples

Example 6.1 The following equations refer to a tension regulating apparatus used in the paper industry. The main input $x = F_R - A$, lever measurement $e = (x - y)/2$, the change in torque provided by the motor, i.e., $T_m = K_m/(1 + \tau_p) e$, roll tension $F_c = T_m/R$, tension in the control spring $y = 2F_c/K$. Draw the individual block diagrams and determine the overall transfer function.

Solution

From the given equations, the block diagrams of the individual blocks can be obtained as follows:

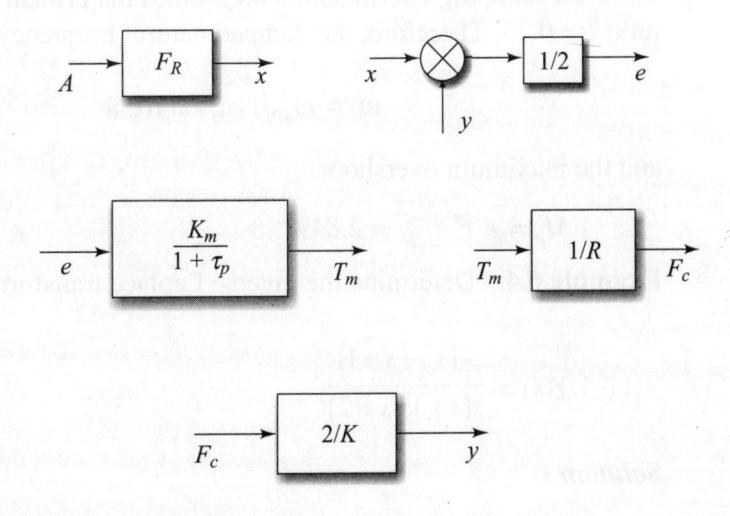

Combining all the individual blocks and by block diagram manipulation, we get the transfer function as

$$\frac{y}{x} = \frac{K_m}{RK(1+z_p)}$$

$$x = F_R A$$

Example 6.2 In a spring damper system, $K = 10$ N/cm and $C = 2$ n/cm sec^{-1}. The spring is suddenly pulled by a force to a height of 5 cm and held at that position. Determine the time taken for the response of the damper to attain 60% of its full value.

Solution

Time constant $\quad T_1 = K/c = 10/2 = 5$

Output response $\quad y(t) = K(1 - e^{t/T_1})$

60% of the input amplitude,

$$y(t) \text{ at } 60\% = 0.6 \times K = 0.6 \times 5 = 3 \text{ cm}$$

Hence the time required for the response to reach 60% of amplitude is 4.58 seconds.

Example 6.3 A servo system for the position control of a rotary table mass is stabilized by viscous friction damping, which is 75% of that needed for critical damping. The undamped natural frequency of the system is 16 Hz. Determine the damped natural frequency and maximum overshoot.

Solution

Since the damping coefficient is 0.75 times the critical damping, the damping ratio $\xi = 0.75$. Therefore, the damped natural frequency

$$\omega_d = \omega_n\sqrt{1-\xi^2} = 10.58$$

and the maximum overshoot

$$M_p = e^{-\varsigma\pi/\sqrt{1-\varsigma^2}} = 2.84\%$$

Example 6.4 Determine the inverse Laplace transform of $Y(s)$, if

$$Y(s) = \frac{-(s^2-s-1)}{s(s+1)(s+2)}$$

Solution

The partial fraction expansion of $Y(s)$ is

$$Y(s) = \frac{C_1}{s} + \frac{C_2}{s+1} + \frac{C_3}{s+2} = \frac{1}{2}$$

$$C_1 = \frac{-(s^2-s-1)}{(s+1)(s+2)} \text{ at } s = 0 = -1$$

$$C_2 = \frac{-(s^2-s-1)}{s(s+2)} \text{ at } s = -1 = \frac{-1}{2}$$

$$C_3 = \frac{-(s^2-s-1)}{s(s+2)} \text{ at } s = -2$$

Hence,

$$y(t) = \frac{1}{2} - e^{-t} - \frac{1}{2}e^{-t}$$

Example 6.5 Obtain the transition matrix for the state matrix

$$A = \begin{bmatrix} 0 & 1 \\ 2 & 3 \end{bmatrix}, \quad B = \begin{bmatrix} 1 & 0 \\ 0 & 1 \end{bmatrix}, \quad C = \begin{bmatrix} 1 & 0 \\ 0 & 1 \end{bmatrix}$$

Solution

$$W_0(s) = \text{adj}\,(sI - A) \begin{bmatrix} s-3 & 1 \\ 2 & s \end{bmatrix}$$

$$H_0(s) = \det(sI - A) = s^2 - 3s - 2$$

Transition matrix

$$\phi(t) = L^{-1} \frac{1}{s^2 - 3s - 2} \begin{bmatrix} s-3 & 1 \\ 2 & s \end{bmatrix}$$

Exercises

6.1 What are the advantages of a closed-loop system?

6.2 Write brief notes on system types.

6.3 Obtain the transfer functions of the zeroth, first, and second order mechanical systems.

6.4 Obtain the transfer functions of the zeroth, first, and second order electrical systems.

6.5 Explain the importance of the roots of the characteristic equation considering a second order system.

6.6 Explain the advantages and disadvantages of different types of controllers.

6.7 Obtain the state space matrices from the transfer functions given below:

(a) $\dfrac{Y(s)}{X(s)} = \dfrac{-4s+3}{s^2+6s+2}$

(b) $\dfrac{Y(s)}{X(s)} = \dfrac{-4s^2+1}{s^3+6s^2+2s+5}$

(c) $\dfrac{Y(s)}{X(s)} = \dfrac{3s^2+6}{s^3+6s^2+2s+5}$

6.8 Obtain the transfer functions from the state space matrices given below:

(a) $A = \begin{bmatrix} 2 & 9 \\ -1 & 4 \end{bmatrix}$, $B = \begin{bmatrix} 4 \\ -3 \end{bmatrix}$, $C = \begin{bmatrix} 2 & -6 \end{bmatrix}$

(b) $A = \begin{bmatrix} -6 & 1 & 0 \\ -9 & 0 & 1 \\ 0 & 0 & 1 \end{bmatrix}$, $B = \begin{bmatrix} 1 \\ 2 \\ 1 \end{bmatrix}$, $C = \begin{bmatrix} 1 & 0 & 0 \\ 0 & 1 & 0 \end{bmatrix}$

(c) $A = \begin{bmatrix} -2 & 1 \\ -3 & 0 \end{bmatrix}$, $B = \begin{bmatrix} 4 \\ 5 \end{bmatrix}$, $C = \begin{bmatrix} 1 & 0 \end{bmatrix}$

6.9 Prove the following, if ϕ is the transition matrix:

(a) $\phi(0) = I$

(b) $\phi(t_1 + t_2) = \phi(t_2)\phi(t_1)$

(c) $[\phi(t)]^n = \phi(nt)$

6.10 Obtain the computer algorithm for a PI controller.

6.11 Find the Z transform transfer function for the following cases:

(a) $y(n+3) + 3y(n+2) - y(n+1) + y(n) = u(n+2) + u(n+1) - 4u(n)$

(b) $y(n+3) = 0.5y(n+2) - y(n+1) + 0.125y(n) = 10\,u(n+3)$

(c) $y(n+2) - 5y(n+1) - 6y(n) = 2u(n)$

6.12 Find the inverse Z transform for the following:

(a) $\dfrac{Y(s)}{X(s)} = \dfrac{4}{Z+3}$

(b) $\dfrac{Y(s)}{X(s)} = \dfrac{z}{2z-1}$

6.13 Determine whether the discrete systems given below are stable or not.

(a) $\dfrac{Y(z)}{U(z)} = \dfrac{-3z^2 + 1}{4z^2 + 2z - 1}$

(b) $\dfrac{Y(z)}{U(z)} = \dfrac{5(z - 0.5)}{z^3 - 2.8z^2 + 1.75z - 0.3}$

Chapter 7

Motion Control Devices

Motion control devices are the elements of mechatronic systems, which are responsible for transforming the output of a microprocessor or control system into a controlling action on a machine or device. For example, motor control devices transform electrical signal output from a controller into a linear or rotary motion. Motion control devices are called servo systems. Hydraulic, pneumatic, and electrical systems can be used as servo systems. Hydraulic systems are well suited for large systems in which power requirements are high. The cost of hydraulic drives is not proportional to the power requirements and, thus, they are expensive for small- and medium-sized mechatronic systems. However, the speed of the hydraulic system is very low compared to that of a pneumatic or an electrical system. Pneumatic systems provide fast and high load-carrying mechatronic systems. The manufacturing of a pneumatic system is expensive, since the accuracy requirements are high and the working medium is compressed air. Electrical actuators have a very high speed of operation without any delay. However, the system is bulky for high-power applications as compared to hydraulic or pneumatic systems.

A rigid body can have very complex motion, which might seem difficult to describe. The motion of any rigid body can be considered to be a combination of translational and rotational motion. By considering three-dimensional (3D) space, a translational motion can be considered to be the movement that can be resolved into components along one or more of the three axes of the 3D space. Similarly, rotation can be considered as a motion that has components rotating about one or more of the axes. A cyclic motion is a combination of translational and rotational motion. Three axes' translation and three rotations give six degrees of freedom. If a joint is constrained to move along a line, then it is a translation with one degree of freedom. A constraint is needed for each degree of freedom that is to be prevented from occurring, provided there is no redundant constraint; then the number of the degrees of freedom would be six minus the number of constraint(s). However, the redundant constraint often occurs. So the basic rule for constraint on a single rigid body is

Number of constraint(s) = Number of degrees of freedom − Number of redundancy

Thus, if a body is required to be fixed, i.e., needs no degree of freedom, no redundant constraints are introduced and the number of constraints required is 6.

The principle of least constraint(s) is used to design devices. This principle states that in fixing a body or guiding it to a particular type of motion, the minimum number of constraints should be used, that is, there should not be redundancy. This is often referred to as the kinematic design. For example, to have a shaft that rotates only about one axis with no translation motion, the number of degrees of freedom should be reduced to 1. Thus, the minimum number of constraints to achieve this is 5 and any more constraints than this will lead to redundancy. If the shaft is mounted on one end of a ball bearing and other end is at roller bearing, the bearing pair together prevents translation and rotation at right angle to the shaft axis. The ball bearing prevents translation along the x axis of the shaft, whereas roller bearing allows motion along the axis of the shaft. Thus, there is a total of five constraints; which give one degree of freedom, rotation about axis of the shaft. If there is a ball bearing at each end of the shaft, then bearings could have prevented translation along the axis of the shaft, thus, there would have been redundancy. Such redundancy can cause damage. If ball bearings are used at both ends of the shaft, then in order to prevent redundancy, one of the bearing's outer race is not fixed in its housing so that it could slide in the axial direction.

Actuators are structures that transmit and support load. A joint is a connection between two or more links at their nodes and allows some motion between the connected links. Levers, cranks, connecting rods, pistons, sliders, pulleys, belts, and shafts are all examples of links. A sequence of joints and links is known as a *kinematics chain*. For a kinematics chain to transmit motion, one link must be fixed. The movement of one link produces predictable relative movement of other links in the chain. For mechatronics system actuation, one can use hydraulic, pneumatic, or electrical drives with kinematic chains.

7.1 Hydraulic and Pneumatic Actuators

The hydraulic actuation is powered by fluids. Fluids usually are pressurized oils. The operation of hydraulic actuators is generally similar, except in their ability to contain the pressure of the fluid. Hydraulic systems operate in a pressure range between 60 bars and 200 bars. The main constituents of a hydraulic system are the power supply unit, hydraulic fluid, direction control valve, linear and rotary actuators, and interaction components. The power supply unit is the most important component in a hydraulic pump. The pump drives the hydraulic fluid from a reservoir (tank) and delivers it through a system of lines in the hydraulic installation against the offering resistance. Pressure should not build up in the flowing liquid which encounters a resistance. An oil filtration unit is also often contained in the power supply section. Impurities are often introduced into a system as a result of mechanical wear. For this reason, filters are installed in the hydraulic circuit to remove impurities in the form of dirt particles from the hydraulic fluids. Water and gases in oil can also act as disrupting factors, so special measures must be taken to remove them. Heaters and coolers are installed for conditioning the hydraulic fluids. The hydraulic fluid is the working medium that transfers the generated energy from the power supply unit to the drive section. Hydraulic fluids have a wide range of characteristics. Therefore, care needs to taken to choose a fluid with characteristics that suit the application. Hydraulic fluids on a mineral oil base are generally used. Such fluids are called hydraulic oils.

Valves are the devices for controlling the energy flow. They can control and regulate the direction, pressure, and rate of the hydraulic fluid flow. Four types of valves are commonly used: direction control valves, pressure valves, flow control valves, and non-return valves. Direction control valves control the direction of flow of hydraulic fluid and thus the direction of motion and the positioning of the working components. Direction control valves may be actuated manually, mechanically, electrically, pneumatically, or hydraulically. They convert and amplify signals and form an interface between the power control section and the signal control section. When labelling direction control valves, it is necessary to specify the number of ports, followed by the number of switching position. Direction control (DC) valves have at least two switching positions and two ports. Such a valve is designated as a 2/2 way valve.

Pressure valves are used to influence the pressure in the complete hydraulic system. There are three main types of pressure control valves:

(1) Pressure regulating valves used to maintain a constant operating pressure in a circuit.

(2) Pressure sequence valves used to sense the pressure of an external line and give a signal when it reaches some preset value.

(3) Pressure limiting valves used as safety devices to limit the pressure in a circuit to below some safe value.

The flow control valve interacts with pressure valves to control the flow rate. They both make it possible to control or regulate the speed of motion of the power components. If the flow rate is constant, the division of flow must take place. This is generally achieved through the interaction of the flow control valve with the pressure valve. Non-return valves block the flow in one direction and permit free flow in the other direction. As there must not be any leakage from the closed direction, these valves always have a poppet design. In the case of a non-return valve, a distinction is made between ordinary non-return valves and piloted non-return valves. In the case of piloted non-return valves, flow in the blocked direction can be released by a signal. Figure 7.1 shows schematic diagrams of various control valves.

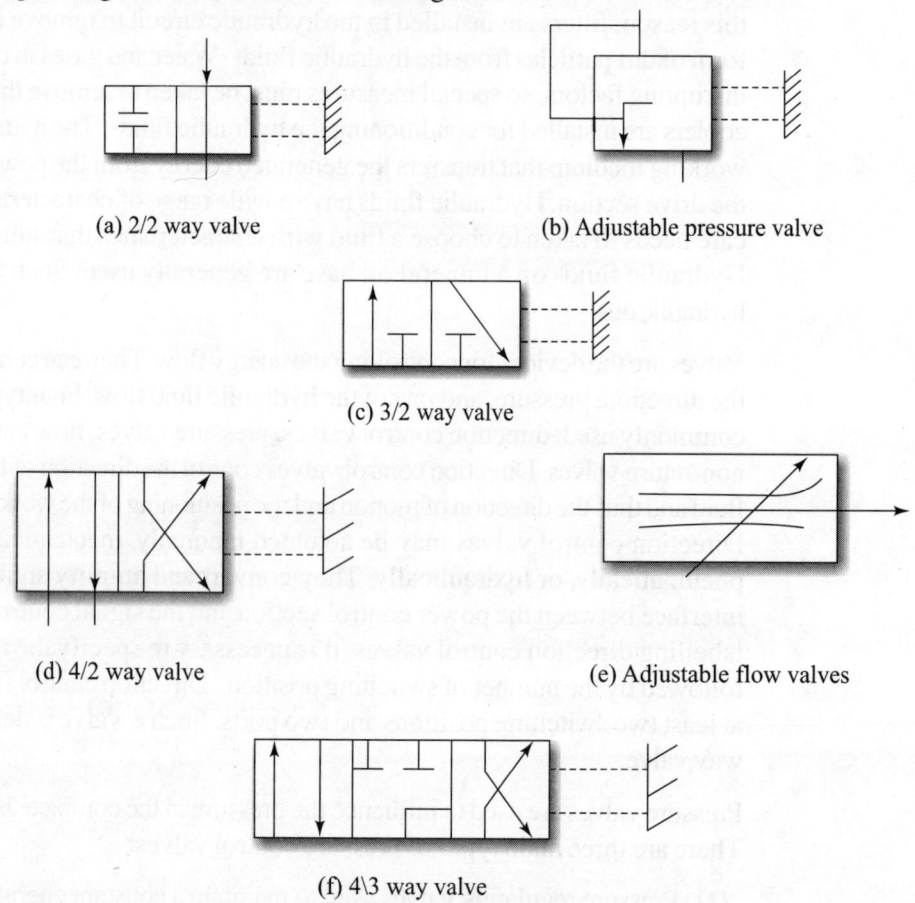

(a) 2/2 way valve

(b) Adjustable pressure valve

(c) 3/2 way valve

(d) 4/2 way valve

(e) Adjustable flow valves

(f) 4\3 way valve

Fig. 7.1 Control valves

Cylinders are drive components that convert hydraulic power into mechanical power. They generate linear movements while passing on the surface of a movable piston. In single acting cylinders, the fluid pressure can be applied to either side of the piston, with the result that the drive movement is produced only in one direction. The return stroke of the piston is affected by an external force or by return speed. In double acting cylinders, the fluid pressure can be applied to either side of the piston, with the result that the drive movements are produced in two directions. Like cylinders, hydraulic motors are drive components controlled by valves. They too convert hydraulic power into mechanical power, but the difference is that they generate rotary or swivel movements instead of linear movement. Figure 7.2 shows the schematic diagrams of a single acting cylinder and a rotary actuator. Two parameters are of particular interest while discussing actuators, namely, the velocity of actuation and the force of actuator. The first depends on the flow rate, where as the latter depends on the input fluid pressure. A number of different alternatives such as pipes, filters, etc. are used for interaction of hydraulic components.

Force $F = \dfrac{P}{(\pi/4)D^2}$

Speed $V = \dfrac{Q}{(\pi/4)D^2}$

Force $F = P(R - r)b$

Torque $T = P(R - r)b\left(\dfrac{R + r}{2}\right) = Pb\left(\dfrac{R^2 + r^2}{2}\right)$

Angular velocity $\omega = Q\dfrac{2}{b(R^2 + r^2)}$

P—supply pressure
Q—fluid flow
D—diameter of the piston

R—vane radius
r—hub radius
b—width of the vane

(a) Single acting cylinder

(b) Vane or rotary actuator

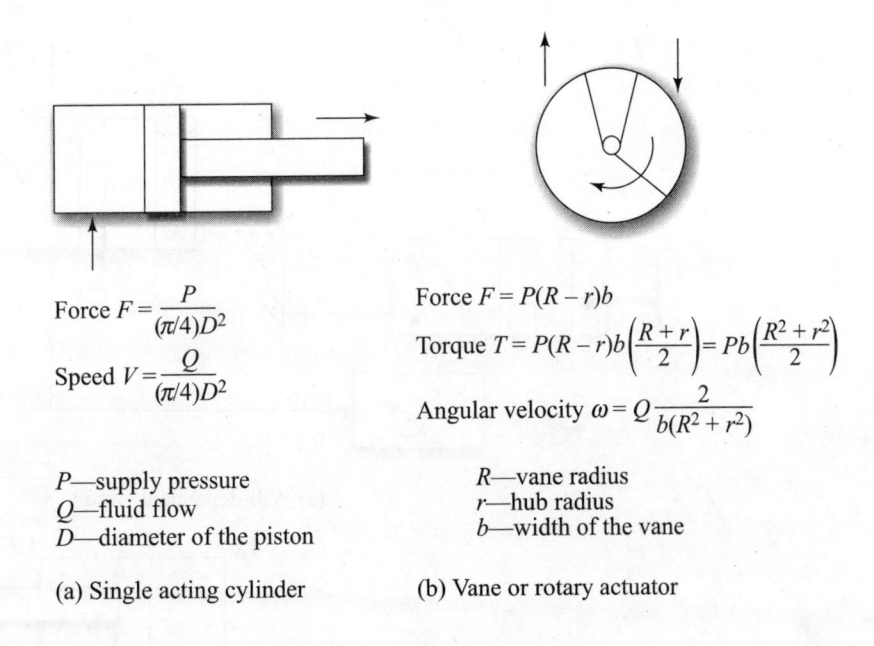

Fig. 7.2 Hydraulic actuator

Many mechatronic systems employ the hydraulic cylinder as an actuating element and require a sequence of extensions and retractions of the cylinder to occur. The sequence of operation can be designed in three ways, namely, will-, travel-, and time-dependent sequence operations. In will-dependent sequence operation, actuation performs according to the will of the operator. In travel-dependent sequence operation, the initial actuation of the sequence is by pushing the start button. Afterwards, the travel of the joint piston actuates the operation of the second piston, and the travel of the second piston is used to actuate another operation. This completes the cycle of operation. The cycle can be started by pushing the start button again. If one wants the system to run continuously, then the last movement in the sequence would have to trigger the first movement. In time-dependent operation, each sequence of operation starts with a time delay, which is incorporated into the sequence circuit diagram. Time-dependent sequence operations are not used for simple sequence circuits. A combination of travel-, will-, and-time dependent sequence circuits is used in some mechatronic systems. Figure 7.3 illustrates the three types of commonly used sequence circuits.

156   Introduction to Mechatronics

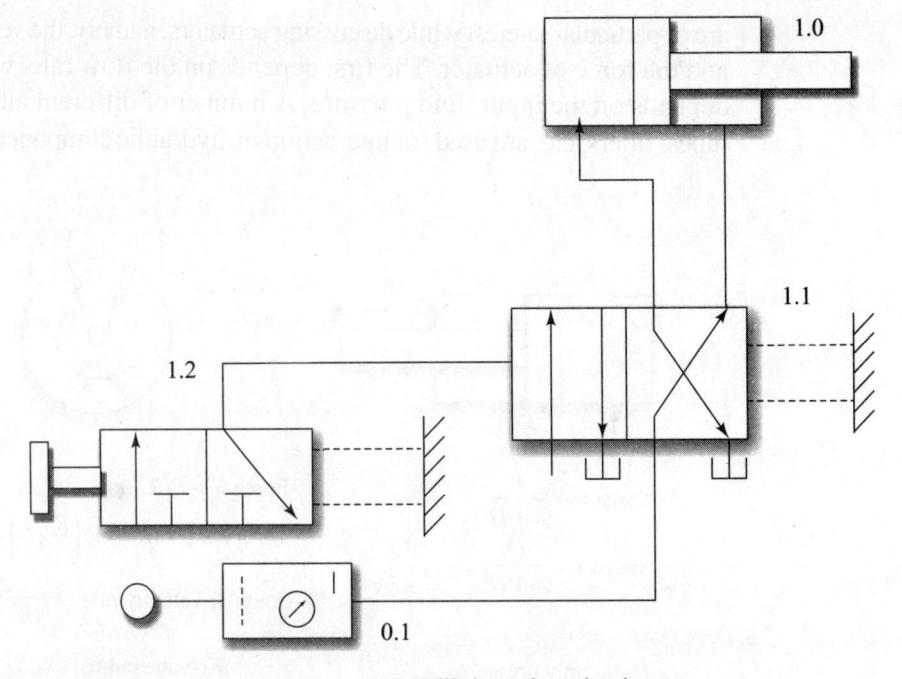

(a) Will-dependent circuit

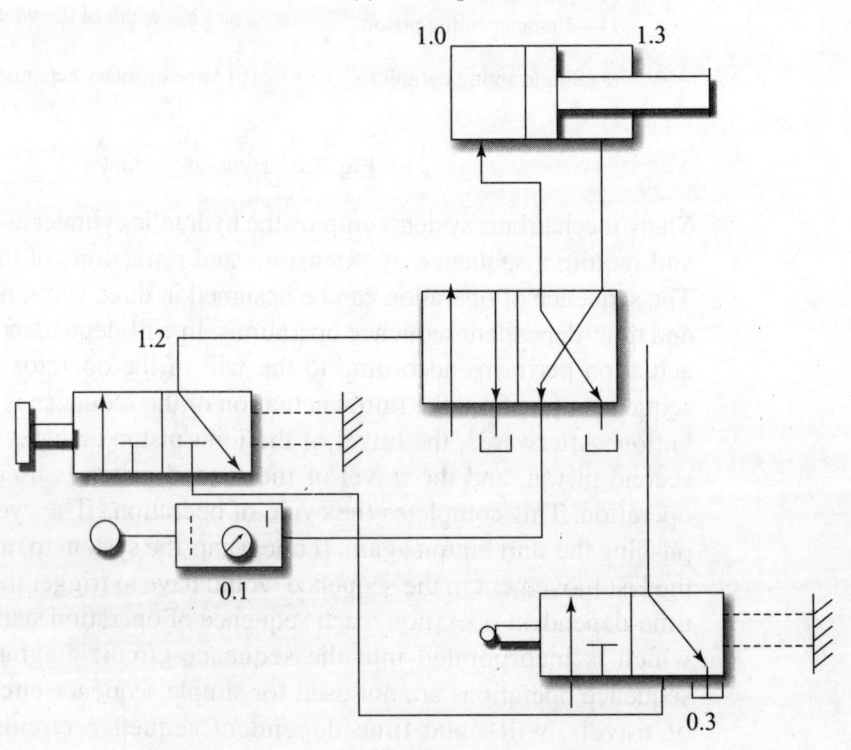

(b) Travel-dependent circuit

(c) Time-dependent circuit

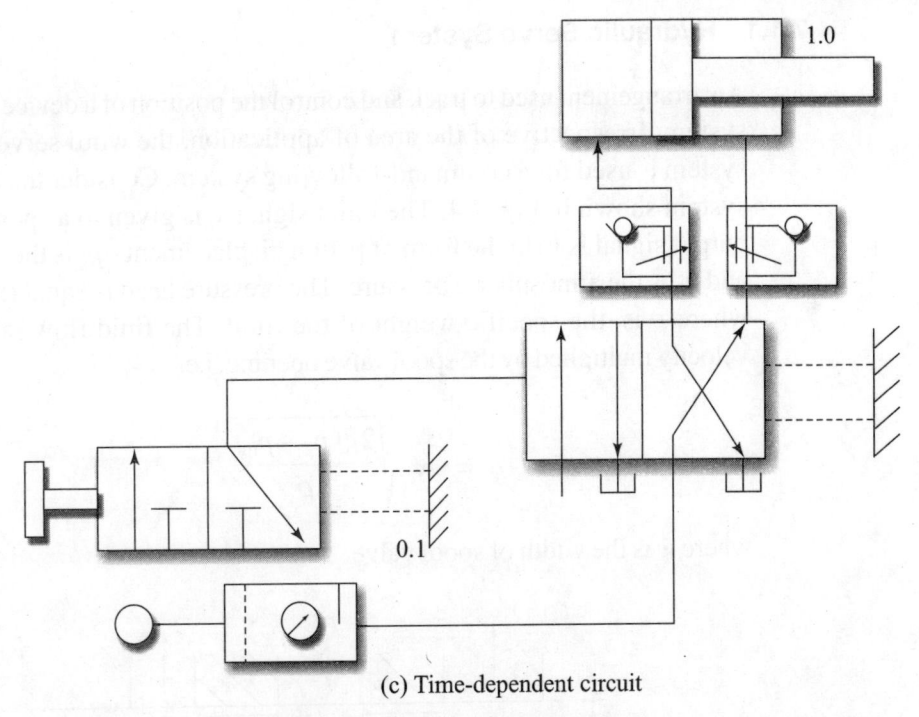

Fig. 7.3 Types of sequence circuits

Some advantages of hydraulic systems are as follows:

1. High load-carrying capacity
2. Low actuator inertia
3. Simple field design of system
4. High flexibility
5. Indefinite stocking capacity
6. High speed at high load
7. Very good strength

Some disadvantages of hydraulic system are as follows:

1. High cost of servo system
2. Need for high resolution feedback
3. Effect of oil temperature on performance
4. Non-availability of small actuators
5. Leakage
6. Low sensitivity
7. Difficulties in maintenance
8. Requirement of skilled workers to connect the system
9. Less natural frequency when actuator is in full value

7.1.1 Hydraulic Servo System

An arrangement used to track and control the position of a device is called a servo system. Irrespective of the area of application, the word servomotor or servo system is used for a command-following system. Consider the hydraulic servo system shown in Fig. 7.4. The input signal x_i is given to a spool valve and the output signal is y (in the form of piston displacement). p_s is the supply pressure and p_0 is the atmospheric pressure. The pressure head is equal to $(p_s - p_0)/\rho = h$, where ρ is the specific weight of the fluid. The fluid flow rate q is equal to velocity multiplied by the spool valve opening, i.e.,

$$q = wx_i \sqrt{\frac{2g(p_s - p_0)}{\rho}}$$

where w is the width of spool valve.

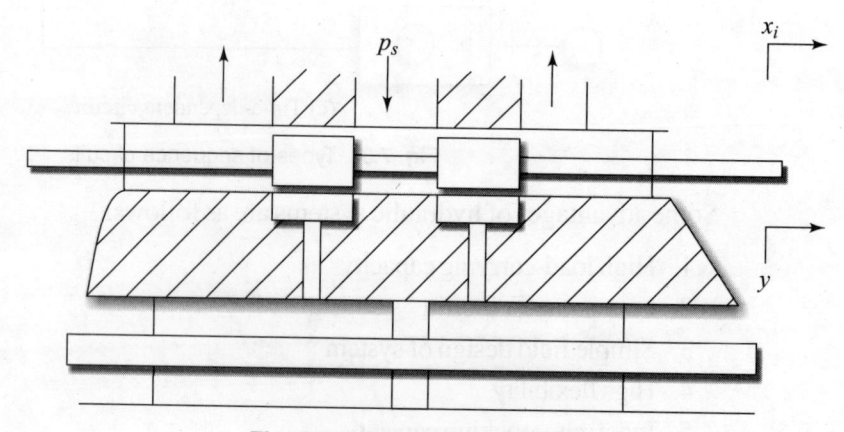

Fig. 7.4 Hydraulic servo system

Due to the flow rate, the piston moves forward, and hence the rate of fluid displaced by the piston is

$$q = A\frac{dy}{dt}$$

where A is the area of the piston. Equating, we get

$$A\frac{dy}{dt} = \{wx_i \sqrt{\frac{2g(p_s - p_0)}{\rho}}$$

Taking Laplace transform,

$$\frac{Y(s)}{X(s)} = \frac{K_v}{s}$$

where $K_v = W\sqrt{2g(p_s - p_0)/\rho}\ A$ is known as the velocity constant. The hydraulic servo system acts as an integral system.

7.1.2 Pneumatic System

Pneumatic systems have been used for some considerable time for carrying out simplest mechanical tasks. In more recent times, such systems have played an important role in the development of pneumatic technology for automation through mechatronic systems. A pneumatic system can be broken down into a number of levels representing hardware and signal flow. Except the energy supply system, all other levels can be represented as in a hydraulic system. The air supply for a particular pneumatic application should be sufficient and of adequate quality. The air is compressed to 1/7th of its volume with the help of an air compressor and is delivered to an air distribution system in factory. To ensure proper quality of the air, air service equipment is utilized to prepare the air before it is applied to a control system. As a rule, all pneumatic components are designed for a maximum operating pressure of 8–10 bars. But it is recommended to operate the same between 5 and 6 bars for economic use. An air receiver is fitted to reduce pressure fluctuations. In normal operations, a compressor is fitted with the receiver when required, and the receiver is available as a reserve at all the times. This helps reduce the switching cycle of the compressor. If oil is required for a pneumatic system, then there should be a separate oil meter using air service unit. An air service unit is a combination of compressed-air filter, regulator, and lubricator. The compressed-air filter performs the job of filtering all contamination from the compressed air flowing through it as well as water which has already condensed. The purpose of the regulator is to keep the operating pressure virtually constant regardless of any fluctuation in line pressure and air consumption. Different types of sequence operations can be designed using sequence circuits associated with hydraulic systems. Pneumatic actuators are very useful and have the following advantages:

1. Compressed air is readily available in most factories.
2. Compressed air can be stored and conveyed easily to larger distances.
3. Compressed air need not be returned to sump.
4. Compressed air is clean.
5. Operation is fast and offers high load-carrying capacity.
6. Digital and logical switches can be prepared using fluidic circuits.
7. Pneumatic elements are simple and reliable.

7.1.3 Diaphragms and Bellows Actuator

A wide variety of flexible metallic elements can be used as pneumatic actuators. Flat diaphragms are widely used for micro-actuators. Using two diaphragms side by side, a pneumatic capsule is formed. Corrugated diaphragms are capsules having large displacement actuation. Central deflection due to pneumatic air supply is used for actuation. Very large actuation can be obtained by using bellows. The bellows and diaphragms are used relatively at low pressures. The bellows and diaphragm serve the same function. If bellows and diaphragms are used for cyclic loading, then their fatigue strength is important. Usually high carbon steel, phosphor bronze, or beryllium copper is used for bellows or diaphragms. Figure 7.5 illustrates the working principle behind small displacement pneumatic actuators.

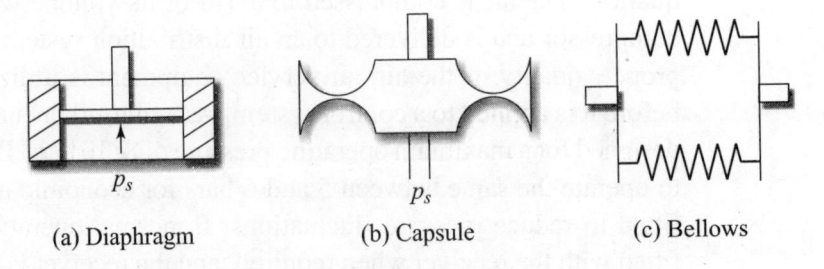

p_s p_s

(a) Diaphragm   (b) Capsule   (c) Bellows

Fig. 7.5 Small displacement pneumatic actuators

7.2 Electrical Actuators

Electrical systems are used as actuators in many mechanical systems. There are primarily three different classes of electrical systems, namely, switching devices, solenoid type devices, and drive systems. The switching devices include mechanical switches and solid-state switches. Relays and limiting switches are examples of mechanical switches. Diodes, thyristors, and transistors are the control signal switches for on-and-off electrical devices. There are a number of mechanical switches which can be activated by lever operators, roller operators, or cam operators to give signal. The method of actuation of a hydraulic or pneumatic directional control (DC) valve is dependent on the task requirement. For travel-dependent actuation, mechanical action is used. The type of actuation may be mechanical, hydraulic, pneumatic, electrical, or a combination of any of these. In the solenoid type devices, current through solenoid is used to actuate a soft iron core. Direct current, alternate current, and stepper motors are drive systems. We will now discuss these in some more detail in the following subsections.

7.2.1 Mechanical Switches

The electrical relay is an example of the mechanical switch which is used in a control system as an actuator. It offers a simple on/off switching action in response to a control signal. Figure 7.6(a) illustrates the mechanics behind the working of a relay. When current flows through the coil of wire, a magnetic field is produced. This magnetic field pulls a movable arm, called the armature, which forces the contacts to open or close. Usually, there are two sets of contacts, with one being opened and the other closed by the actuator. This action can then be used to supply current to a motor or an electric heater in mechanical systems. The relay may be used to control the action of pneumatic valves, which in turn control the movement of the piston. Time relays are controlled relays that facilitate a delayed switching action. The time delay is usually adjustable and can be initiated after current flow through the relay coil ceases.

(a) Relay switch

(b) Lever, roller, and cam-operated switches

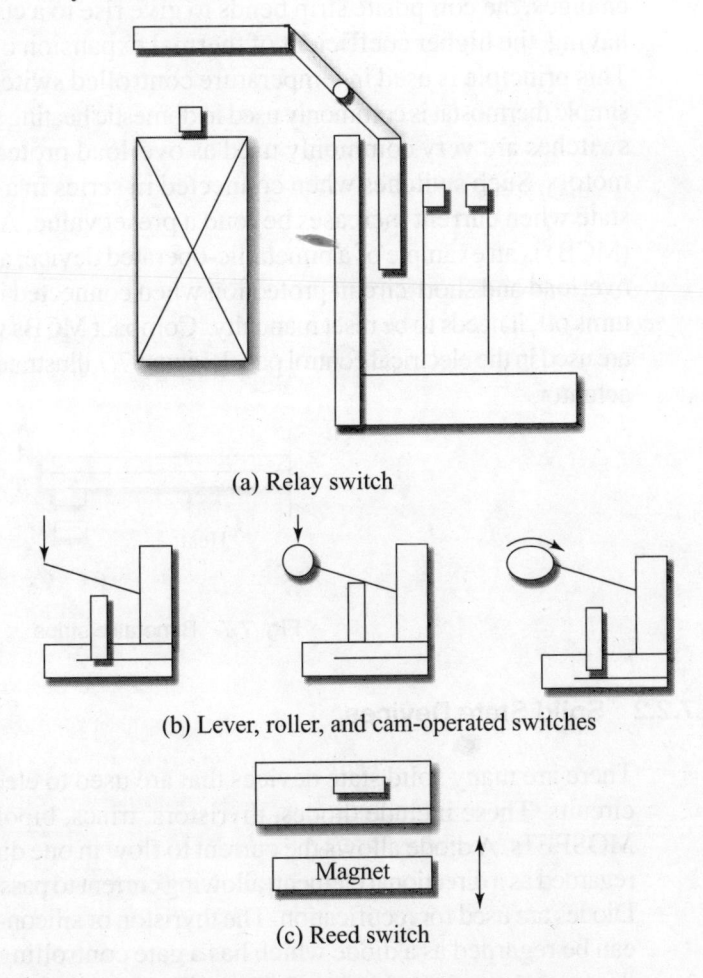

Magnet

(c) Reed switch

Fig. 7.6 Mechanisms of switches

A reed switch consists of two magnetic switches in a cut-out sealed in a glass tube. When a magnet is brought close to the switch, the magnet leads are attracted to each other and close the switch contacts [Fig. 7.6 (c)]. Such a switch is widely used for ensuring the closure of doors. It is also used with switching devices as a tachometer, which involves rotation of a toothed wheel past the reed switch. If one of the teeth has a magnet attached, then every time it passes the switch, it will momentarily close the contact and hence produce a current/voltage pulse in the associated electrical circuit.

Micro-switches are small electrical switches that require physical contact and a small operating force to close the contacts. Figure 7.6(b) illustrates the working of lever, roller, and cam-operated switches.

Bimetallic strips consist of two different metallic strips bonded together. The metals used have different thermal expansion coefficients. So when the temperature changes, the composite strip bends to give rise to a curved shape with the metal having the higher coefficient of thermal expansion on the outside of the curve. This principle is used in temperature controlled switch systems. For example, a simple thermostat is commonly used in domestic heating systems. Bimetallic thermal switches are very commonly used as overload protection devices, for running motors. Such switches when connected in series in a power circuit go to an off state when current increases beyond a preset value. A miniature circuit breaker (MCB) is an example of a bimetallic-operated device, and it usually provides both overload and short-circuit protection when connected in a circuit. When an MCB turns off, it needs to be reset manually. Compact MCBs with auxiliary strip contacts are used in the electrical control panel. Figure 7.7 illustrates the principle of bimetallic actuator.

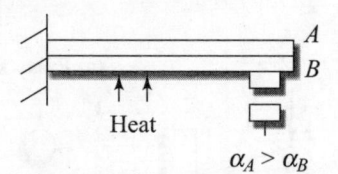

$\alpha_A > \alpha_B$

Fig. 7.7 Bimetallic strips

7.2.2 Solid State Devices

There are many solid state devices that are used to electronically control switch circuits. These include diodes, thyristors, triacs, bipolar transistors, and power MOSFETs. A diode allows the current to flow in one direction only. It can thus be regarded as a directional element, allowing current to pass only when forward biased. Diodes are used for rectification. The thyristor, or silicon-controlled rectifier (SCR), can be regarded as a diode which has a gate controlling conductor under which a diode can be switched on. An SCR is a PNPN device having four layers of P and N types as shown in Fig. 7.8. There are three terminals, called the anode, cathode, and gate. The SCR works like a controlled rectifier, in which the conductor can be controlled by a triggering signal at the gate terminal. The SCR is fixed or turned on by a pulse of current flowing through its gate. The SCR remains open until it triggers the gate. Then the SCR latches and remains closed even if the triggering pulse at the gate is disconnected. The only way to turn off the SCR is by decreasing the anode current nearly to zero. The triac is similar to the thyristor and is equivalent to a pair of thyristors connected in reverse and parallel on the same chip. The triac can be turned on in either the forward or reverse direction. Figure 7.8 illustrates the basic principle of thyristor and triac, as well as their characteristics.

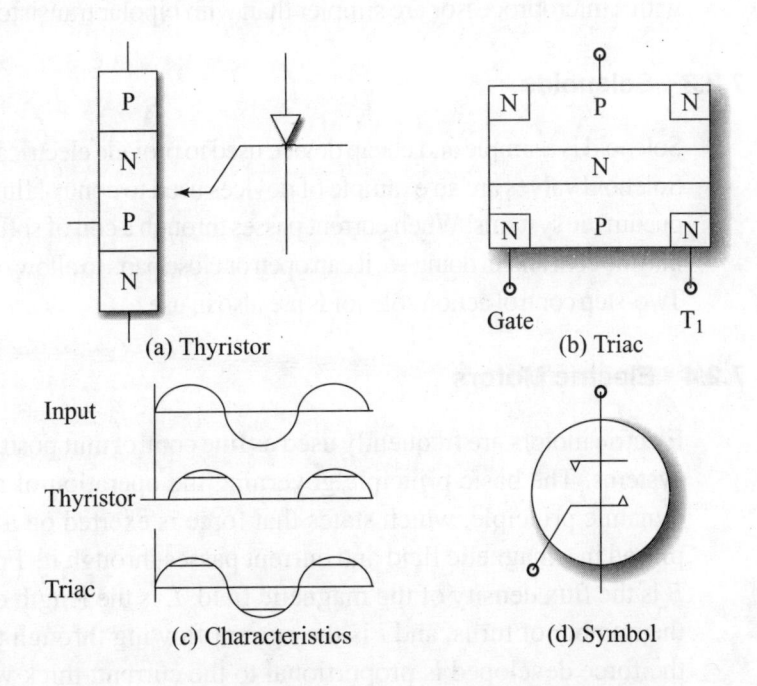

Fig. 7.8 Thyristor and triac characteristics

Bipolar transistors come in two forms, namely, NPN and PNP forms. For the NPN transistors, the main current flows in at the collector and out at the emitter. As against this, the PNP transistor has the main current flowing in at the emitter and flowing out at the collector. However, the control signal is applied to the base in both the types. When a small NPN transistor is combined with a large PNP transistor, the resultant is a large NPN transistor with a large amplification factor. Similarly, a small PNP transistor in conjunction with a large NPN transistor results in a single, large PNP transistor. Such a combination of transistors that enables a high current to be switched with a small input current is termed as a *Darlington pair*. These switches are available as a single chip. Bipolar transistor switching is implemented by basic current. Bipolar transistors allow a higher frequency of switching than what is possible with a thyristor. However, the power handling capability of bipolar transistors is less than that of thyristors.

Metal oxide field effect transistors (MOSFETs) come in two types, namely, N-channel and P-channel. The main difference between the use of a MOSFET and a bipolar transistor for switching is that in MOSFET no current flows into the gate to exercise control. The gate voltage is the controlling signal, thus the drive circuit can be amplified in that there is no need to be concerned about the size of the current. With MOSFETs, very high frequency switching is possible, up to 1 MHz, and interfaces with a microprocessor are simpler than with bipolar transistors.

7.2.3 Solenoids

Solenoid is a simple and cheap device used to provide electrically operated actuators. Solenoid valves are an example of devices used to control fluid flow in hydraulic or pneumatic systems. When current passes through a coil of soft iron, the core is pulled into the coil, and in doing so, it can open or close parts to allow or stop the flow of fluid. Two-step control action solenoids are also in use.

7.2.4 Electric Motors

Electric motors are frequently used as fine control unit positional or speed control systems. The basic principle governing the operation of a motor is the elctro-dynamic principle, which states that force is exerted on a conductor when it is placed in a magnetic field and current passes through it. Force $F = BLNI$, where B is the flux density of the magnetic field, L is the length of the conductor, N is the number of turns, and I is the current flowing through the conductor. Since the force developed is proportional to the current, thick wires have to be used for high power electric motors. The second law of electrodynamics states that when a conductor moves in a magnetic field, an emf, which is equal to the rate at which the magnetic flux shifts through by the conductor changes, is induced. The induced emf $e = BLNV$, where V is the velocity at which the conductor is moving. The second law of electrodynamic is used in electric generators.

The advantages and disadvantages of electric motors are as follows.

Advantages

1. High performance control can be achieved at lower cost.
2. A wide range of motor sizes and capacity is available.

3. Accuracy and repeatability of electric drives are excellent.
4. Electric motors can provide the best resolution.
5. Electric actuators are smaller, hence require less floor space.
6. Temperature stability for overload up to twice of the minimal load for small durations is possible.

Disadvantages

1. Limited choice of gear box.
2. Power not as much as provided by a hydraulic system.
3. Need for energy peak loading duty cycle to avoid thermal load on the motor.
4. Motor and gear box inertia is high.
5. To reverse, a special controlled circuit is required.

Motors are classified into two main categories, namely, direct current (DC) and alternating current (AC) motors. DC motors are used in most of the modern control systems which act as closed-loop systems. Stepper motors are used for many mechatronic systems which work as open-loop systems.

In the conventional DC motor, coils of wire are mounted in slots on a cylinder of magnetic material called armature. The armature is mounted on a bearing and is free to rotate. It is mounted in the magnetic field produced by field poles. The magnetic field may be produced by a permanent magnet or an electromagnet. The field may be self-excited or separately excited. Each end of the armature coil is connected to the adjacent segments of a segmented ring called commutator. Electric contacts to the segments are made through the carbon contact called brushes. As the armature rotates, the commutator reverses the current in each coil. As it moves between field poles, the direction of rotation of the DC motor can be reversed by reversing either the armature current or the field current. In the permanent magnet DC motors, the permanent magnet gives a constant value of magnetic flux density. The torque developed $T = K_t I$, where K_t is known as the torque constant and is equal to BLN. Since the armature coil is rotating in the magnetic field, electromagnetic induction occurs and back emf is induced. The back emf $V_B = K_B \omega$, where K_B is the back emf constant and ω is the angular velocity. Hence, the net voltage acting in the armature winding is $V - V_B$, where V is the applied voltage. Hence, current in the armature is equal to $(V - V_B)/R$ and torque $T = K_t(V - V_B)/R$. The construction features and torque–speed characteristic of a four-pole DC motor are shown in Fig. 7.9.

(a) Construction features

(b) Torque–speed characteristic

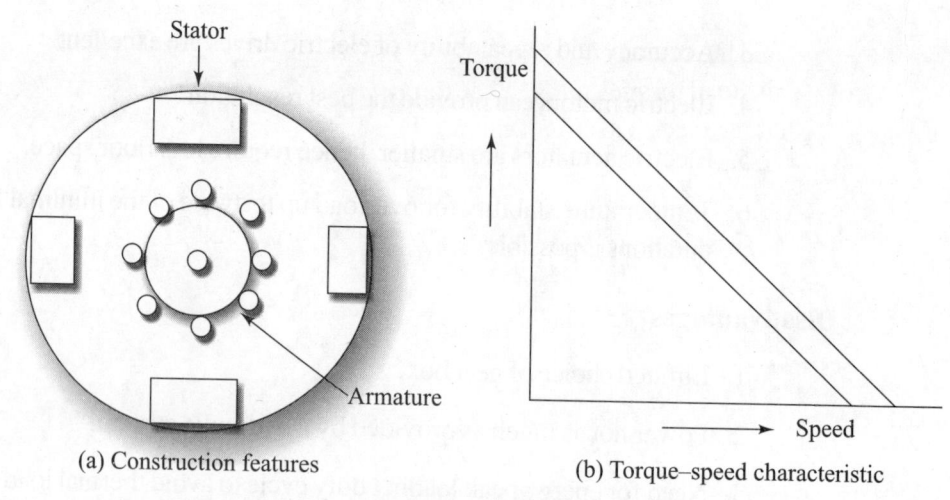

Fig. 7.9 DC motor

DC motors with field coil excitation are classified as series, shunt, compound, or separately excited according to how the field winding and armature winding are connected. The series wound DC motor exerts the highest starting torque and has the greatest no-load speed. Reversing the polarity of the supply to the coil has no effect on the direction of rotation of the motor. With high loads, there is a danger that the series wound motor runs at a high speed. The shunt wound motor is good for speed regulation. Because of their constant speed regardless of the load, shunt wound motors are used widely. Separately excited motors have separate controls for the armature and field current. For the field-coil DC motor, speed can be changed by varying either the armature current or the field current. Generally, the armature current is varied. Thus, speed control can be obtained by controlling the applied voltage to the armature. With an alternating current supply, the thyristor circuit can be used to control the average voltage applied to the armature.

7.3 DC Servomotor

In servo applications a DC motor is required to reduce rapid acceleration from standstill. The physical requirements of such a motor are low inertia and high starting torque. Low inertia is attained with a reduced armature diameter, with a consequent increase in the armature length such that the desired power output is achieved. The symbolic representation of an armature controlled DC motor is shown in Fig. 7.10. In this set-up, an electrical DC source supplies a constant current I_f to the field winding. The armature circuit consists of an armature resistance R_a and armature inductance L_a, both are due to the total armature winding. V is the applied armature voltage and V_B is the back emf on the mechanical side, i.e., the motor rotor. The attached load can be treated as inertia I

and the viscous friction coefficient is represented as c. The disturbance torque T_D is the load torque and T_M is the torque developed by the motor. The transfer function of the electrical circuit can be obtained using Kirchhoff's law from the following equation:

$$L_a\left(\frac{di_a}{dt}\right) + R_a i_a + V_B = V$$

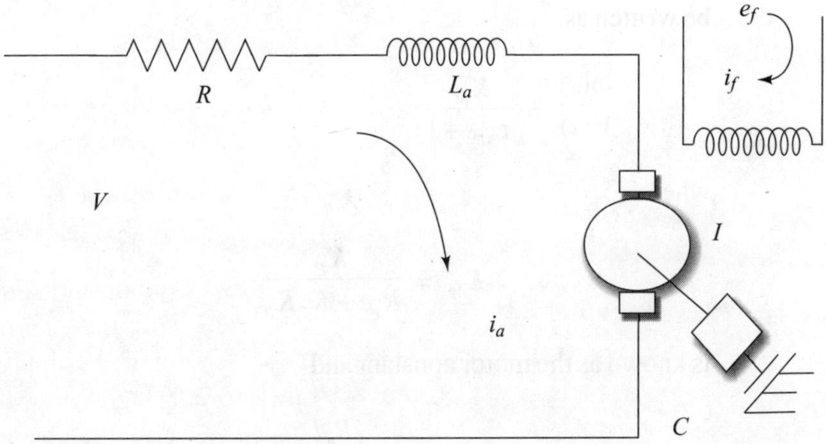

Fig. 7.10 Armature control DC motor

The torque equation is

$$I\left(\frac{d\omega}{dt}\right) + c\omega + T_D = T_M$$

Taking Laplace transform and assuming the initial condition to be zero, we get

$$T_M(s) = K_T I_a(s)$$

$$E_b(s) = K_b \omega(s)$$

$$(L_a s + R_a) I(s) = V - V_B$$

$$(Is + c)\omega(s) = T_M(s) - T_D(s)$$

By manipulation,

$$\frac{\omega(s)}{V(s)} = \frac{K_T}{(L_a s + R_a)(Is + c) + K_T K_B}$$

The inductance L_a in the armature circuit is usually small and may be neglected. If L_a is neglected, the transfer function reduces to

$$\frac{\omega(s)}{V(s)} = \frac{K_T/R_a}{Is+c+\dfrac{K_1 K_B}{R_a}}$$

The back emf constant K_B represents an added term to the viscous friction coefficient c. Therefore, the back emf effect is equivalent to electric friction, which tends to improve the stability of the DC motor. By manipulating the transfer function, it may be written as

$$\frac{\omega(s)}{V(s)} = \frac{K_M}{\tau_M s+1}$$

where

$$K_M = \frac{K_T}{R_a c + K_T K_B}$$

is known as the motor constant and

$$\tau_m = \frac{R_a I}{R_a c + K_T K_B}$$

is known as the motor time constant. When the motor is used to control the shaft position θ,

$$\frac{\theta(s)}{V(s)} = \frac{K_M}{s(\tau_m s+1)}$$

which is the most unstable system. Hence a good compensator must be used if the DC motor is to be used as a position control device.

The commutator type DC motor has the following advantages:

1. Large inertia
2. High power
3. High reliability
4. Low cost
5. High inductance

Its demerits are as follows:

1. Stable control is difficult.
2. Cogging effect—when each slot of the armature comes close to the magnetic field, there will be stalling.

Pan-cake DC motors provide less cogging effect since the armature manufacturing adopts the printed circuit board manufacturing technology. It has a very low inductance capacity compared to the conventional DC motor.

If one needs to use the DC motor as a servo mechanism, a DC motor tacho feedback can be used. The tachogenerator generates a voltage V_f proportional to the actual motor speed but in the opposite polarity. A schematic diagram of the DC servomotor and tachogenerator is shown in Fig. 7.11. Voltage coming in between the armature is $V_a = V - V_f$, where V_a is the voltage across the armature and V is the supply voltage. When the motor is not rotating, $V_a = V$. If $V_f > V_a$, the supply voltage becomes negative. Hence, the motor reduces its speed. Similarly, if the actual speed of the motor reduces below the set value, then V_f is lesser than V_a in magnitude. Then the difference is a positive voltage. The positive voltage when applied to the armature increases the actual speed. This process continues and the speed is regulated within the band. If the parameter of the system is optimized properly, it is possible to reduce the differential to the minimum so that the set speed is maintained irrespective of the load condition.

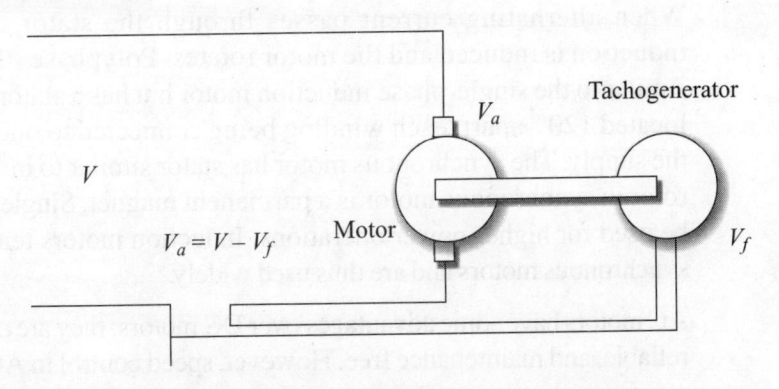

Fig. 7.11 Servo mechanism for constant speed

7.4 Brushless Permanent Magnet DC Motor

A problem with the DC motor is that it requires a commutator and brushes to periodically reverse the current through the each armature coil. The brushes make sliding contact with the commutator and, as a consequence, sparks jump between the two and the brushes suffer wear. The brushes thus have to be periodically changed and the commutator requires resurfacing. To avoid this problem, brushless motors have been designed.

Essentially, a brushless motor consists of a sequence of stator coils and a permanent magnet rotor. The current carrying conductor in a magnetic field experiences a force. As a consequence of the reaction force, the magnet will also experience an opposite and equal force. In the case of the brushless DC motor, the carrying conductors are fixed and the magnet moves. The rotor is a ferrite or ceramic permanent magnet. The current to the stator coil is electrically switched by transistors in sequence around the coils. The switch is controlled by the position of the rotor so that force is always acting on the magnet, causing it to rotate in the same direction. Hall sensors are used to sense the position of the rotor and initiate switching by transistors. The sensors are positioned around the stator.

Brushless permanent magnet DC motors are used in situations where high performance coupled with reliability and low maintenance is essential. Brushless permanent magnet DC motors are quiet and capable of high speeds.

7.5 AC Servomotor

The alternating current (AC) motors can be classified into two groups: single phase and polyphase. Each group is further subdivided into inductive and synchronous motors. The single-phase induction motor consists of a squirrel cage rotor in which alluminium or copper bars fit into slots to form electrical circuit. There is no external connection to the rotor. The stator has a set of windings. When alternating current passes through the stator, an electromagnetic induction is induced and the motor rotates. Polyphase (three phase) motor is similar to the single-phase induction motor but has a stator with three windings located 120° apart, each winding being connected to one of the three lines of the supply. The synchronous motor has stator similar to induction motor but the rotor in synchronous motor is a parmanent magnet. Single-phase motors tend to be used for higher power operations. Induction motors tend to be cheaper than synchronous motors and are thus used widely.

AC motors have some advantages over DC motors; they are cheaper, more rugged, reliable, and maintenance free. However, speed control in AC motors is generally more complex than in DC motors. As a consequence, a speed controlled DC drive generally works out cheaper than a speed controlled AC motor. The speed control of AC motors is based on the provision of variable frequency supply because the speed of such motors is determined by the frequency of the supply. The torque developed by the AC motor is constant when the ratio of the applied stator voltage to frequency is constant. There are two methods to maintain a constant torque at different speeds when the frequency is varied and the voltage applied to the stator also varies. In the first method the AC is first rectified to give a DC with the help of a converter and then inverted back to AC again, but with a frequency that can be selected. Another method is often used for operating slow speed and uses the cycle converter. This converts AC at one frequency directly to AC at another frequency without the intermediate DC conversion.

The characteristic features of AC servomotors are as follows:

1. High power density with low weight
2. Low rotor inertia
3. Constant continuous torque and constant overload capacity over the full speed range
4. No need of additional cooling of the motor

7.6 Stepper Motor

The stepper motor is a device that produces rotation through equal angles, also called steps, for each digital pulse supplied to it as input. If with such a motor one pulse produces rotation through 9 deg, then 60 pulses will produce a rotation through 360 deg. There are two types of pulses that are used as the input: unipolar and bipolar pulses. Figure 7.12(a) shows the difference between unipolar and bipolar pulses.

Stepper motor

(b) Variable reluctance stepper

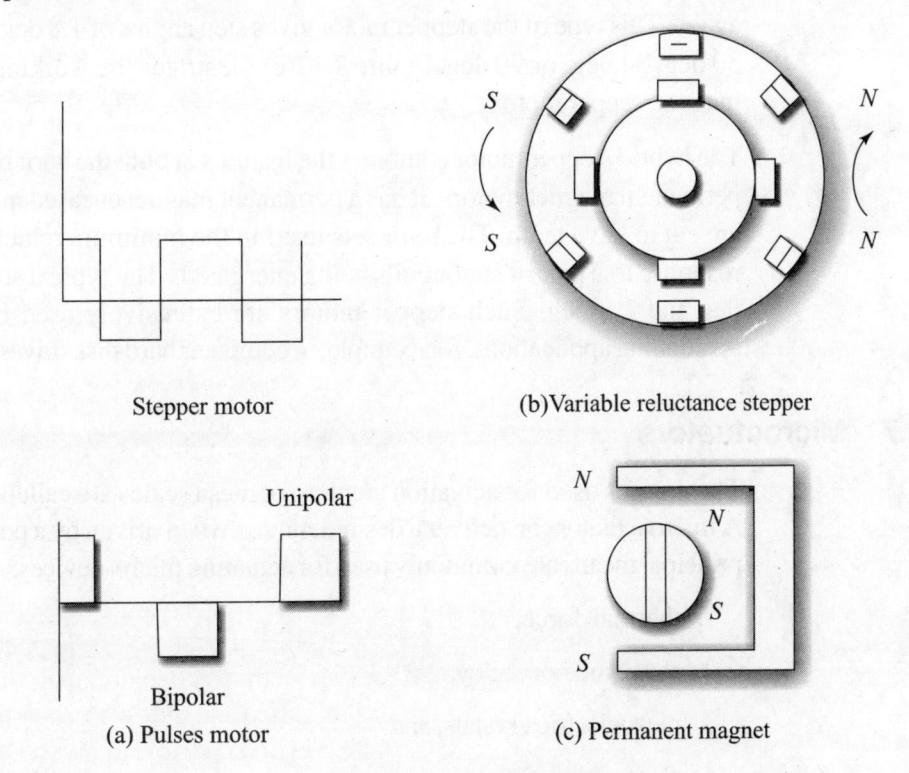

Unipolar

Bipolar

(a) Pulses motor

(c) Permanent magnet

Fig. 7.12 Stepper motor

There are a number of forms of stepper motors. The most commonly used form is the variable reluctant stepper motor. In this type of the stepper motor the rotor is made of soft steel. It is cylindrical with four poles. A cylindrical state has few poles more than the stator. When an opposite pair of windings have current flowing through them, a magnetic field is produced with lines of force which pass from the stator poles through the nearest set of poles on the rotor. These lines of force can be considered to be like elastic threads and are always trying to shorten themselves. Hence, the rotor will move until the rotor and stator poles line up. This state is termed the position of minimum reluctance. This form of stepper motor generally gives a step angle of 7.5 or 15 deg. Figure 7.12(b) shows the schematic diagram of the variable reluctance stepper motor.

The permanent magnet stepper motor has a stator with poles. Each pole is wound with field windings, with the coil of opposite pairs of poles being in series. Current is supplied from a DC source as a pulse train. The rotor is a permanent magnet and when a pair of stator poles has current switched on, the rotor will move to line up with it. Thus, for the given current, the rotor moves to the first 45 deg position, and when polarities are reversed, the rotor will move further 45 deg in order to line up again. This type of the stepper motor gives step angles of 1.8 deg, 7.5 deg, 15 deg, 30 deg, 34 deg, or 90 deg. Figure 7.12(c) illustrates the working of a permanent magnet stepper motor.

The hybrid stepper motor combines the features of both the variable reluctance and permanent magnet motors. It has a permanent magnet encased in iron caps, which are cut to have teeth. The rotor sets itself in the minimum reluctance position in response to a pair of stator coils being energized. The typical step angles are 0.9 deg and 1.8 deg. Such stepper motors are extensively used in high-accuracy positioning applications, for example, in computer hard disk drives.

7.7 Microctuators

The devices used for actuation at micro or mesa scales are called micro actuators. A microactuator can deliver a desired motion when driven by a power source. Four principal means are commonly used for actuating micro-devices:

1. thermal force,
2. shape memory alloy,
3. piezoelectric crystals, and
4. electrostatic forces.

Electromagnetic actuators are widely used in devices at macro scales. However, these are rarely used in micro-devices because of the lack of the sufficient space in micro-devices to accommodate a coil inductance for generating sufficient magnetic field for actuation power. The obvious weakness of the electrostatic actuator is the inherent low magnitude. However, not much force is required to actuate most micro devices; instead electrostatic actuation is used.

Bimetallic strips are actuators based on the phenomenon that forces a strip to bend when it is heated or cooled from some initial reference temperature. The bimetallic strip principle has been used to produce several micro actuators such as micro-clamps or valves.

Shape memory alloys (SMAs) can produce microactuators more accurately and effectively. Nitinol or TiNi alloys act as shape memory alloys. These alloys tend to return to their original shape at a preset temperature. Figure 7.13 illustrates the working principle of a microactuator using SMA. An SMA strip is originally in a bent shape at a designed preset temperature. The strip is set straight at room temperature. However, heating a beam with the attached SMA strip promotes the strip memory to return to its original bent shape. The deformation of the SMA strip causes the attached section of the beam to deform, thus causing microactuation of the beam.

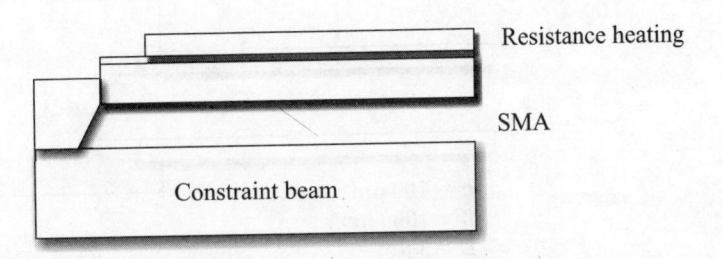

Fig. 7.13 Microactuation using SMA

Certain crystals such as quartz deform with the application of electric voltage. This phenomenon is known as piezoelectric actuation. Piezoelectric actuation is used in micro-positioning mechanisms and micro-clamps. Piezocrystals with electrodes can be mounted on a silicon cantilever beam for microactuation.

Electrostatic forces are used as driving forces for many actuators. Accurate assessment of electrostatic forces is an essential part of the design of many micro-motors and actuators. Figure 7.14 represents two charged plates separated by a dielectric material with a gap d. The plates become electrically charged when an electromotive force or voltage is applied to them. This induces capacitance in the charged plates. The induced capacitance in the charged plates can be expressed as

$$C = \frac{\varepsilon_r \varepsilon_0 WL}{d}$$

where W is the width of the plate, L is the length of the plate, ε_r is the relative permeability, ε_0 is the permeability of air, and d is the distance between the plates. The energy associated with the electric potential can be expressed as

$$U = -\frac{1}{2}CV^2 = \varepsilon_r \varepsilon_0 \frac{WLV^2}{2d}$$

The associated electrostatic force that is normal to the plates in the direction of d is

$$F_d = -\frac{dU}{da} = \varepsilon_r \varepsilon_0 \frac{WLV^2}{d^2}$$

Similarly, the force along the direction of width $F_W = \varepsilon_r \varepsilon_0 LV^2/d$ and the force along the length direction $F_L = \varepsilon_r \varepsilon_0 WV^2/d$. F_W is independent of the width and F_L is independent of the length. These electrostatic forces are prime driving forces of micromotors.

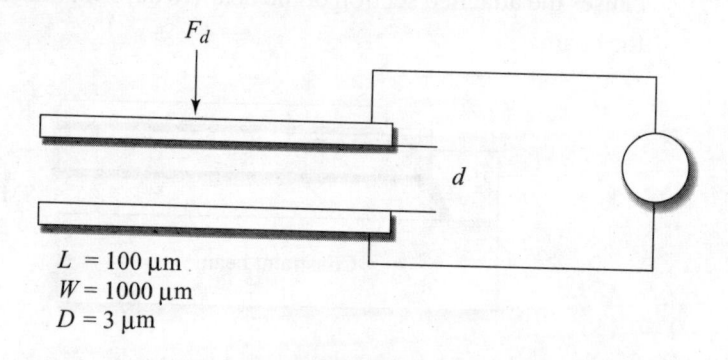

Fig. 7.14 Electrostatic forces in parallel plates

7.7.1 Microgripper

The electrostatic force generated in parallel charged plates can be used as the driving force for gripping an object. The required gripping force in a gripper can be provided by normal forces or by in plane force from a pair of misaligned plates. The arrangement that uses normal gripping forces from parallel plates [Fig. 7.15 (a)] appears to be simple in practice. The major disadvantage of this arrangement is that excessive space is required for occupying the electrode in the microgripper. Consequently, it is rarely used. The other arrangement, shown in

Fig. 7.15(b), uses multiple pairs of misaligned plates. This arrangement is commonly used for micro-devices. This arrangement is frequently referred to as the comb drive. The gripper action at the tip of the gripper is initiated by applying voltage across the plate attached to the drive arms and closing of arms. The electrostatic force generated by the pairs of misaligned plates tends to align them. This causes the drive arms to bend, which in turn closes the extension arms for gripping. Microgrippers can be adapted as micromanipulators in micro-manufacturing processes or microsurgeries. The length of the gripper produced is 400 μm. It has a tip opening of 10 μm.

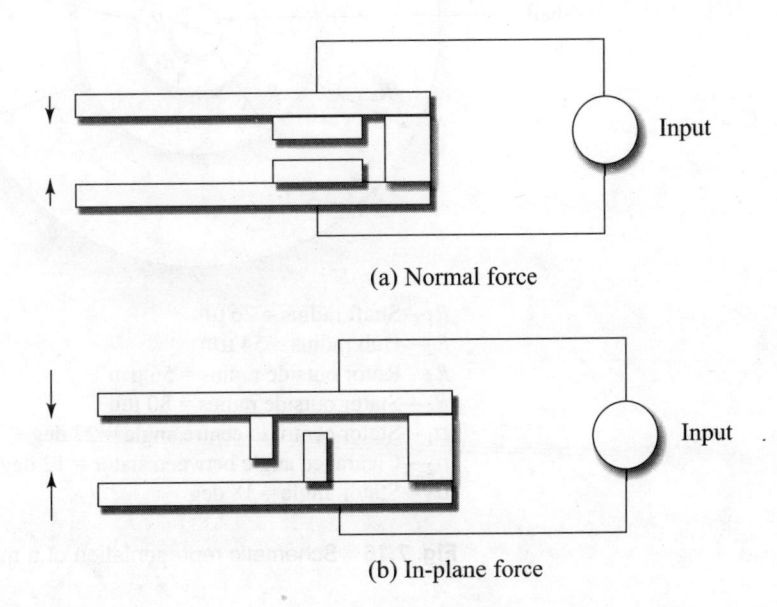

(a) Normal force

(b) In-plane force

Fig. 7.15 Microgripper

7.7.2 Micromotor

A micromotor driven by electrostatic force is shown in Fig. 7.16. As can be seen in the figure, electrodes in rotor poles and stator poles are mismatched in such a way that they will generate an electrostatic driving force due to the misalignment of the energized pairs of electrodes. The ratio of the poles in the stator to those in the rotor is 3:2. The air gap between rotor poles and stator poles can be as small as 2 μm. The outside diameter of the stator poles is 100 μm, whereas the length of the rotor poles is about 20 to 25 μm.

One serious problem that is encountered by engineers in designing and manufacturing of the micro-rotary motor is the wear and lubrication of bearings. Typically, the motor rotates at over 10000 RPM. Much effort is needed in dealing with the wear problem. Consequently, micro-tribology, which deals with friction, wear, and lubrication, has become a critical research area in micro-technology.

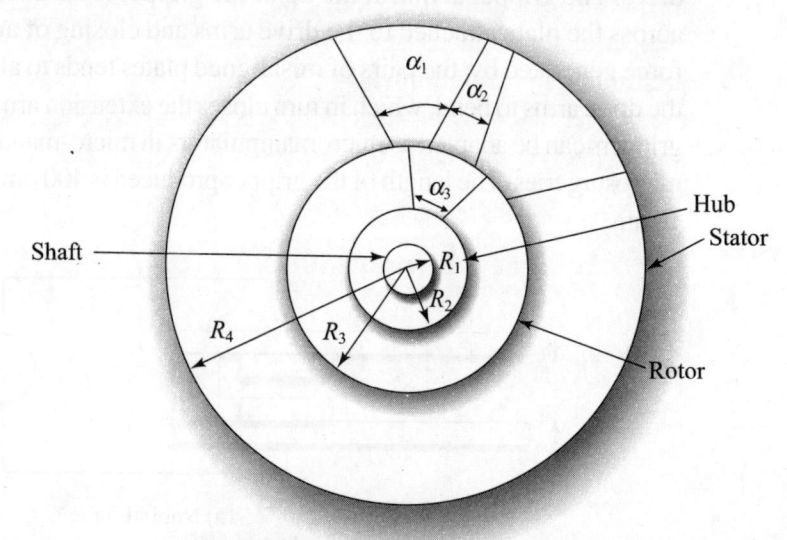

R_1—Shaft radius = 26 µm
R_2—Hub radius = 54 µm
R_3—Rotor outside radius = 56 µm
R_4—Stator outside radius = 80 µm
α_1—Stator centre to centre angle = 27 deg
α_2—Clearance angle between stator = 12 deg
α_3—Stator angle = 18 deg

Fig. 7.16 Schematic representation of a motor

7.8 Drive Selection and Applications

DC servomotors provide excellent speed regulation, high torque, and high efficiency. Therefore, they are ideally suited for control applications. The DC motor can be designed to meet a wide range of power requirements utilized in most small- to medium-size mechatronic systems. Some of the commercial products that use electric motors are

- Vacuum cleaner
- Electric saw
- Fan
- Hair drier
- Disk drive

- Washer
- Electric drill
- Furnace blower
- Electric razor
- Toys

Stepper motors are not appropriate drives for heavy-duty mechatronic systems. The allowable speed of a stepper motor is a function of its load torque. But load torque strongly depends on the system load required. An excessive load on the stepper motor might cause a subsequent loss of steps. In addition, stepper motors are limited in resolution and tend to be noisy. For all of these reasons, stepper motors are seldom used in mechatronic systems which require a high torque. Stepper motors are very popular due to their functional capabilities such as the following:

1. Precision incremental movements in steps.
2. Repeatability and precision in positioning.
3. Reliable design, since brush, gear box, etc. are not needed.

The disadvantage of the stepper motor is that resonance can occur at high speeds. The applications of stepper motor are found in dot matrix printers, open-loop NC equipment, scanners, disk drives, machine tools, XY recorders, and robotics.

Hydraulic or pneumatic devices are well suited for large mechatronic systems where power requirements are high. The cost of the hydraulic drive is not proportional to the power required, and thus is expensive for small- and medium-size mechatronic systems. Some of the applications of hydraulic or pneumatic systems are in robotics, earth excavators, automatic mine diggers, etc.

Presently microactuators are widely used in many applications, which include the following:

1. Nanometrology
2. Optical device alignment
3. Micro-fan and micro-pump
4. Autofocusing system
5. Dick spin stand
6. Bio-mimetic robot
7. Micro-lithography
8. CD-ROM lead contact
9. Nano-positioning

Figure 7.17 gives some qualitative rather than quantitative selection guidelines for choosing actuators.

178 Introduction to Mechatronics

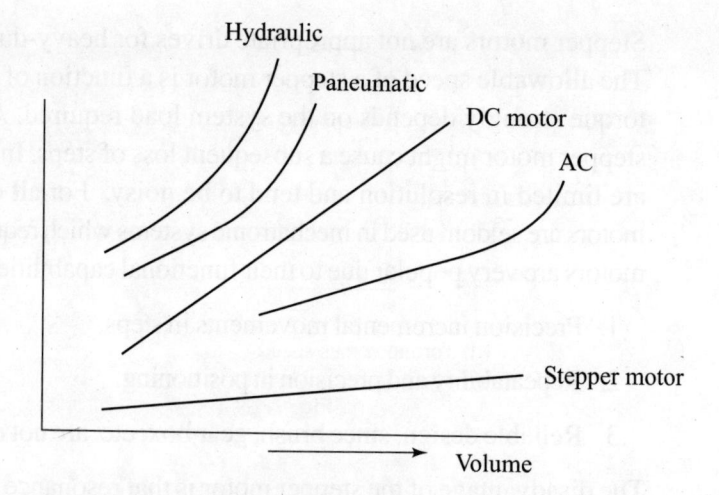

(a) Torque versus volume of actuators

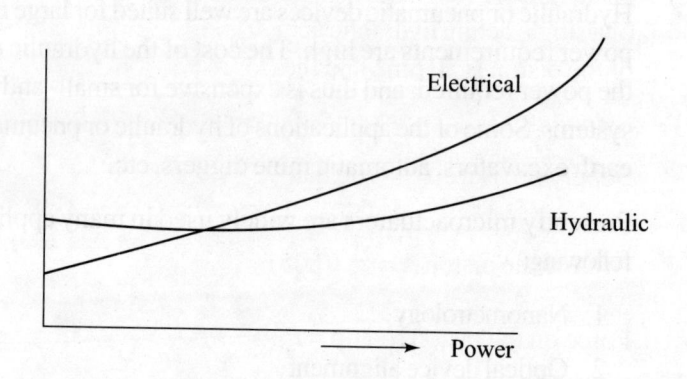

(b) Weight versus power of the electrical and hydraulic actuators

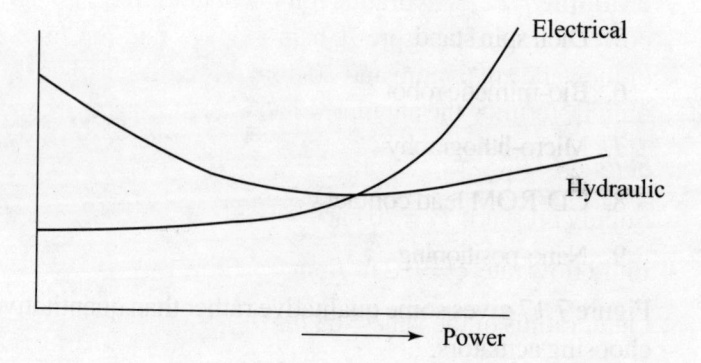

(c) Cost versus power of the electrical and hydraulic actuators

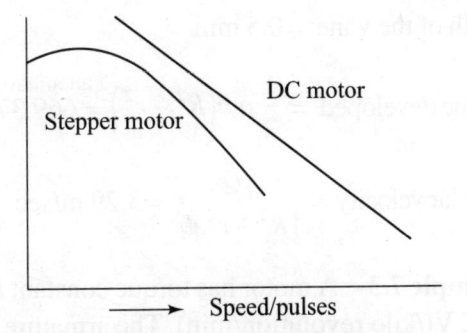

(d) Torque versus speed characteristic of stepper and DC motors

Fig. 7.17 Selection of drives

Illustrative Examples

Example 7.1 What are the holding force and velocity of the movement of piston of a single acting hydraulic actuator when the fluid pressure is 100 bar, the diameter of the piston is 50 mm, and the flow rate is 0.3 m³/min?

Solution

Supply pressure $P_s = 100$ bar

Area of the piston $A = \dfrac{\pi}{4}(0.05)^2 = 0.00196$

Force on the piston rod tip $= P_s A = 1960$ N

Therefore,

Velocity of the piston movement $= Q/A = 8$ m/sec

Example 7.2 A hydraulic rotor actuator is used for a twist joint with a hydraulic power source of pressure 70 bars and flow rate 2 m³/min. The outer and inner radii of the vane are 65 mm and 20 mm, respectively. The thickness of the vane is 0.5 mm. Determine the angular velocity and torque that can be generated.

Solution

Supply pressure $P_s = 70$ bar

Fluid flow rate $Q = 0.2$ m³/min

Outer radius of the vane $= 65$ mm

Inner radius of the vane $= 20$ mm

Width of the vane = 0.5 mm

Torque developed = $\frac{1}{2} p_s b (R^2 - r^2) = 669.37$ N-cm

Angular velocity = $\frac{2Q}{(R^2 - r^2)b} = 3.29$ m/sec

Example 7.3 A motor has torque constant $K_t = 100$ Nm A and voltage constant of 12 V(/kilo revolution/min). The armature resistance is 2 Ω. If 24 V voltage is applied to the terminal, what would be (a) the torque at the rotor (b) speed at zero load, and (c) torque at 100 revolution/min. Plot the result as a speed versus torque graph.

Solution

At zero RPM, the back emf developed

$$V_B = 0$$

The armature current

$$I = V/R_a = 24/2 = 12$$

Therefore, generated torque

$$K_t I = 10 \times 12 = 120 \text{ N-m}$$

At no load, the input voltage is equal to the back emf,

$$24 = 12 \text{ V/(kilo revolution/min)} \times \omega$$

Hence, the angular speed

$$\omega = 2 \text{ kilo revolution/min} = 2000 \text{ RPM}$$

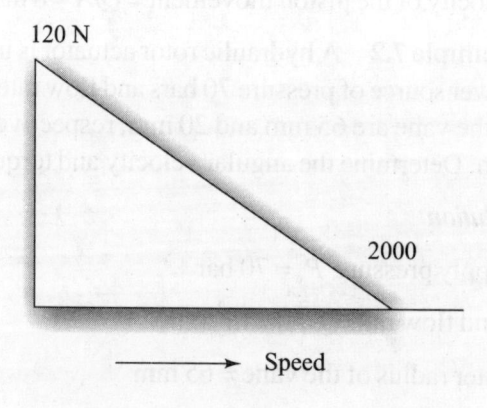

Fig. 7.18 Speed–torque characteristic

At 1000 RPM, the back emf is 12 V. Therefore,

Armature current = $\dfrac{24-12}{2} = 6$ A

Torque developed = $6 \times 10 = 60$ N-m

Example 7.4 A stepper motor is to be used to drive a linear axis of a mechatronic system. The motor output shaft is connected to a screw thread with a 30 mm pitch. It is desired to control each axis at 0.5 mm. What is the corresponding step angle?

Solution

Pitch of the screw thread, $p = 30$ mm

Linear resolution required = 0.5 mm

Pitch of one revolution of screw thread = 30 mm

Therefore, the number of steps required is

$30/0.5 = 60$ steps

So, the step angle required is

$360/60 = 6$ deg

Example 7.5 Ingredients are fed continuously to a paint-drying furnace by a conveyor. To minimize heat loss through the furnace door, it should be opened only for the duration of time necessary to allow the ingredients to pass through. The hydraulic circuit should be designed so that the door can be safely held in position for a long period of time without dropping down.

Solution

Install a de-lockable non-return valve into the supply line to the piston rod side of the cylinder after the directional control valve to ensure that the non-return valve closes immediately when the door stops. The output of the directional control valve to the tank must be depressed by A, B, T, connected and P closed.

The interaction of the non-return valve with a 4/3 way valve guarantees that the de-lockable non-return valve is loaded. As a result the door can be held open for a long time without dropping down when the 4/3 valve is switched off in the mid-position. When the piston rod retracts, that is, when the door opens, the de-lockable non-return valve is unseated via a control line. The non-return valve opens when the directional control valve reverses the cylinder. The required circuit diagram is shown in Fig. 7.19.

182 Introduction to Mechatronics

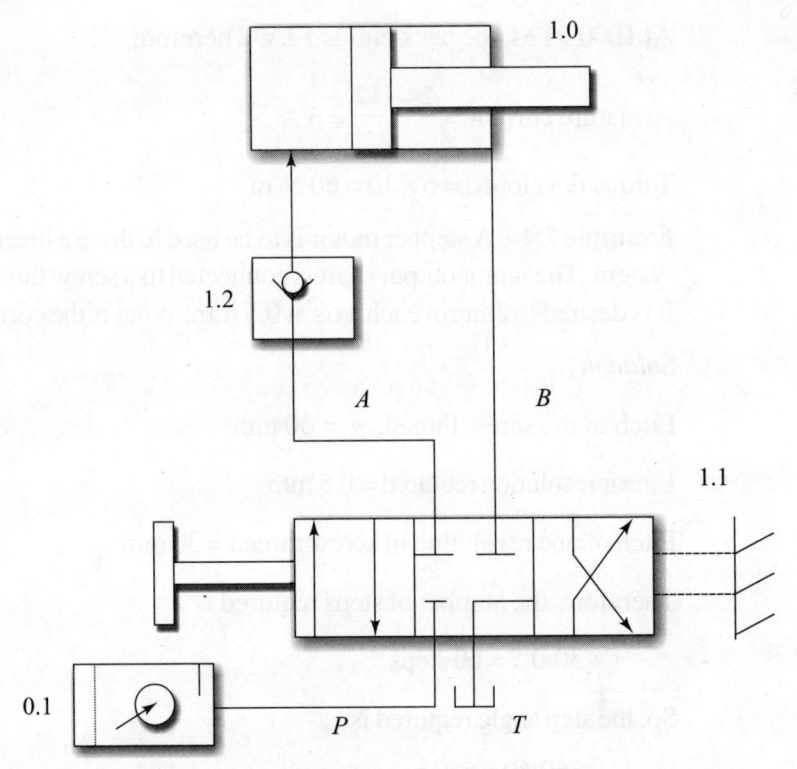

Fig. 7.19 Circuit diagram for opening a door in furnace

Example 7.6 A differential cylinder is on one side of the piston area $A = 10$ cm^2 and the annular piston surface area on other side is $A_p = 5$ cm^2. The stroke length s is 10 cm. The pump delivers 10 L/min and the maximum pressure is 100 bar. Determine the speed of advance (pitch) and return stroke. Determine the time taken to extend and retract the piston and the maximum force which can be attained on extending and retracting. Figure 7.20 shows the hydraulic cylinder.

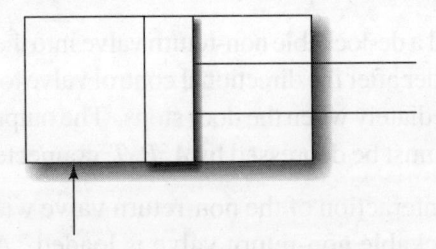

Fig. 7.20 Hydraulic cylinder

Solution

$$\text{Velocity during advance} = \frac{Q}{A} = \frac{10000}{10} = 10 \, \text{m/sec}$$

Velocity during retraction $= \dfrac{Q}{A_p} = \dfrac{10000}{5} = 20$ m/sec

Time for advance $= s/v = 100/1000 = 6$ sec

Time for retraction $= s/v = 100/1000 = 3$ sec

Force during advance $= P \times A = 1000$ N $\times 10 = 10000$ N

Force during retard $= P \times A_P = 1000$ N $\times 5 = 5000$ N

Example 7.7 Calculate normal electrostatic force exerted on a 1 mm × 1 mm plate of $\varepsilon_r = 1.0$ for air dielectric and $\varepsilon_0 = 8.85$ pF/m or 8.85×10^{-12} C^2/N m^2 for material.

Solution

Width of the plate = 1 mm

Length of the plate = 1 mm

Dielectric constant for air = 1.0

Dielectric constant for material = 8.85×10^{-12} C^2/N-m^2

Normal electrostatic force developed

$$F_d = \varepsilon_r \varepsilon_0 WLV^2/2d = -1.106 \; 10^{-6} \, V^2 \, N$$

Thus, 11 mN force is generated for a voltage of 100 V applied to the plate.

Exercises

7.1 What is a servo system? Explain with a suitable sketch the working of a hydraulic servo system.

7.2 What is the velocity of the piston and the force generated by it if the fluid pressure is 50 bar inside the cylinder? The piston diameter is 80 mm and flow rate is 8 cm^3/min.

7.3 A hydraulic rotary servo actuator is to be used for a twist joint with a hydraulic power system. The outer and the inner radii of the vane are 80 mm and 20 mm, respectively. The width of each vane is 10 mm. Determine the angular velocity and torque generated if the supply pressure is 50 bar and the flow rate is 8 cm^3/min.

7.4 List out the symbols of different elements used in hydraulic and pneumatic systems.

7.5 What are the advantages and disadvantages of a pneumatic power system and a hydraulic power system?

7.6 Differentiate between thyristor and triac.

7.7 Explain SCR and MCB.

7.8 Obtain the transfer function of a DC motor with load.

7.9 What is a brushless DC motor?

7.10 Sketch and explain the working of an AC servomotor.

7.11 Sketch and explain the working of a magnetic stepper motor.

7.12 What is micro-actuation? Explain the principles used for micro-actuation.

7.13 Sketch and explain the working of a microgripper.

7.14 Explain the principle and working of a micromotor.

Chapter 8

Sensors and Transducers

A transducer is a device, usually electrical, electronic, or electro-mechanical, that converts one type of energy into another for various purposes including measurement or information transfer. In a broader sense, a transducer is sometimes defined as any device that converts a signal from one form into another. The term 'sensor' is often used in place of the transducer. A sensor is defined as an element which when subjected to some physical change experiences a relative change. Transducers/sensors may act as passive or active devices. A sensor in which the output energy is supplied entirely or almost entirely by its input signals is called a passive element. An active element has an auxiliary source of power that supplies a major part of the output power. There may or may not be a conversion of energy from one form to another.

The treatment of the instrument performance characteristics is generally broken down into two subareas: static characteristics and dynamic characteristics. The static characteristics are the values given when steady state conditions occur. The dynamic characteristics refer to the behaviour between the time that the input value changes and the time required given by a transducer to settle down to steady state values. Accuracy, precision, threshold, resolution, hystersis, dead band, sensitivity, non-linearity, range or span, and errors are examples of the static performance characteristic parameters. The response time, time constant, settling time, peak time, rise time are examples of the dynamic performance characteristics. For better functioning of mechatronics, sensors or transducers, both static and dynamic parameters are very important. Sensors or transducers are used in mechatronics for the following purposes:

1. To provide position, velocity, and acceleration information of the measuring element in a system which provides feedback information
2. To act as protective mechanism for a system

3. To help eliminate mechanically complex and expensive feeding and sorting devices
4. To provide identification and indication of the presence of different components
5. To provide real time information concerning the nature of the task being performed

8.1 Static Performance Characteristics

This section gives the definitions of the static performance characteristic parameters (see Fig. 8.1).

Accuracy is defined as the proximity of the measured value to the true value. It is expressed as a percentage of the full range output or full-scale deflection.

Precision or repeatability is used to describe the ability of the instrument to give the same output for repeated application(s) of the same input value. Accurate sensors are always precise, but vice versa may not be true.

Sensitivity is defined as the ratio of the incremental change in output to the incremental change in input. More sensitive the instrument, better the readability.

Threshold is the minimum input signal required by the instrument to start functioning from its initial position.

Resolution is the smallest increment that can be measured in any range of the sensor.

Hysteresis A sensor can give different outputs for the same value of quantity being measured depending on whether that value is reached by a continuously increasing or decreasing change. This effect is called hysteresis.

Hysteresis error is the maximum difference in the input or output for the increasing and decreasing input value.

Dead band or **dead space** of a transducer is the range of the input value for which there is no output.

Drift is the change in output that occurs over a time. It may be expressed as a percentage of the full range output. The term 'zero drift' is used for the changes that occur in the output when there is zero input.

Error is the difference between the actual results of the measurement and the time value of the quantity being measured. The error may be a systemic error or random error. Systemic errors may arise due to calibration errors, computing errors, procedure errors, etc. Systemic errors are known even before the use of the sensor. Random errors are due to ambient change, loading error, etc. and are analysed through statistical error analysis.

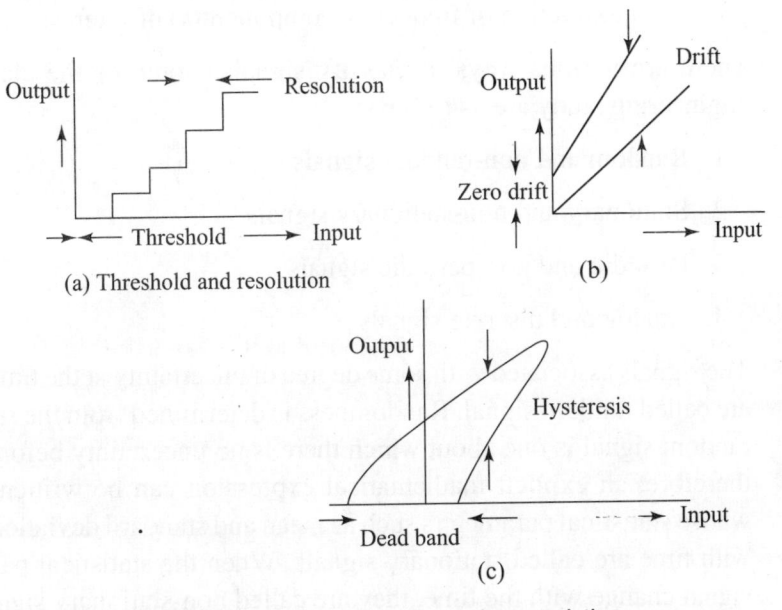

Fig. 8.1 Static performance characteristics

8.2 Dynamic Performance Characteristics

Static performance characteristic parameters are constant in time whereas dynamic performance characteristics vary with time. The dynamic characteristics of a sensor or transducer can be classified as steady state, periodic, and non-repetitive or transient. The transient characteristics have been discussed in detail in Chapter 6.

Fundamentally, the source of a signal is the transducer or sensor, which helps to describe the variation in the process parameters. The dynamic performance characteristic indicates the process parameter and is also called the process variable. Examples of natural process parameters are the ambient temperature, humidity level, intensity of light inside the room, etc. Some of the engineering process parameters are the rotational speed of an electric motor, transient temperature of a typical ball-bearing system of a rotary shaft, pressure in a pneumatic actuator, field current in a DC motor, etc. All these signals are functions of time. On the horizontal axis, the signal originates at some reference point of time whereas the strength or amplitude originates on the vertical axis. Each component within the signal is characterised by three parameters, namely, amplitude, frequency, and phase. Each parameter has a bearing on the shape of the signal. For practical use, one requires a signal of single frequency component or group frequency components. This requires filtering, which is defined as the process of extraction of frequency component(s) of interest.

There are various ways to classify signals. Some of the classifications for engineering study are as follows:

1. Random and non-random signals
2. Stationary and non-stationary signals
3. Periodic and non-periodic signals
4. Analog and discrete signals

The signals associated with some degree of uncertainty at the time of appearance are called random signal. Randomness is determined from the statistics. A non-random signal is one about which there is no uncertainty before it occurs and, therefore, an explicit mathematical expression can be written for it. Signals whose statistical parameters such as mean and standard deviation do not change with time are called stationary signals. When the statistical parameters of the signal change with the time, they are called non-stationary signals. A signal is said to be periodic if its amplitude is retreated over a fixed interval as the time elapses. Non-periodic signals do not have this characteristic. Analog signals are a continuous function of time whereas discrete signals are non-continuous. A periodic analog signal with period T can be represented as $v(t) = v(t + T)$ for all values of t, where T is called the time period. A periodic signal may be a simple sinusoidal signal or more complex signal, as shown in Fig. 8.2.

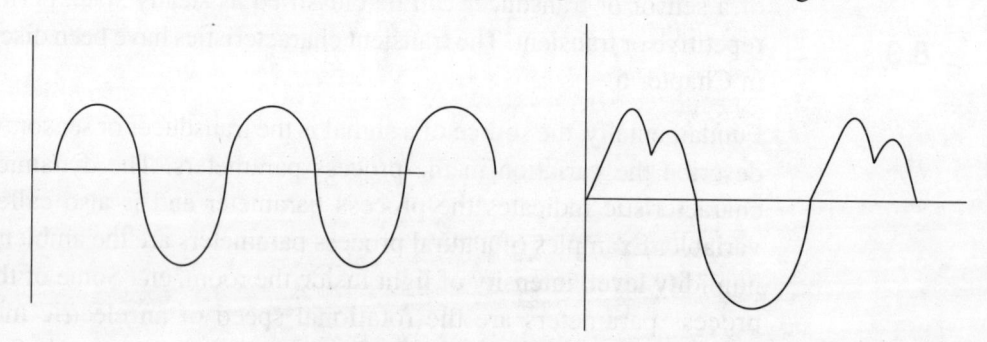

Fig. 8.2 Periodic signals

A sinusoidal periodic signal can be expressed as $v(t) = A \sin \omega t$, where A is the peak amplitude and $\omega = 2\pi f$, f being the frequency association. A periodic signal of period T can be expressed as a summation of an infinite number of trigonometric functions, such as sine and cosine functions. As long as the signal has a finite number of maxima and minima and is discontinuous within T, the periodic signal can be expressed as

$$v(t) = \frac{a_0}{2} + \sum_{n=1}^{\infty} a_n \cos\frac{2\pi nf}{T} + \sum_{n=1}^{\infty} b_n \sin\frac{2\pi nf}{T}$$

The above equation is called a Fourier series expansion of f. The component of the time periodic signal is called fundamental frequency. a_0 is the derivative component of the signal, and a_n and b_n are the coefficients of the trigonometric functions. The constants can be obtained by using the following equations:

$$a_0 = \frac{2}{T}\int_{t_0}^{t_0+T} v(t)dt$$

$$a_n = \frac{2}{T}\int_{t_0}^{t_0+T} v(t)\cos\frac{2\pi nf}{T}dt$$

$$b_n = \frac{2}{T}\int_{t_0}^{t_0+T} v(t)\sin\frac{2\pi nf}{T}dt$$

The frequency counts of any periodic signal can be immediately obtained from the time domain equations of the Fourier series. Fourier transform (FT) is a mathematical tool that transforms a time domain signal into a frequency domain signal. A time domain signal can also be extracted from a frequency domain signal. Mathematically,

$$FT[v(t)] = v(f) = \int_{-\infty}^{+\infty} v(t)e^{-\pi ft}dt$$

$$L^{-1}F(t)v(t) = v(t) = \int_{-\infty}^{+\infty} v(t)e^{2\pi ft}dt$$

8.3 Internal Sensors

Feedback sensors provided in mechatronic systems are called internal sensors. Internal sensor devices are specifically chosen to suit the need for a particular task. Internal sensors are needed for the feedback of position, velocity or acceleration of various elements in the mechantronic systems. Internal sensors may be classified as potentiometers, tachometers, resolvers, optical encoders, Hall effect sensors, and Moire's fringes.

8.3.1 Potentiometer

A potentiometer consists of a resistance element and a wiper. It is provided with three terminals; one at each end of the resistance element and one for the wiper. The output signal of the potentiometer is the resistance between one end of the resistance wire and the wiper terminal. According to the type of resistance element, potentiometers can be divided into three classes: wire wound, cement, and plastic potentiometers. The wire wound potentiometer is usually a nichrome wire made of 75 per cent nickel and 25 per cent chromium. Its resistance is not more than 0.055 Ω for a temperature change from 25°C to 125°C. The linearity of the wire wound potentiometer is within ±0.003 per cent. Cement potentiometers are usually used in electronic circuits. Plastic potentiometers are also called conductive plastic potentiometers. Conductive plastics are made up of epoxy polyester and other resins which are blended with carbon powder to make them conductive. Linearity can be held at 0.025 per cent. The device gives one million cycles of operation. Figure 8.3 illustrates the schematic diagram of a potentiometer and its characteristics.

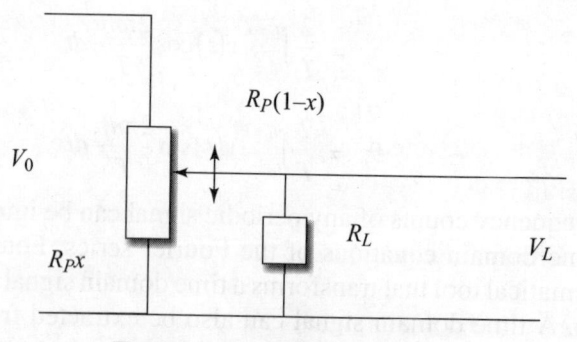

Fig. 8.3 Potentiometer with load

The potential difference across the load V_L is directly proportional to only the input voltage V_0 if the load resistance is infinite. For finite loads, the relationship between the output and slide position is non-linear. The potentiometer resistance $R_P(1-x)$ is in parallel with the load resistance R_L. The combined resistance is given by $R_L x R_P/(R_L + xR_P)$.

Total resistance = $R_P(1-x) + \dfrac{R_L x R_P}{R_L + x R_P}$

Hence

$$\dfrac{V_L}{V_0} = \dfrac{xR_L R_P / (R_L + R_P)}{R_P(1-x) + xR_L R_P / (R_L + xR_P)} = \dfrac{x}{\dfrac{R_P}{R_L} x(1-x) + 1}$$

If the load is of infinite resistance, then $V_L = V_0$. Hence, the error due to load having finite resistance is equal to

$$(xV_L - V_L) = xV_0 - \frac{xV_0}{\frac{R_P}{R_L}x(1-x)+1} = V_0 \frac{R_P}{R_L}(x^2 - x^3)$$

For instance, a potentiometer of resistance 500 Ω and displacement of the slide being half of its maximum, with a load resistance 10 kΩ and supply voltage of 4 V gives an error equal to

$$4\frac{500}{10000}(0.5^2 - 0.5^3) = 0.025 \text{ V}$$

Usually, a potentiometer has very low resistance.

8.3.2 Tachometer

A tachogenerator is used to measure angular velocity. A variable-reluctance tachogenerator consists of a toothed wheel of ferromagnetic material attached to the rotary shaft. A pick-up coil is wound on a permanent magnet. As the wheel rotates, the teeth move past the coil and the air gap between the coil and the ferromagnetic material changes. Since the air gap periodically changes, the flux linked to the pick-up coil changes. The resulting cyclic changes in the linked flux produce an alternating emf in the coil. Flux change $\Phi = \Phi_0 + \Phi_s \cos \omega t$, where Φ_0 is the mean value of flux and Φ_s is the amplitude of the flux variation. The induced emf $e = N(-d\Phi/dt) = N\Phi_s n\omega \sin \omega t$, where n is the number of teeth in the wheel and N is the number of turns in the coil. The maximum value of the induced emf, $E_{max} = N\Phi_s n\omega$. The pulse count gives the angular velocity and the maximum value of emf gives a measure of the angular velocity.

Another form of tachogenerator is essentially an AC or a DC generator. The amplitude or frequency of the alternating output emf can be used as a measure of angular velocity of the rotor. The output may be rectified to give a DC voltage, which is proportional to the angular velocity. Non-linearity for such sensors is typically of the order of ±0.15 per cent of the full scale range and sensors are typically used for rotation up to 1000 rpm. Similarly, the amplitude of the output voltage of a DC motor can be used as a measure of the angular velocity.

8.3.3 Resolver

In a synchro-resolver, two stator windings are positioned at right angles to each other as shown in Fig. 8.4. A resolver resolves the voltage into its two components: sine and cosine of the angle made by the rotor. One rotor winding is excited with the AC signal of constant amplitude with a frequency of 60–400 cycle/sec. When the rotor rotates through and angle θ_i from the null position, it gives AC signal outputs proportional to $\sin \theta_i$ and $\cos \theta_i$. A typical high accuracy resolver has an excitation voltage of 26 V at the maximum operating frequency of 400 cycles/sec. The connection to the rotor is made by the brushes and slip rings. A digitizer converts the input signal and the output from two stator windings into a digital signal, which gives the exact rotation of the rotor. Resolvers are rugged and operate on a wide temperature range. They can operate at very low speeds. The main disadvantage of the resolver is the requirement of oscillator and digital converter. These requirements make it costly as well.

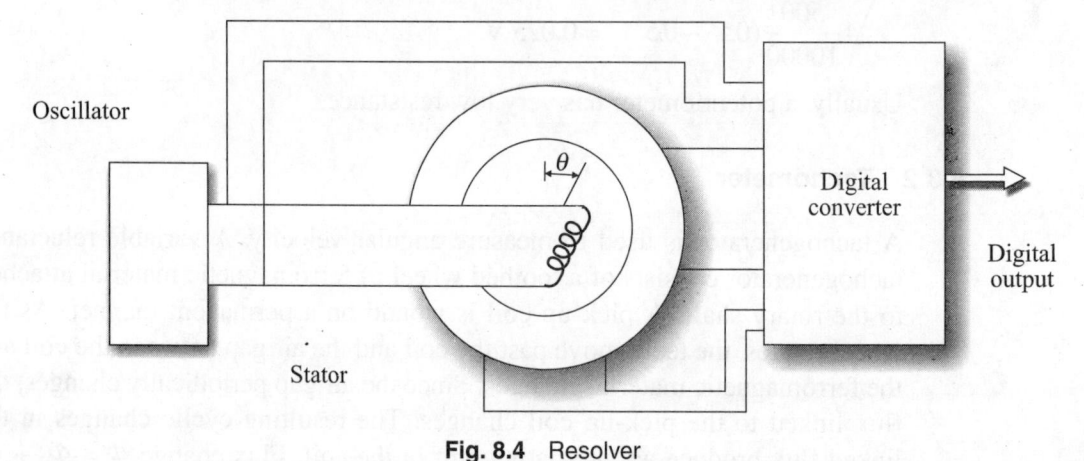

Fig. 8.4 Resolver

8.3.4 Optical Encoder

An encoder is a device that provides digital output as a result of a linear or angular displacement. Based on the position of the encoder, it is grouped into two categories, namely, incremental and absolute encoders. The incremental encoder gives the direct change in rotation from the same datum position whereas the absolute encoder gives the actual angular position. Figure 8.5(a) shows the basic form of an incremental encoder used for the measurement of angular displacement. In this set-up a beam of light passes through slots in a disk and is detected by a suitable light sensor. When the disk is rotated, the sensor produces a pulsed output, with the number of pulses being proportional to the angle through which the disk rotates. Thus the angular position of the disk in the shaft rotation can be determined. In practice, concentric tracks with these sensors are used. The inner track has just one hole and is used to locate the home position of the disk. The other two tracks are used to determine the direction of rotation of the shaft. This is achieved by off-setting the two track holes into one and a half width of the hole. Figure 8.5(b) shows the basic form of an absolute encoder for the measurement of angular displacement. This encoder gives an output in the form of a binary number of several digits and each digit represents a particular angular position.

(a) Incremental encoder (b) Normal binary

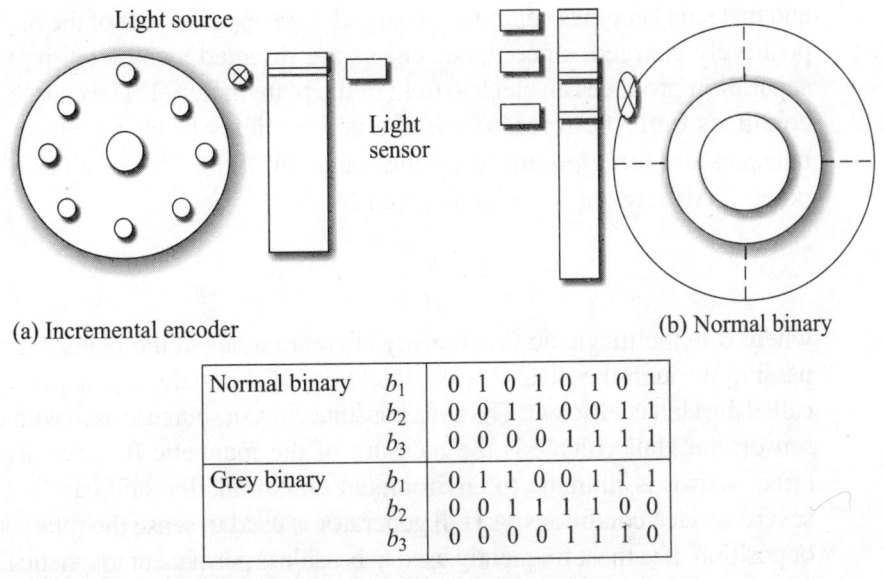

Normal binary	b_1	0 1 0 1 0 1 0 1
	b_2	0 0 1 1 0 0 1 1
	b_3	0 0 0 0 1 1 1 1
Grey binary	b_1	0 1 1 0 0 1 1 1
	b_2	0 0 1 1 1 1 0 0
	b_3	0 0 0 0 1 1 1 0

(c) Truth table for grey binary and normal binary codes

Fig. 8.5 Optical encoder

The rotary disk has three concentric circles of slots and three sensors to detect the light pulses. The slots are arranged in such a way that the sequential output from the sensor is a number in the binary code. The number of bits in the binary number will be equal to the number of tracks. The normal form of the binary code is generally not used, because changing from one binary number to the next can result in more than one bit change. This may lead to false counting if there is any misalignment. To overcome this difficulty grey codes are generally used. With this only one bit changes in moving from one number to the next. The grey code bit, $b_i = b_i \oplus b_{i-1}$, where the right side bits correspond to the normal form of the binary bit. Figure 8.5(c) shows the truth table of the normal binary and grey binary codes.

8.3.5 Hall Sensor

When a beam of charged particles passes through a magnetic field, force acts on the particles and the beam is deflected from its straight-line path. A current flowing through a conductor is like a beam of moving charged particles and, thus, can be deflected by a magnetic field. This phenomenon was discovered by E.R. Hall in 1879 and was accordingly called Hall effect. Let us consider a conductive plate with a magnetic field applied at right angles to its plane. A current is passed though this plate. As a consequence of the magnetic field, the moving electrons in the stream of current are deflected to one side of the plate and that side becomes negatively charged. The opposite side of the plate becomes positively charged, since the electrons are directed away from it. This charge separation produces an electric field in the plate material. The charge separation continues until the force on the charged particles from the electric field just balances the force produced by the magnetic field. The result is a transverse potential difference V, which is given by

$$V = \frac{K_H B I}{t}$$

where B is the magnetic flux density at right angles to the plate, I is the current passing through the plate, t is the thickness of the plate, and K_H is the constant called the Hall coefficient. Thus, if a constant current source is used with a particular sensor, the Hall voltage is the measure of the magnetic flux density. The Hall effect sensor is immune to environment contamination and can be used under severe service conditions. A Hall generator is used to sense the rotary shaft speed or position. It is most frequently used in brushless permanent magnetic DC motors. Figure 8.6 illustrates the working principle of the Hall sensor.

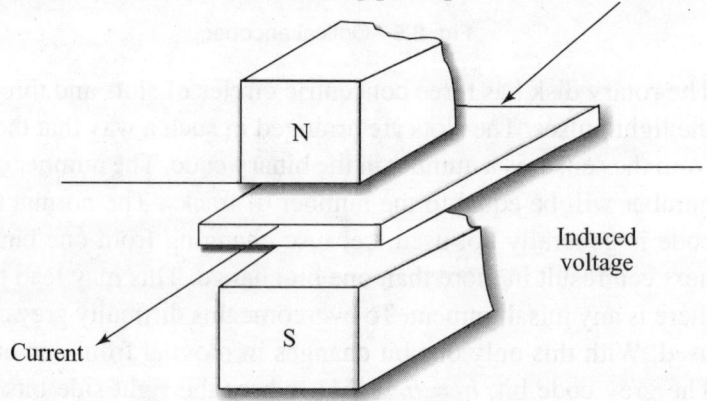

Fig. 8.6 Hall sensor

The Hall sensor uses a small, semi-conducting chip of indium arsenide or indium arsenide oxide. The chip is not more than 0.025 mm thick. It is usually enclosed in a plastic envelope. Two of its opposite edges are connected to a constant DC voltage source. The other two edges are connected to an amplifier to transmit the Hall voltage, which is generated as soon as the chip is exposed to the magnetic field.

8.3.6 Moire's Fringe Sensor

The linear or angular position of any link of a mechatronic system can be detected by using Moire's fringe grating. In the linear grating, a precision scale is engraved with closely spaced parallel lines (approximately 2800 to 3000 lines/cm is fixed) and a plate moves over the scale. The moving plate has the same number of grating as in the main scale. A light source and photocell detector is fixed on either side of the stationary element of the mechatronic system. As the plate rotates, bright and dark fringes are formed due to diffraction of light. When a transparent region is exposed to the light source, a pulse is registered in the photocell. By knowing the pitch of the engraved lines on the linear grating and by counting the number of pulses, the movement of the arm can be established.

In the radial Moire's fringe grating, two radial gratings are positioned adjacent to one another. One grating is fixed to the rotational element and the other grating is fixed to the frame. If the disk centre is offset, bright and dark fringe patterns are produced when the shaft rotates. This pattern is called Moire's fringes, which can be sensed by the use of a photocell. Figure 8.7 shows the working principle of Moire's fringes for linear and rotary motion measurements.

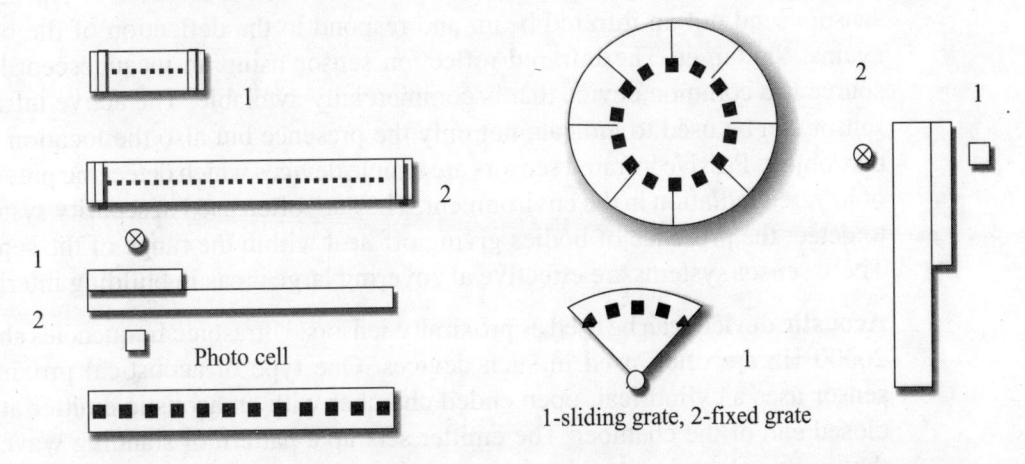

(a) Moire's fringes for linear motion measurement (b) Moire's fringes for rotary motion measurement

1-sliding grate, 2-fixed grate

Fig. 8.7 Moire's fringe sensors

8.4 External Sensors

External sensors are peripheral devices used in mechatronic systems. External sensors are used in mechatronic systems for the following purposes:

- Safety monitoring
- Work cell control
- Part inspection and quality control
- Gaining information about the system

There are many types of external sensors such as proximity sensors, range finders, tactile sensors, machine vision sensors, and force sensors. We will discuss these in the following sub-sections.

8.4.1 Proximity Sensors

Proximity sensors are the devices that are used to locate objects in close proximity. The closeness of the object, for which the sensor can sense its presence, depends on the device used for the purpose. The distance can be anywhere between several millimetres to several metres. A variety of technologies are available for designing proximity and range sensors. These technologies include optical devices, acoustics, electrical field techniques, and pneumatic pressure drop methods.

Optical proximity sensors can be designed to suit operations in the visible or invisible (infrared) light. Infrared sensors may be active or passive. The active sensors send out an infrared beam and respond to the deflection of the beam against the target. The infrared reflection sensor using an incandescent light source is a common device that is commercially available. The active infrared sensor can be used to indicate not only the presence but also the location of a part/object. Passive infrared sensors are simple devices which detect the presence of infrared radiation in the environment. They are often used in security systems to detect the presence of bodies giving off heat within the range of the sensor. These sensor systems are effective at covering large areas in building interiors.

Acoustic devices can be used as proximity sensors. Ultrasonic frequencies above 20000 Hz are often used in such devices. One type of acoustical proximity sensor uses a cylindrical, open ended chamber with an acoustic emitter at the closed end of the chamber. The emitter sets up a pattern of standing waves in the cavity, which is altered by the presence of an object near the open end. A microphone located in the well of the chamber is used to sense the change in the sound pattern. This kind of sensors can also be used as range finders.

There are various forms of switches that can be activated by the proximity of an object. The proximity sensors in such switches sense the object and give an output which may indicate either an on state or an off state. The microswitch is a small electrical switch and requires a small physical contact and a small operating force to close the contacts.

Pneumatic sensors involve the use of compressed air. In this device the proximity of an object is transformed into a change in air pressure. Figure 8.8(a) shows the basic form of such a sensor. Low pressure is allowed to escape through a port in front of the sensor. This displaces the air in the absence of any closeby object, thereby reducing the pressure in the vicinity of the sensor output port.

If there is a closeby object, the air cannot so readily escape and the result is that pressure increases in the sensor output port. The output pressure from the sensor, thus, depends on the proximity of the object. Such sensors are used for the measurement of displacements of a fraction of a millimetre in range which typically is 2–12 mm.

Proximity sensors involving the use of electrical field are commercially available. Two types of sensors in this category are *eddy current proximity sensors and magnetic field sensors.* If a coil is fed with an alternating current input, an alternating magnetic field is produced. If there is a metal object in close proximity to this alternating magnetic field, then eddy currents are induced in it. The eddy currents themselves produce magnetic field. This distorts the magnetic field responsible for their production. As a result, the impedance of the coil changes the amplitude of the alternating current. At some preset level, this change can be used to trigger a switch. Figure 8.8(b) shows the basic form of the eddy current proximity sensor. It is used for detecting non-magnetic but conductive materials. These sensors have the advantage of being relatively inexpensive, smaller size, high reliability, and high sensitivity to small displacement.

Magnetic field proximity sensors are relatively simple and can be made using a permanent magnet. The magnet can be made a part of the object being detected or can be part of the sensor device. It can only be used for the detection of metal objects and is best for ferrous metal.

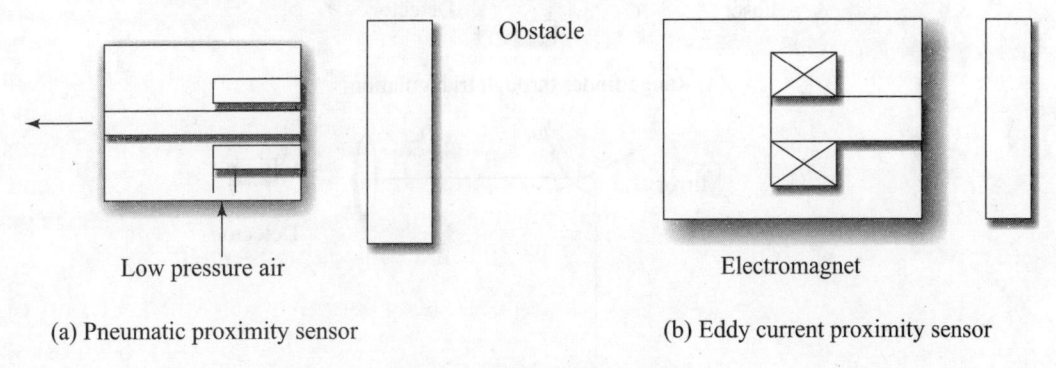

(a) Pneumatic proximity sensor

(b) Eddy current proximity sensor

Fig. 8.8 Proximity sensors

8.4.2 Range Finder

Range sensing is concerned with detecting how near or how far a component is from the sensing position. Range sensing (also called distance sensing) works on the noncontact analog technique. Short range sensing may be accomplished with the electrical capacitance, inductive or magnetic technique. They can be made active with a range of few millimetres to 400 mm.

Long range sensing usually involves transmitting energy waves and a receiver situated on the sensing device receives the reflected waves to determine the distance. The time taken for energy waves to reach the receiver is used to find out the distance between the sensing device and the object. Continuous ranging can be achieved by transmitting energy waves towards the target and detecting them when these are reflected back. This principle is well established in radar and sonar techniques. The use of laser for range sensing for long distance ranging has been used for long and is now a well established technique for long range sensing.

Figure 8.9(a) illustrates the simple configuration of a laser-based triangular range finder. In this case, a single mode He-Ne laser is used. The laser beam impinges on the centre of the mirror that rotates at a fixed speed so that the given plane is

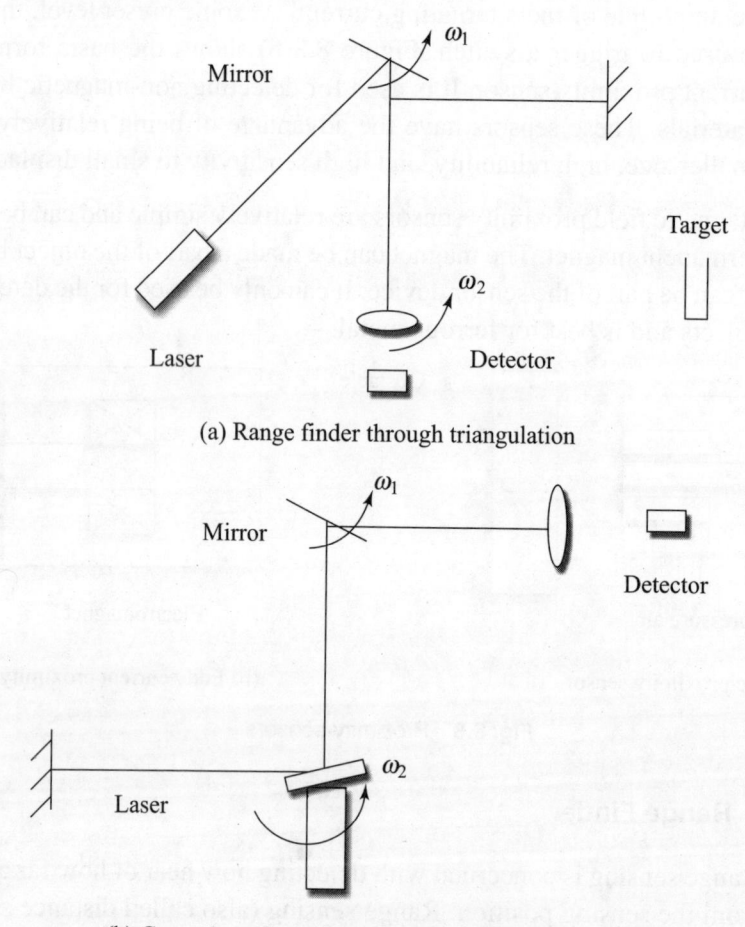

(a) Range finder through triangulation

(b) General configuration of the laser beam triangulation

Fig. 8.9 Range finder

continuously scanned. The distance between the arm of rotation and the photodetector constitutes the base line of the range finder. The development of the solid state laser diode and the ease of interfacing the optical sensor to the microcomputer have led to the design and implementation of a compact and accurate optical range finder. An ideal range finder should meet the following requirements:

1. The sensor must be capable of gauging the distances between 1 cm to 1 m to meet with a resolution of 1 mm.
2. The size of the sensor should be of the order of few cubic inches, so that it can be mounted on any mechatronics system.
3. The sensor should be relatively insensitive to the degree of effectiveness of the object.

Figure 8.9(b) illustrates the general configuration of the laser beam triangulation. In this set-up, a narrow collimated ray is directed towards the target with the help of a set of collimated lens. The beam impinges on the cube of the mirror that can rotate at a fixed range, so that the given plane is continuously scanned. A position-sensitive photodetector is placed in the scanning plane; the distance between the axis of rotation of the mirror and centre of the photodetector constitutes the baseline of the range finder.

8.4.3 Tactile Sensors

Tactile sensing means sensing through touch. In its widest context, tactile sensing can be taken to mean any sensing wherein a contact is involved. The simplest form of such a device is a square or rectangular array in which sensors are arranged in rows and columns. They are commonly called matured sensors. Each individual sensor is activated when brought into contact with the object. By detecting the active sensors or by analysing the magnitude of the output signal of the sensors, the input to the component being touched can be determined. This input is then compared with the previously stored input information to determine the size and shape of the component. Basically, there are three designs available for tactile sensors. These are the conductive skin, large scale integrated (LSI) circuit tactile sensor, and magnetic resistive skin.

Figure 8.10 illustrates the artificial skin or conductive membrane which was developed at the MIT laboratory. There are 256 tactile sensors in a space of the size of the finger tip. Each sensor has an area of 0.0163 cm^2 and is a multi-layer construction of different materials. The top layer is a sheet of elastomer conductive silicon rubber impregnated within a graphite or silver layer of approximately 0.25 mm thickness. The bottom layer is a printed circuit board (PCB), etched into fine parallel lines so that it conducts in only one direction. The sensitivity of the sensor depends on the construction of the middle layer, which consists of a fine mesh of either nylon or non-conductive paint. Wires are connected to the PCB at the edge of the array. The array is scanned by applying voltage to one row at a time. When an object is placed on the conductive membrane, the conductive membrane pierces through the mesh and short circuits the conductance of the wire and hence the current flow through the wire varies. Measurement of the current flow in each row of the wire identifies the object.

Fine mesh

Conductive membrane

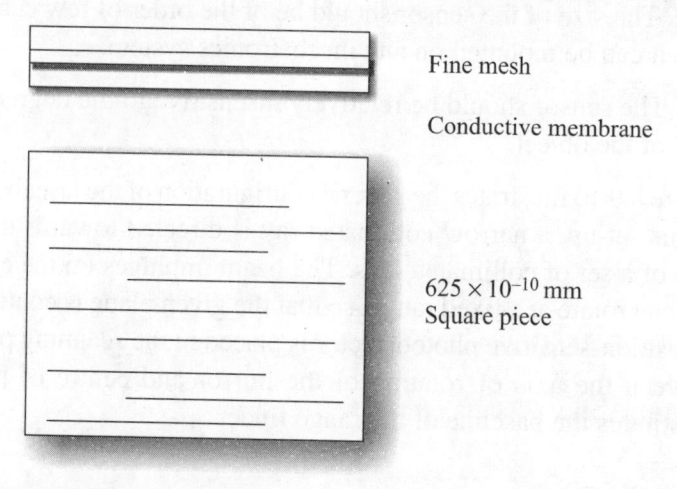

625×10^{-10} mm Square piece

Fig. 8.10 Artificial skin

Scientists at the NASA Jet Propulsion Laboratory have been working on the design of large scale integrated (LSI) circuits as tactile sensors. This type of sensor combines the function of transduction computation and communication. The basic concept of the LSI circuit design is shown in Fig. 8.11. The LSI wafer concept uses an array of tactile sensors with sensing materials connected directly to the computing element of the wafer. Such an arrangement reduces the rows of sensing data and sends the resulting signal to control the computer. A number of functions are thus combined. The signal bandwidth, number of connection wires, weights, and power consumption are reduced, while a high tactile resolution is achieved. The grasping force, touch pattern, contact area, and slipper can be extracted from the reading. The exposed surface of the LSI wafer contains an array of pairs of electrodes covered by a sheet of electrically conductive rubber. When the rubber sheet is deformed by touch, its resistivity varies locally. The current passing through the sheet between the pair of electrodes is indicated by the local contact pressure. An assembled electrode pair is the computational element consisting of an analog comparator, an adding register, and accumulators.

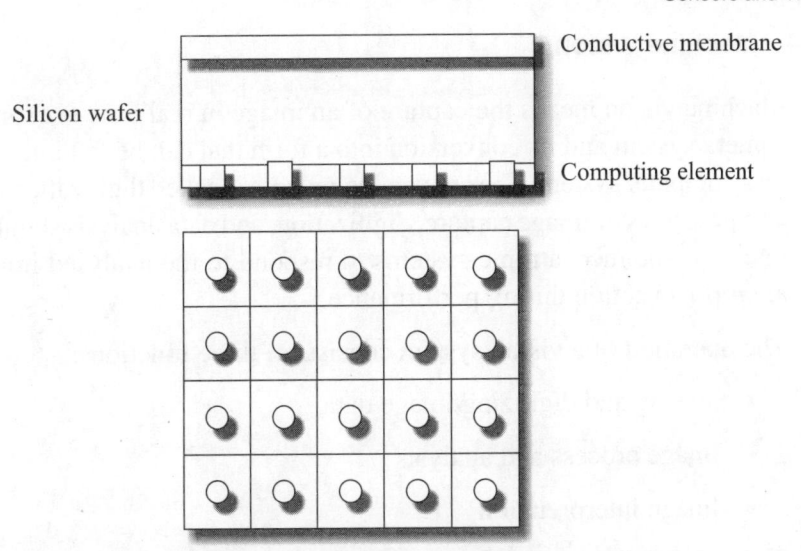

Fig. 8.11 Basic concept of LSI tactile sensor

Figure 8.12 shows the design of magneto-resistive skin. Magneto-resistive skin works on the ability of a material to change its electrical conductivity under the influence of the magnetic field. The magneto-resistive materials use an alloy of nickel and iron, called para alloys. In this case, it consists of 81% of Ni and 19% of iron. The resistance of para alloy is 50 Ω. Magneto-resistive elements are etched on a substrate of aluminium oxide connected by gold struts and thick film edge conductance. An array of magneto-resistive elements is arranged on a thin film of rubber. Typically, an open cell spring rubber strip is about 0.00625 mm wide in size. A copper strip provides the magnetic field. The magnetic field varies due to magneto-resistive, para alloy elements located underneath. The para alloy element changes its resistivity, which is proportional to its distance from the copper strip. A multiplicity circuit measures the change in resistivity of the element continuously and identifies the object.

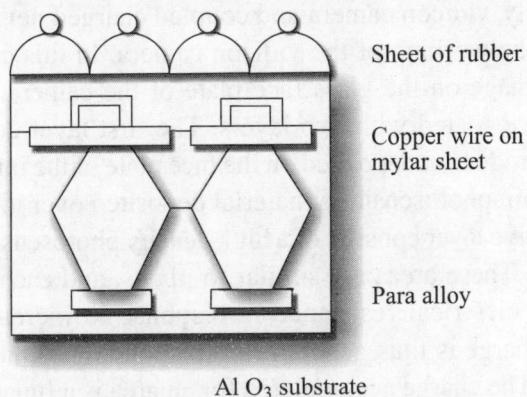

Fig. 8.12 Magneto-resistive tactile sensor

8.4.4 Machine Vision

Machine vision means the capture of an image in real time via some form of a camera system and its conversion into a form that can be fed into and analysed by a computer system. The conversion process is called digitization of the image. The processes of image capture, digitization, and data analysis should be enough to enable the mechatronic systems to respond to the analysed image and take appropriate action during performance.

The operation of a vision system consists of three functions:

- Sensing and digitizing image data
- Image process and analysis
- Image interpretation

Sensing and digitizing functions involve the input of the vision data by means of a camera focused on the screen of interest. Special lighting techniques are frequently used to obtain the image of sufficient contrast for later processing. The image received by a camera is typically digitized and stored in computer memory. The digital vision is called a frame grabber. These devices are capable of digitizing images at a rate of 30 frames/sec. A frame consists of a matrix of data representing projection(s) of the scene sensed by the camera. The elements of the matrix are called picture elements or pixels. The number of pixels is determined by a sampling process performed on each image frame. A single pixel is the projection of the small portion of the scene, which reduces to a single value. Usually 236×236 pixels will be there in an area of 1 cm^2. The value from a single pixel is a measure of high intensity for that element of the scan. The pixel's intensity is converted into a digital value.

Image capture is achieved through the vision camera. Two types of cameras used, namely, vidicon camera and coupled charged device (CCD). Figure 8.13 illustrates the working of the vidicon camera. In this camera system, the lens forms an image on the glass face plate of the camera. The face plate has its inner surface coated with two layers. The first layer consists of a transparent signal electrode film deposited on the face plate of the inner surface. The second layer is a thin photosensitive material deposited over the conducting film. The photosensitive layer consists of a high density photosensitive material in a small layer area. These areas are similar to pixels, and each rear pixel generates a decreasing electrical resistance in response to increasing illumination. An electrical charge is thus generated corresponding to the image formed on the face plate. The charge accumulated for an area is a function of the intensity for impinging light over the span feed time. Once light-sensitive scan is built up, the charge is read out to produce the vedicon signal. This is accomplished by scanning the photosensitive layer by an electron beam. The scanning is controlled by a deflection coil mounted along the length of the tube. For accumulated positive charge, the electron beam deposits enough electrons to neutralize the charge. An equal number of electrons flow to cause current to flow at the video signal electrode. The magnitude of the current signal in each pixel is proportional to the intensity of light and the pixel size. The current is then directed through a load resistor, which develops a signal voltage, which is further amplified and analysed. Raster scanning eliminates the need to consider time at each area by making scan time equal for all areas. Only the line intensity of the impinging light is considered. Raster scanning is typically done by scanning the electrode beam from left to right and from top to bottom. The output of the camera is a continuous voltage signal for each line scanned. The voltage signal for each scanned line is subsequently sampled and quantified, resulting in a series of version processes for the complete screen. This finally gives two-dimensional arrays of picture elements. Typically a single pixel is quantified between 6 and 8 bit by the analog-to-digital converter.

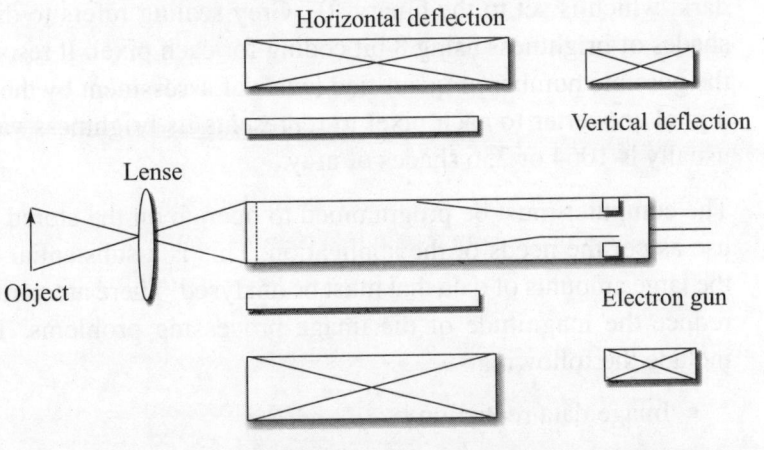

Fig. 8.13 Vidicon camera

The coupled charge device (CCD) is a solid state camera. In this technology the image is projected by a camera on a CCD, which detects, stores and read outs the accumulated charges generated by the light on each portion of the image. The light detection occurs through the absorption of light on a photoconductive substrate of silicon. Charges accumulate under positive controlled electrodes in isolated wells due to voltage applied to the controlled electrodes. Each isolated well represents one pixel and can be transferred to an output storage resistor by varying the voltage on the metal control electrode. Figure 8.14 illustrates the working of a CCD.

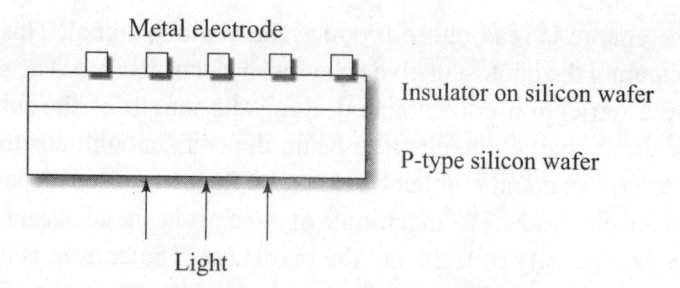

Fig. 8.14 Coupled charge device

In most vision systems video signal is digitized by an analog-to-digital converter and stored in fine buffer for subsequent computer analysis. The resolution of the device is determined by the sampling rate of the analog-to-digital converter. The frame grabber is an image storing and computational device which stores a given pixel array. A digital representation of the image means that the image information is stored sequentially as a binary bit pattern in computer memory. The grey level is a measure of the pixel brightness. The simplex level of each individual bit can be interpreted as either bright, which is set to binary '1', or dark, which is set to the binary '0'. Grey scaling refers to digitization of 256 shades of brightness using 8 bit coding for each pixel. It resolves to determine the possible number of quantified levels of assessment by the video analog-to-digital converter to each pixel to represents its brightness value. The number usually is 1064 or 256 shades of grey.

The computer must be programmed to operate on the stored digital image for use as per the needs of the application. This is a substantial task, considering the large amounts of data that must be analysed. There are various techniques to reduce the magnitude of the image processing problems. These techniques include the following:

- Image data reduction
- Segmentation
- Feature extraction
- Object recognition

In the image data reduction technique, the objective is to reduce the volume of data. Two schemes have found common use, namely, digital conversion and windowing. Digital conversion reduces the number of grey levels used by the machine vision system. Windowing involves uses only a portion of the total image stored in the frame buffer for image processing and analysis. This portion is called the window. In this technique the image is processed in a portion-by-portion manner.

Segmentation is a general term and applies to various methods of data reduction. The objective of segmentation is to grab areas of an image having similar features, singular edges, areas, etc. There are many ways to segment an image such as thresholding, region growing, and edge detection. Thresholding is a binary conversion technique in which each pixel is converted into a binary value, either black or white. Region growing is a collection of segmentation techniques in which pixels are grouped in regions, called grid elements, based on attribute similarities. This technique creates runs of 1's and 0's and is often used as the first pass analysis to partition the image into identification segments or blobs. For a simple image such as a dark blob on a light background, the run technique can provide useful information. For more complex images this technique may not provide an adequate portion of an image into a set of useful regions. Such regions might contain pixels that are connected to each other and have similar attributes. Edge detection considers the intensity change that occurs in the pixels of the boundary or edge of a part.

Feature extraction is usually accomplished by means of features that uniquely characterize the object. Some features of the object that can be used in machine vision include the area, diameter, perimeter, etc. A feature is a parameter that permits ease of comparison and identification.

Identification is accomplished using the extract feature information. The object recognition algorithm must be powerful enough to uniquely identify the object. The object recognition method may be implemented in two different ways:

1. Template match technique
2. Structural technique

The template match technique is a subset of the more general statistical pattern recognition technique that serves to classify the object in an image into a predetermined category. The basic objective in template matching is to match its object with a stored pattern feature set defined, called model template. This technique is suited to applications that do not need a large number of model templates. The structural technique of pattern recognition considers the relationships between features and edges of an object. It can be computationally time consuming for complete pattern recognition. It is often more appropriate for searching a simple region or edge within an image. These simpler regions can then be used to extract the required features.

8.4.5 Force Sensors

A spring balance is an example of a force sensor in which a force or weight is applied to cause displacement. The resultant displacement gives the measure of the force applied. The force applied can be static or dynamic. Force sensors are analog in operation and are sensitive to the direction of force applied. There are six types of forces that may require sensing. These are given below.

- Tensile force
- Compressive force
- Shear force
- Torsional force
- Bending force
- Frictional force

There are many techniques used in sensing these forces. Some of these techniques are direct and others indirect. The selection of a particular technique depends on the type and magnitude of the force and mode of its application.

A commonly used form of the force-measuring transducer is based on the use of electrical resistance and strain gauge monitor. Application of force produces strain in some members when they are stretched or compressed. This strain causes them to bend. This arrangement is generally referred to as a load cell. Figure 8.15 shows an example of such a cell. It consists of a cylindrical tube. Strain gauges are attached to this tube when force is applied to compress it. Then the strain gauges give a resistance change, which is a measure of the strain and hence the applied force. Typically such load cells are used for forces up to 10 MN. Strain gauge load cells are used for small forces varying from 0 to 5N up to 0 to 50 kN. Uncertainties in the sensor are typically a non-linearity error of about ±0.03% of full range, hysteresis error ±0.02% of full range and repeatability error ±0.02% of full range.

An unknown force may be measured by

1. Balancing against the known gravitational force on a standard mass
2. Measuring the acceleration of a body of known mass to which the unknown force is applied
3. Balancing against a magnetic force developed by the interaction of a current carrying coil and magnet
4. Transducing the force to a pressure and measuring the pressure
5. Applying the force to some elastic member and measuring the resulting deflection

Torque transmission to a rotating shaft generally involves both the source of power and a sink. Torque-measuring devices are called dynamometers. These devices may be generative type, absorption dynamometer, and transmission dynamometer. The transmission dynamometer is based on the strain gauge arrangement.

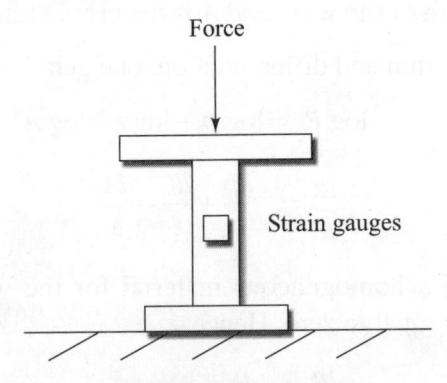

Fig. 8.15 Load sensor

8.5 Microsensors

The rapid advances in micro-fabrication technologies have led to the development of many microsensors. Microsensors are the most widely used mechatronic devices today. A smart sensor unit performs the following functions: automatic calibration, interference signal reduction, parasite effect compensation, offset correction, and self testing. All these functions make the sensor unit an intelligent micro-system.

Many types of microsensors are used to perform various functions in a variety of industries. These are generally used to measure physical quantities. We will discuss and explain the working principle of some of the microsensors in the following sections.

8.5.1 Strain Gauges

Strain gauges are resistive pick ups used as transducers. The resistance of the gauge changes in accordance with the strain, which in turn can be used for the measurement of input signals such as load or stress. The strain gauge consists of a metallic filament of approximately 0.03 mm diameter, which is sandwiched between thin layers of paper or epoxy resins. When load is applied to the surface of the elastic member where the strain gauges are mounted, a change in the length of the wire results. This change leads to a change in the resistance. This change in resistance can be measured with the help of a conventional Wheatstone bridge circuit. The change in electrical resistance is linear with strain. Gauge factor is the ratio of fractional change in electrical resistance to the fractional change in length (strain). The gauge factor for a strain gauge depends only on the material of the gauge wire. The resistance R of the wire can be obtained from the relationship $R = \rho \ell /A$, where ρ is the specific resistance of the material, ℓ is the length of the wire, and A is the cross-section area of the wire.

Taking logarithm and differentiation, one gets

$$\log R = \log \rho + \log \ell - \log A$$

$$\frac{dR}{R} = \frac{d\rho}{\rho} + \frac{d\ell}{\ell} - \frac{dA}{A}$$

Considering a homogeneous material for the wire, the change in specific resistance is equal to zero. Hence,

$$\frac{dR}{R} = \frac{d\ell}{\ell}(1+v)$$

where v is Poisson's ratio and is defined as the ratio of the transverse strain to the axial strain with a negative sign. Hence the gauge factor $G = (1 + v)$, which is constant for a particular wire material. Then, strain

$$\varepsilon = \frac{d\ell}{\ell} = \frac{1}{G}\frac{dR}{R}$$

which clearly indicates that the change in resistance is directly proportional to the change in strain.

Strain gauges are used for the measurement of displacement, pressure, load, force, torque, and stress. As an example, we can take a micro accelerometer constructed with suspended mass from a cantilever plate. The displacement of mass due to acceleration can be correlated with the change in the resistance of the strain gauges mounted on the root of the cantilever plate. Figure 8.16 illustrates the working of the microaccelerometer.

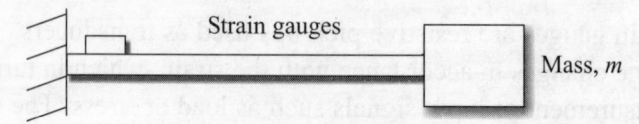

Fig. 8.16 Microaccelerometer

Micro-pressure sensors are widely used in automotive and aerospace industries. The main component of a microsensor is a thin diaphragm, which gets deflected (or stressed) when pressure or force acts on it. The amount of deflection or stress induced in the diaphragm deformation is then converted into an electrical signal output through transduction. There are generally two types of pressure sensors: absolute pressure and gauge pressure sensors. The absolute pressure sensor has an evacuated cavity on one side of the diaphragm. The measured pressure is the absolute value with vacuum as the reference pressure. In the absolute pressure type, nano-evacuation is necessary. Figure 8.17 illustrates a micro-pressure sensor. The top view of the silicon die shows four piezoresistors planted beneath the surface of the silicon die. These piezoresistors convert the stresses/deflection induced in the silicon diaphragm by the applied pressure into a charge of electrical resistance, which is then converted into voltage output by a Wheatstone bridge circuit. The piezoresistors are essentially miniaturized semiconductor strain gauges which can produce the charge of electrical resistance induced by the mechanical stresses output from the Wheatstone bridge.

$$V_o = V_i \left(\frac{R_1}{R_1 + R_4} - \frac{R_3}{R_2 + R_3} \right)$$

where V_i and V_o are the input and output voltages of the Wheatstone bridge, respectively.

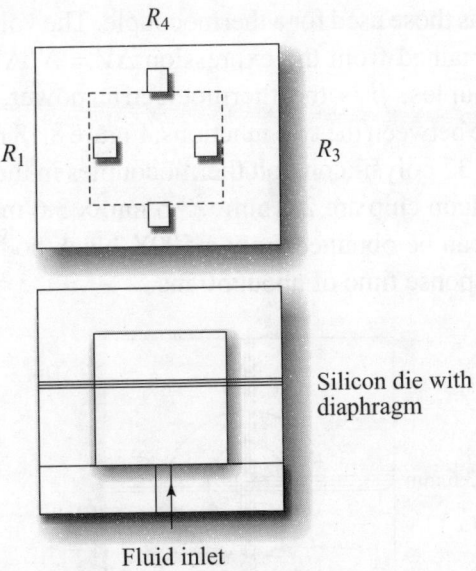

Fig. 8.17 Micro-pressure sensor

8.5.2 Thermocouples

Thermocouples are employed as temperature sensing elements. Basic principle of a thermocouple is seeback effect. When two different metals are joined together and the temperatures at the two junctions are different, then a potential difference occurs across the junctions. This phenomenon is known as the Seebeck effect. The reverse of this phenomenon is called the Peltier effect. When a voltage is applied to this kind of circuit, the temperature at one junction is lower and at the other junction is higher than the ambient temperature. The principle of the Peltier effect is used for micro-space air conditioning. The Thomson effect states that there is a temperature gradient in a conductor when current is passed through the conductor. The principles of the Seeback effect, Peltier effect, and Thomson effect are effectively used in microsystems application. A thermocouple used for temperature measurement is an example of the application of the Seeback effect. Since there is a temperature difference between the two junctions of a thermocouple, the value of emf, E, depends on the two metals concerned and the temperature T of the junctions. Mathematically, $E = aT + bT^2$, where a and b are constants. Some commonly used thermocouple materials are iron/constantam, chromel/alumal, copper/constantam, and platinum/platinum-rhodium alloys.

Micro-thermopile is a more realistic solution for miniaturized heat sensing. It operates with both hot and cold junctions. In this case, the thermocouple is in parallel and the voltage output is in series. Materials used for a thermopile are the same as those used for a thermocouple. The voltage output from a thermopile can be obtained from the expression $\Delta V = N\beta\Delta T$, where N is the number of thermocouples, β is the thermoelectric power, and ΔT is the temperature difference between the two junctions. Figure 8.18 represents a micro-thermopile with total 32 polysilicon volt thermocouples in the thermopile. The dimensions of this silicon chip are 2.6 mm × 3.6 mm × 3.6 mm. A typical output signal of 100 mV can be obtained from a 500K black body source of $Q = 0.29$ W/cm^2 with a response time of about 50 ms.

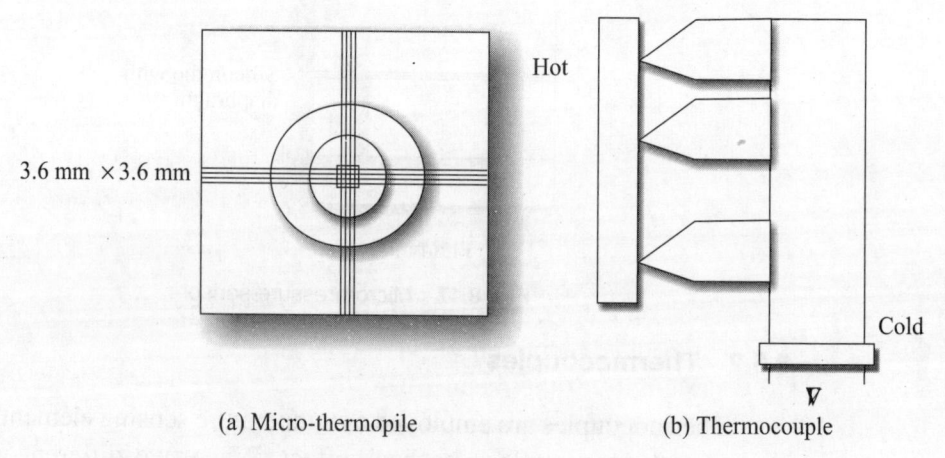

(a) Micro-thermopile   (b) Thermocouple

Fig. 8.18 Schematic diagrams of micro-thermopile and thermocouple

8.5.3 Acoustic Wave Sensor

The principal application of an acoustic wave sensor is to measure chemical composition in a gas. Each sensor generates acoustic waves by converting mechanical energy into electrical acoustic waves. These devices are also used to activate or regulate the fluid flow through micro-fluidic systems. The actuation energy for this type of sensors is provided by piezoelectric and magnetostriction mechanisms. The former mechanism is a more popular method for generating acoustic waves. Piezoelectricity is a common means for transducing mechanical energy into electrical energy and vice versa.

8.5.4 Biomedical Sensors and Biosensors

Biosensors present great challenges to engineers as the designing and manufacturing of this type of sensors requires considerable knowledge and experience in molecular biology as well as physical chemistry. Basically two types of sensors are used in biomedical applications:

- Biomedical sensors
- Biosensors

Biomedical sensors are used to detect biological substances. As against this class of sensors, biosensors are measuring device that use biological element(s) such as antibodies or enzymes for sensing activities. Biomedical sensors and biosensors are discussed in detail in Chapter 11.

8.5.5 Chemical Sensors

Chemical sensors are used to identify chemical compounds such as various gas species. The working principle behind these sensors is very simple. We know that many materials are sensitive to chemical attacks. For example, most materials are vulnerable to oxidation when exposed to air for a long time. A significant oxide layer built up over a metal surface can change material properties, such as the electrical resistance, of the original metal. Material sensitivity to specific chemicals is used as a basic principle for many chemical sensors. Three types of chemical sensors are in use. They are chemi-resistors, chemi-capacitors, and metal-oxide gas sensors. Organic polymers are used with an embedded metal sensor. These polymers can cause a change in the electrical conductivity of the metal when it is exposed to certain gases. For example, a special polymer called phthalocyamine is used with copper to sense ammonia and nitrogen dioxide gases. Some polymers can also be used as dielectric materials in a capacitor. The exposure of these polymers to certain gases alters the dielectric constant of the material, which in turn changes the capacitance between the metal electrodes. For example, polyphenylacetylene (PPA) is used to sense gas species such as CO, CO_2, N_2, CH_4, etc. Chemi-mechanical sensors use certain materials (e.g., polymers) that change shape when exposed to chemicals including moisture. One may detect the presence of such chemicals by measuring the change in the dimensions of the materials. Metal-oxide gas sensors work on a principle which is similar to that of chemi-resistor sensors. Several semiconducting materials such as SnO_2 change their electrical resistance after absorbing certain gases. The process is faster if heat is applied. To enhance the reactivity of the measured gases and the transduction semiconducting materials, metallic catalysts are deposited on the surface of the sensor. Such deposition can speed up the reaction and, hence, increases the sensitivity of the sensor. Figure 8.19 illustrates the principle of a metal-oxide gas sensor.

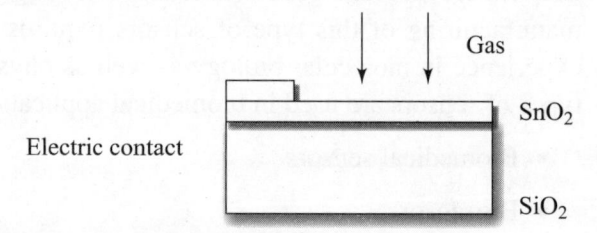

Fig. 8.19 Metal-oxide gas sensor

8.5.6 Optical Sensors

Optical sensors that can convert optical signals into electronic output have been developed and utilized in many consumer products such as television. Micro-optical sensors have been developed to sense the intensity of light. Solid state materials that provide stray photon-electron interactions are used as sensing materials. Selection of materials for optical sensors is principally based on quantum efficiency, that is, on a material's ability to generate electron-hole pairs from input photons. Semiconducting materials such as silicon and gallium arsenic are commonly used for optical sensors. Alkali metals such as cesium, lithium, sodium, potassium, and rubidium are also used. Figure 8.20 illustrates the working of a photovoltaic generator.

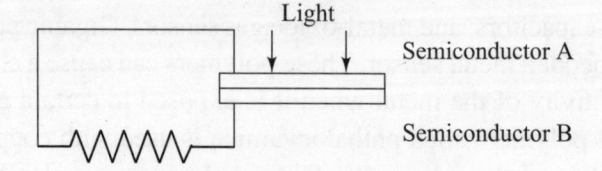

Fig. 8.20 Photovoltaic generator

Illustrative Examples

Example 8.1 A simple pressure-time function consists of two harmonics terms as $F = 100\sin 80t + 50\cos(160t - \pi/4)$. Analyse the relationship of harmonics.

Solution

An inspection of the equation shows that the circular frequency of the fundamental has a value of 80 rad/sec or $80/2\pi = 12.7$ Hz. The period for the pressure variation is $T = 1/12.7 = 0.0785$ sec. The second term has a frequency twice that of the fundamental, as indicated by the circular frequency of 160 rad/sec. It also lags the fundamental by 1/8 cycle of $\pi/4$ rad. In addition, the equation indicates that the amplitude of the fundamental is 50.

Example 8.2 What is the probable error at any point of the scale for a one-percent-accuracy indicator of range 1000°C? If the lower scale starts from 200°C, what is the range of the instrument? What is the percentage of error when the indicator reads 700°C?

Solution

$$\text{Accuracy in percentage} = \frac{\text{true value} - \text{measured value}}{\text{span}} = \frac{\text{error}}{\text{span}}$$

Hence,

Error = span × 1/100 = 1000/100 = 10°C

Error can be either negative or positive, hence the error is ±10°C.

When the lower scale is 200°C, the range of the instrument is equal to the minimum scale and span, i.e.,

200 + 1000 = 1200

When the meter reads 700°C, the probabilistic error is ± 10°C.

Hence, the percentage of error

$$= \pm \frac{10}{700} 100 = \pm 1.43\%$$

Example 8.3 Suppose a strain gauge has resistance $R = 120\ \Omega$, gauge factor $G = 2.1$. When a shunt resistance $R_s = 100\ \text{k}\Omega$ is connected across the gauge, what is the equivalent strain?

Solution

When a shunt resistance is connected,

Equivalent resistance $= \dfrac{RR_s}{R+R_s}$

since resistances are in parallel. Hence, change in resistance

$$dR = R - \dfrac{RR_s}{R+R_s} = \dfrac{R^2}{R+R_s}$$

Therefore, the equivalent strain

$$\varepsilon_e = \dfrac{1}{G}\dfrac{dR}{R} = \dfrac{1}{2.1}\dfrac{120}{(100000+120)} = 570 \text{ microstrain}$$

Example 8.4 Calculate the force required (in newton) to accelerate a body of mass 2 kg to 5 m/sec^2.

Solution

According to Newton's law of motion,

$$\text{Force} = \text{mass} \times \text{acceleration}$$
$$= 20 \text{ kg} \times 5 \text{ m/sec}^2 = 100 \text{ newton}$$

Example 8.5 Assume the width and thickness of a beam to be 10 μm and 50 μm, respectively. The beam is made of silicon with Young's modulus, 19000 MPa, and 10 mg weight is attached to the lever. Determine the equivalent spring constant and natural frequency of the beam element in a micro-accelerometer.

Solution

Referring to Fig. 8.12, the moment of inertia of the beam cross section

$$I = \dfrac{bt^3}{12} = \dfrac{(10\times 10^{-6})(50\times 10^{-6})}{12} = 0.1042\times 10^{-18}$$

The spring constant

$$k = \dfrac{w}{\delta} = \dfrac{3EI}{l^3}$$

where δ is the deflection, E is Young's modulus, and l is the length of the beam. Hence

$$k = \dfrac{3(190000\times 10^{-6})(0.1042\times 10^{-18})}{(100\times 10^{-6})^3} = 59.39 \text{ N/m}$$

The natural frequency of the cantilever beam

$$= \sqrt{\dfrac{k}{m}} = \sqrt{\dfrac{59.39}{10^{-5}}} = 2437 \text{ rad/sec}$$

Exercises

8.1 Explain, with an example, the static characteristic parameter of a sensor.

8.2 Differentiate between internal and external sensors.

8.3 What is a resolver? Give its applications.

8.4 Differentiate between normal binary code and grey code.

8.5 What is a proximity sensor? Explain the working of an eddy current proximity sensor.

8.6 Explain the working of an optical range finder.

8.7 What is a tactile sensor? Explain the working of a magneto-resistive tactile sensor.

8.8 What are the problems associated with machine vision?

8.9 Explain the working principle of a vidicon camera.

8.10 What are the various techniques used for memory reduction in machine vision. Explain briefly.

8.11 Explain the basic principle used in a microsensor.

Chapter 9

CNC Machines

The term 'mechatronics' was initially coined for the automated manufacturing environment. A mechatronic system consists of a mechanical system and an electronic system such as a microprocessor or computer. The electronic system controls, governs, and regulates the functioning of the mechanical parts of the system. Modern machine tools are excellent examples of mechatronic systems. Nowadays, numerical control machines are an essential part of any manufacturing environment.

The idea of numerical control machines was conceived by John Pearson and his associate Frank Stulen in 1948. Numerical control (NC) is a form of programmable automation in which mechanical actions of a machine tool or other equipment are controlled by a program containing coded alphanumeric data. A prototype of an NC machine successfully performed simultaneous control of three axes motion based on a data-punched binary tape in March 1952. Earlier, such NC machines were called numerical controlled servo-systems. NC machines were used at various aircraft industries in USA between 1958 and 1960. The US Air Force continued to encourage the development and application of NC by sponsoring research at the Massachusetts Institute of Technology (MIT). The research led to the development of the automatic program tool (APT) language in 1958.

An NC system consists of three basic components:
 (i) Program instructions
 (ii) Machine control unit
 (iii) Processing equipment

The program instruction is a list of detached, step-by-step commands that direct the action of the processing equipment. This list of instructions is called *part program*. It is coded on a suitable medium for submission to the machine tool unit. For many years, the common medium was a 25-mm-wide punched tape. This tape used a standard format that could be interpreted by the machine tool unit. Today, the punched tape has largely been replaced by newer storage technologies/devices. These include magnetic tape, diskettes, and electronic transfer of part program from a computer.

In modern NC technology, the machine control unit (MCU) consists of a microprocessor and related control hardware that stores the program of instructions and executes it by sequentially converting each command into mechanical action of the processing equipment. In case the MCU is a computer, the computer numerical control is used to distinguish it from numerical control machines.

Processing equipment is another basic component of an NC system. It accomplishes the processing steps to start and perform useful operations to accomplish a task. The processing equipment comprises a work table, spindle, motors, and control systems. Figure 9.1 illustrates the basic components of an NC system.

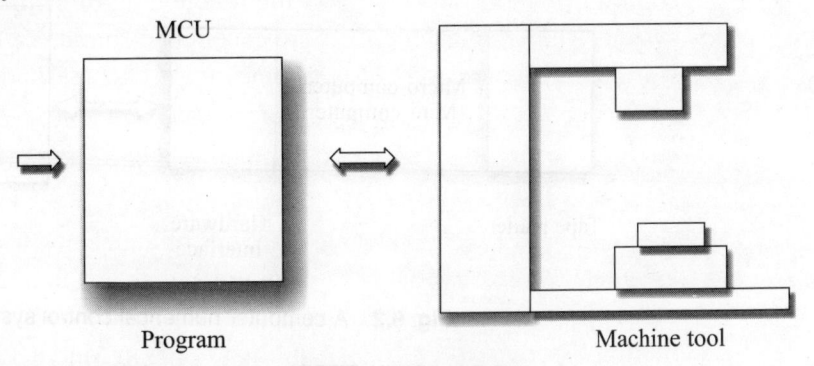

Fig. 9.1 NC machine

In keeping with the trend towards smaller, less expensive computers, a single computer is used for one machine tool. Such NC machines are called computer numerical control (CNC) machines. CNC systems were commercially introduced around the year 1970. One standard computer control unit could be adapted to various types of machine tools by programming the control functions into the computer memory for that particular machine. CNC is an NC system that utilizes the dedicated, stored program computer to perform some or all of the basic numerical control functions. CNC machines use a micro-computer or mini-computer.

There are a number of functions that a CNC machine is designed to perform. Several of these functions could be either impossible or very difficult to accomplish with conventional NC machine. The principal functions of CNC machines are the following:

- Machine tool control
- In-process compensation
- Improved programming and operating features
- Diagnostics

Figure 9.2 illustrates the general configuration of a CNC system. CNC machines have a number of inherent advantages over the conventional NC machines. These are as follows:

- Tape editing at the machine side
- Metric conversion
- Greater flexibility
- User-written program
- Total manufacturing system

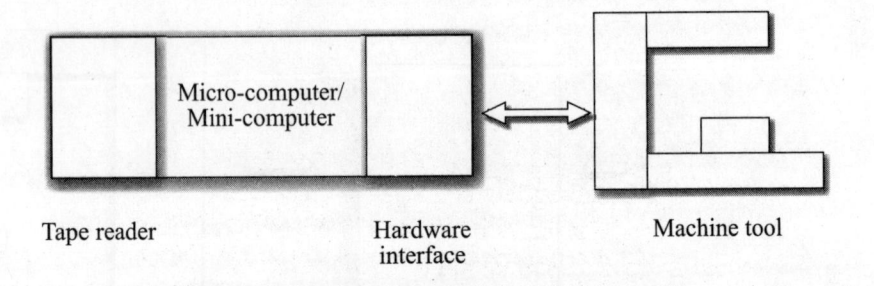

Fig. 9.2 A computer numerical control system

Direct numerical control (DNC) system is defined as a manufacturing system in which a number of machines are controlled by a computer through direct control and in real time. The tape reader is not used in DNC, thus relieving the system of its least reliable component. Instead of using a tape reader, the part program is transmitted to the machine tool directly from the computer memory. In principle, one large computer can be used to control more than 100 separate machines. Figure 9.3 illustrates the configuration of a basic DNC system. A direct numerical control system consists of four basic components:

(i) Central computer

(ii) Bulk memory, which stores the NC program

(iii) Telecommunication line

(iv) Machine tools

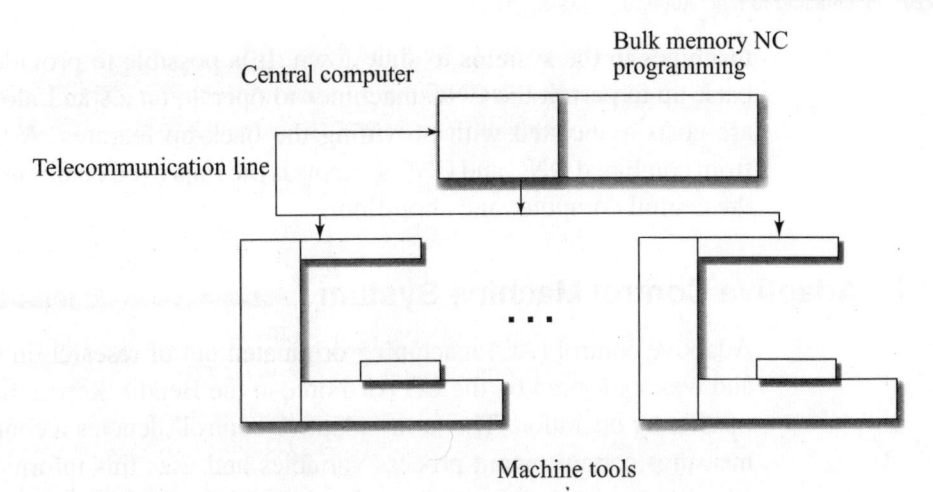

Fig. 9.3 Direct numerical control machine

There are several functions which a DNC system is designed to perform. These functions are unique to DNC and could not be accomplished with either CNC or conventional NC system. The principal functions of DNC are as follows:

- NC without punched tape
- NC part program storage
- Data collection, processing, and reporting
- Communication

Just as a CNC machine has certain advantages over a conventional NC system, there are also advantages associated with the use of DNC. These are as follows:

- Elimination of punched tapes and tape reader
- Creation of computational capability and flexibility
- Convenient storage of part programs in computer files
- Program can be stored as a label
- Reporting of shop performance
- Provides the framework for the evolution of the future computer automated factory

The combination of DNC and CNC provides the opportunity to add new capabilities and refine existing ones in computerized manufacturing systems. The DNC computer downloads the program directly to the CNC computer memory. A second advantage created by combining CNC with DNC is redundancy. If the central DNC computer fails, this will not necessarily cause the individual machines in the systems to shut down. It is possible to provide the necessary back-up to permit the CNC machines to operate on a stand-alone basis. There are costs associated with providing the back-up features. A third advantage from combined DNC and CNC systems is the improved communication between the central computer and shop floor.

9.1 Adaptive Control Machine System

Adaptive control (AC) machining originated out of research in the early 1960s and was sponsored by the US Air Force at the Bendix Research Laboratory for machining operation. The term 'adaptive control' denotes a control system that measures certain output process variables and uses this information to control speed and/or feed. Some of the process variables that have been used in adaptive control machining systems include spindle deflection, force, torque, cutting torque, vibration amplitude, and horse power consumed. The motivation for developing an adaptive machining system lies in trying to operate a process more efficiently.

Adaptive control is not appropriate for every machining situation. In general, the following characteristics can be used to identify situations where adaptive control can be beneficially applied:

(a) The in-process timing consumes a significant portion of machining cycle time.

(b) There are significant sources of variability in the job for which adaptive control can compensate.

(c) The cost of operating the machine tool is high.

(d) The typical jobs are those involving steel, titanium, and high strength alloys.

In the development of AC machining systems, two distinct approaches to the problem can be used. These are

(i) AC optimization (ACO)

(ii) AC constraints (ACC)

In ACO, an index of performance is specified for the system. This performance index is a measure of the overall process performance such as the production rate or cost per unit volume of metal removed. Most of ACO systems attempt to maximize the rate of work material removal to the tool wear rate. The index of performance is a function of the material removal rate divided by the total wear rate. The trouble with this performance index is that the tool wear rate cannot be measured online with the current measurement technology.

The systems developed for actual production are somewhat less sophisticated than the research ACO system. The production AC systems utilize constraint limits imposed on certain measured process variables. These are called adaptive control constraint (ACC) systems.

9.2 CNC Machine Operations

Several modes of operation are usually available in NC systems. The auto mode is the normal mode of operation. Other than the auto mode, NC systems may use the manual or dial-in mode, the jogging mode, or the block-by-block mode. The auto mode permits full automatic operation of the CNC, i.e., it allows continuous execution of the part program. The manual data mode permits the operator to dial in information to the control and thus instructs the machine to follow the machine cycle without the use of a punched tape or stored program. This is useful when setting up soft keys and is used to select information such as spindle speeds, feeds, etc. and a button is used to transfer this information into the control, one block at a time. The cycle start button is used to execute each block of information. For set-up purposes, it is useful in controlling the slide movements. Jogging switches permit this by selecting the direction and appropriate axis and then pressing a button to move the slide. This mode is used for coarse positioning. Fine positioning is carried out by discrete jogging switches. Discrete jogging switches permit the axis to be moved by a known distance, usually by 0.1, 0.01, or 0.001 mm. The axis and the increment of movement are selected and the slides move at the press of a button. These features can be used in setting up and in selecting the depth of a cut. In the block-by-block mode with the tape in position, it is possible for the control to read and execute one block's information at a time. Successive press of the cycle start button permits further blocks of the part program to read and execute block by block.

9.2.1 Compensation and Override

Compensation and feed rate correlation are available in CNC control. Tool compensation is very useful feature in control system and ensures that programming is independent of the tool dimensions. The control system contains a memory in which both the tool length and radius compensation are stored. The tool zero offset or cutter radius compensation features are useful in milling applications, where re-sharpened cutters or cutters of different diameters than the program size can be used without altering the tape. Roughing and finishing cuts can be made from the same tape and modifications to programmed dimensions may also be made. Tool zero offsets are frequently added to lathe controls to compensate for tool wear and to allow different diameters to be machined in order to match the mating parts.

Tool features are similar to tool zero offset, but are applied to the spindle axis only. These are particularly useful in drilling applications, where the varying lengths of drills can be compensated. Feed rate override is often advantageous to the programmed feed rate, especially in cases where adverse cutting conditions are encountered. If the program feed rates are unsuitable, it can be adjusted by a dial usually calibrated in terms of the percentage of the program feed rate; 0 percent to 100 percent is most commonly used.

The sequence number read-out displays each block of information and has a sequence number, for example N23, which means twenty-third block. It is convenient to have the sequence number displayed so that the operator knows which block of the program is being read at a particular time. This facility is very useful when checking tapes or programs as it enables the operator to know the location of any tape error.

The present position read-out display enables the operator to see on electronic displays the present position of one or all axes. This is particularly useful when the machine is being used manually and also when setting up the NC machine.

The CNC controller includes a keyboard for program or tape editing and a cathode ray tube, on which messages are displayed or tool path can be displayed for the operator. The machine tool consists of a work table, spindle, motors, and machine control. It also includes cutting tools, work fixtures, and other auxiliary equipment needed in the machining operation.

9.2.2 Co-ordinate System

A co-ordinate system consists of three mutually perpendicular axes intersecting at a single point called the zero point. The three axes of the co-ordinate system (X, Y, Z) are used to identify the three planes, namely, the XY plane, YZ plane, and ZX plane. Any plane located parallel to one of the planes is called a main plane. A co-ordinate system aids in the location of points in the working zone of a machine. There are two types of co-ordinate systems that are employed by a control system to position the operation tool in relation to the work piece. The Cartesian co-ordinates give paradoxical distance from one point to another. The polar co-ordinates describe the position of a feature by a length measured from a specified point and the angle measured in degrees from a specified datum or reference axis. Most CNC machines operate only in linear dimensions. Only special machining operation requires dimensional information. A second need may be the ability to rotate a feature about a specified location.

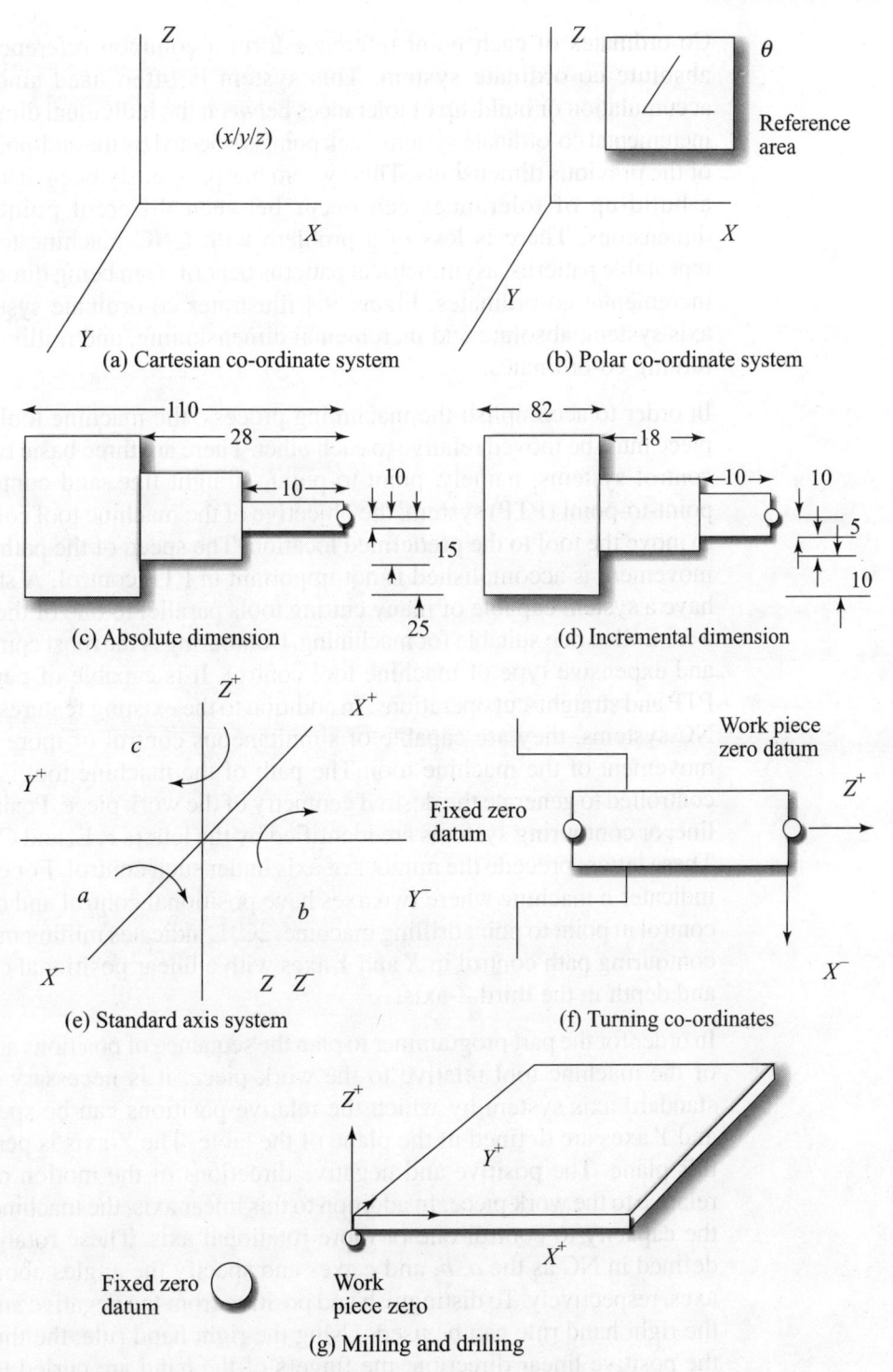

Fig. 9.4 Coordinate systems

(a) Cartesian co-ordinate system
(b) Polar co-ordinate system
(c) Absolute dimension
(d) Incremental dimension
(e) Standard axis system
(f) Turning co-ordinates
(g) Milling and drilling

Co-ordinates of each point reference form a common reference point in the absolute co-ordinate system. This system is often used since there is no accumulation or build-up of tolerances between the individual dimensions. In the incremental co-ordinate system, each point is selected by the end-point co-ordinates of the previous dimensions. This system has previously been discouraged since a build-up of tolerances can occur between different points of featured dimensions. There is less of a problem with CNC machine tools. Common repeatable patterns, asymmetrical patterns benefit from being dimensional, using incremental co-ordinates. Figure 9.4 illustrates co-ordinate systems, standard axis system, absolute and incremental dimensioning, and milling, drilling, and turning co-ordinates.

In order to accomplish the machining process, the machine tool and the work piece must be moved relative to each other. There are three basic types of motion control systems, namely, point-to-point, straight line, and contouring. In the point-to-point (PTP) system, the objective of the machine tool control system is to move the tool to the predefined location. The speed of the path by which this movement is accomplished is not important in PTP control. A straight cut can have a system capable of many cutting tools parallel to one of the major axis at a controlled rate suitable for machining. Contouring is the most complex, flexible, and expensive type of machine tool control. It is capable of performing both PTP and straight-cut operations. In addition to the existing features of contouring NC systems, they are capable of simultaneous control of more than one axis movement of the machine tool. The path of the machine tool is continuously controlled to generate the desired geometry of the work piece. Positional, straight line, or contouring systems are identified by the letters P, L, and C, respectively. These letters precede the number of axis under such control. For example, 2P, L indicates a machine where two axes have positional control and one axis linear control in point to point drilling machine. 2C, L indicates milling machine having contouring path control in X and Y axes with a linear positional control of feed and depth in the third Z-axis.

In order for the part programmer to plan the sequence of positions and movements of the machine tool relative to the work piece, it is necessary to establish a standard axis system by which the relative positions can be specified. The X and Y axes are defined in the plane of the table. The Z-axis is perpendicular to this plane. The positive and negative directions of the motion of the tool are relative to the work piece. In addition to this linear axis, the machine may possess the capacity to control one or more rotational axis. These rotational axes are defined in NC as the a, b, and c axes and specify the angles about X, Y, and Z axes, respectively. To distinguish and positive from the negative angular motion, the right hand rule can be used. Using the right hand rule, the thumb points in the positive linear direction, the fingers of the hand are curled to point in the positive rotational direction.

For a turning operation, two axes are normally all that are required to command the movement of the machine tool relative to the rotating work piece. The Z-axis is the axis of rotation of the work piece and the X-axis defines the radial location of the machine tool. The positive Z is the movement away from the work piece and positive X is to increase the radius.

The work piece datum may be defined as the point, line, or surface from which dimensions are referred. The machine datum is an established position within the programmable area of the movement of the machine about which the machine makes its programmed dimensional moves. It is often termed the zero datum or simply zero. Each axis of the movement on the machine must have an established zero datum point. This point may be fixed by the machine tool manufacturer or may be defined by the user and set by the operator. It follows that the machine datum must be referenced in some way to the work piece datum. Its location will depend on where the component is mounted. Machine datum facilities generally take three forms, namely, fixed zero, zero offset, and datum shift and floating zero. Fixed zero refers to an absolute fixed point on the machine. The machine zero datum is fixed and cannot be repositioned by the operator. When a machine utilizes this datum type, an important principle known as single co-ordinate positioning is established. This means that all programmable dimensions are assumed to be positive in sign from the fixed zero. This simplifies the job of the part programmer. With this system, all programmed moves must be made relative to the fixed datum of the machine. The machine operator needs to fix the component on the table in the specified position accurately, which is more difficult. The zero offset facility allows the zero point to be specified to any desired position within the programmable area of movement. Datum shift permits the convenient location of the work piece on the machine work table, since the machine zero can now be made to correspond with the work piece datum. Part programming is simplified, since the component can be positioned on the work table as per the convenience of the machine tool operator. The floating zero or full floating zero is the most common datum system in operation. Any co-ordinate position may be described in terms of positive or negative dimensions.

9.2.3 CNC Programming Procedure

The part programming often involves carrying out the task of process planning. Process planning is the procedure of deciding what operations are to be done on the component, in what order, and with what tool and work holding facilities.

Both part programming and process planning aspects of manufacture occur after the detailed drawings of the component have been finalized. A part programmer should be able to do the following:

- Plan work holding
- Plan operation sequence
- Decide on tooling
- Tabulate co-ordinate dimensions
- Ascertain part program format
- Code the part program
- Check, review, and modify the part program
- Formulate documentation

The necessary data for producing a part may come from the following:

- Information taken directly from the drawing
- Machining parameters
- Data determined by the part programming
- Information depending on particular NC system

There are two methods of part programming: manual part programming and computer-assisted programming. In manual part programming, the program is developed by means of NC codes; in computer-assisted programming, the program is written in a higher language and the part program containing NC codes is generated by the computer. Basically, there are two types of format:

- Fixed-block format
- Word-address format

In the fixed-block format, instructions contain all words in the same sequence irrespective of the words being the same in the previous block. In each instruction, words are in the same sequence, but each word is preceded by a tab character. If the instructions remain unchanged in the succeeding block, the instructions need not be repeated, but the tab character must be punched in. In both the cases, the identifying letter address need not be enlarged. Presently, the fixed-block format is not used, since punched tapes are already obsolete.

In the word-address format, each word is preceded and identified by its letter address. This format enables instructions that remain unchanged from the preceding blocks. This system speeds up programming and the length of tapes is considerably reduced. This format is adopted by most of the CNC machine control units.

A block is a group of words containing coded instructions for the NC system to execute at a particular movement. Each block is separated from the next block by an end of block (EOB) character. Each word consists of a letter and some number. An example of a block follows:

N0090　G01　X150　Y270　Z-10　F10　S55　T01　D01　M08　EOB

where N0090 is a sequence number or line number, which is used to identify the block. N0090 is called a word, N is the address and 0090 is a numerical value. G01 is a preparatory function, X150 is the geometric *X*-axis instruction, Y270 is the *Y*-axis instruction, Z10 is the Z-axis instruction, F10 is the feed rate instruction, S55 is the spindle speed in the structure, T01 is tool instruction, D01 is the length compensation instruction, and M08 is the miscellaneous instruction.

A preparatory function is a word used to prepare the controller for instructions. Some preparatory function codes are as follows:

G00	Point-to-point positioning
G01	Linear interpolation
G02	Circular interpolation—clockwise
G03	Circular interpolation—anticlockwise
G04	Dwell
G05	Hold
G06	Parabolic interpolation
G08	Acceleration of the feed rate
G09	Deceleration of the feed rate
G10	Linear interpolation similar to G01 except that all dimensions are multiplied by 10(used for incremental dimension)
G11	Linear interpolation similar to G01 except that all dimensions are divided by 10
G13–G16	Used for axis selection for direct control system to operate in a specific axis
G17	*XY* plane selection

G18	ZX plane selection
G19	YZ plane selection
G20	Circular interpolation in clockwise direction for long dimension
G21	Circular interpolation in clockwise direction for short dimension
G30	Circular interpolation in counterclockwise direction for long dimension
G31	Circular interpolation in counterclockwise direction for short dimension
G33	Thread cutting for constant lead
G34	Thread cutting for increasing lead
G35	Thread cutting for decreasing lead
G40	Cutter compensation cancel
G41	Cutter compensation left
G42	Cutter compensation right
G43–G49	Cutter compensation to adjust for difference between actual and programmed cutter radius.
G60–G70	Reserved for positioning for part to part system
G80–G89	A fixed cycle or canned cycle while direct the machine to complete such action
G90	Absolute dimension programming
G91	Incremental dimension programming

Co-ordinates specify the position for the tool. *X*, *Y*, and *Z* are used for the primary motion of the tool and *U*, *V*, and *W* for the secondary motion. *a*, *b*, and *c* specify angular dimensions and *I*, *J*, and *K* specify circular dimensions used for thread cutting.

The F and S functions are used to specify the feed in a machining operation and spindle rotation or cutting speed in RPM respectively. Tool function T is a word needed only for machines with a tool turret or automatic tool changer. The word T specifies which tool is to be used in the operation. The word D is used to activate the cutter radius or length compensation. D0 is used to cancel the tool length offset. The miscellaneous word M is used to specify certain miscellaneous or auxiliary functions which may be available on the machine tool such as coolant on/off, spindle/off, etc. The M function is also known as an auxiliary function. Some standard machine functions are given below:

M00	Program stop, which stops spindle coolant feed command.
M01	Optional stop, which is similar to M00 but it is performed only when the operator has previously pushed a button.
M02	End of the program, which indicates the completion of all the instructions.
M03	Starts spindle rotation in clockwise direction.
M04	Starts spindle rotation in counterclockwise direction.
M05	Stops the spindle and coolant.
M06	Executes the change of tool manually not to include tool selection.
M07	Turns flood coolant on.
M08	Turns mist coolant on.
M09	Automatically shuts the coolant off.
M10	Automatically clamps the slides, work piece, fixtures, spindle, etc.
M11	Unclamp command.
M13	Spindle rotates clockwise and coolant turns on.
M14	Spindle rotates counterclockwise and coolant turns on.
M15	Rapid transverse or feed rotate '+'.
M16	Motion enters + or − direction.
M19	The spindle to stop a predetermined angular position.
M30	End of tape similar to M02 except that rewinding tape or stop. This makes ready for the next work piece.
M31	Interlock by pass which temporarily releases the normal interlock.
M32–M35	The control maintains a constant cutting speed by adjusting the rotation of speed of work pieces.
M40–M45	Gear changers, if used, otherwise unassigned.

9.2.4 Interpolation

In contouring or continuous path systems, the tool is cutting while the axes of motion are many. All axes of motion might move simultaneously, each at a different velocity. When a non-linear path is required, the axial velocity changes even within the segment, so the position of the cutting tool at the end of each segment and the ratio between the axial velocities determine the desired contour of the part. The system must contain a continuous position control loop in addition to velocity controls. Dimensional information is given for each axis and is fed through the data processing unit. The programmed feed rate of the contour, however, has to be processed by the control process unit (CPU) in order to provide the proper velocity commands for each axis. This is done by means of an interpolation which is contained in the CPU of the contouring system. The function of the interpolator is to obtain intermediate points lying between those taken from the drawing. In NC or CNC systems, three types of interpolators exist—linear, circular, and parabolic. The most commonly used ones are linear and circular.

For linear interpolation, G01 is used, which gives the displacement in straight to the point where co-ordinates have been given in the same block. The same goes for points in the subsequent blocks so long as G01 is not replaced by a preparatory function of the same group. This displacement is at the programmed feed rate and, therefore, fit for machining.

The function G02 or G03 is used to predict the displacement of the tool delay circular arc. In G03 the tool movement is counterclockwise, while G02 has clockwise movement. After G02 or G03, the coordinates X, Y, and Z of the end position of the arc are programmed. The coordinates of the centre of the arc are programmed with i, j, and k. Instead of the co-ordinates of the centre, the arc can also be defined by programming the radius, R. Figure 9.5 illustrates the interpolation commands.

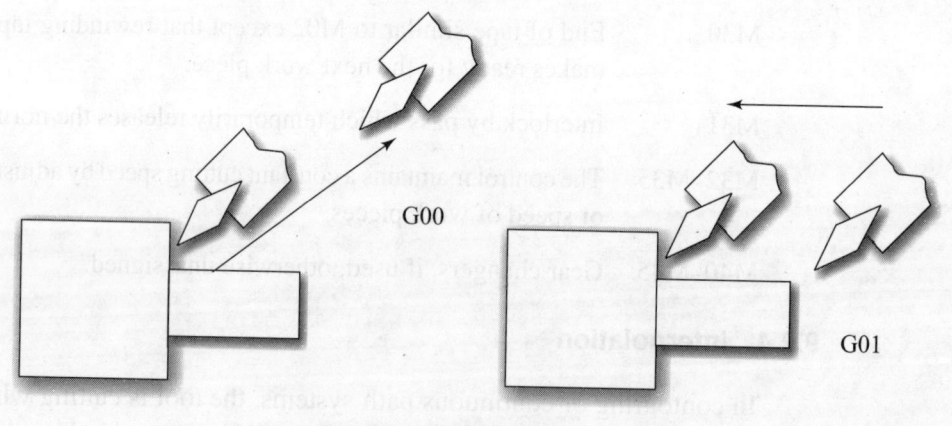

(a) Linear interpolation

CNC Machines 231

(b) Circular interpolation clockwise

A = Start point
E = End point
O = Center point

(c) Circular interpolation counterclockwise

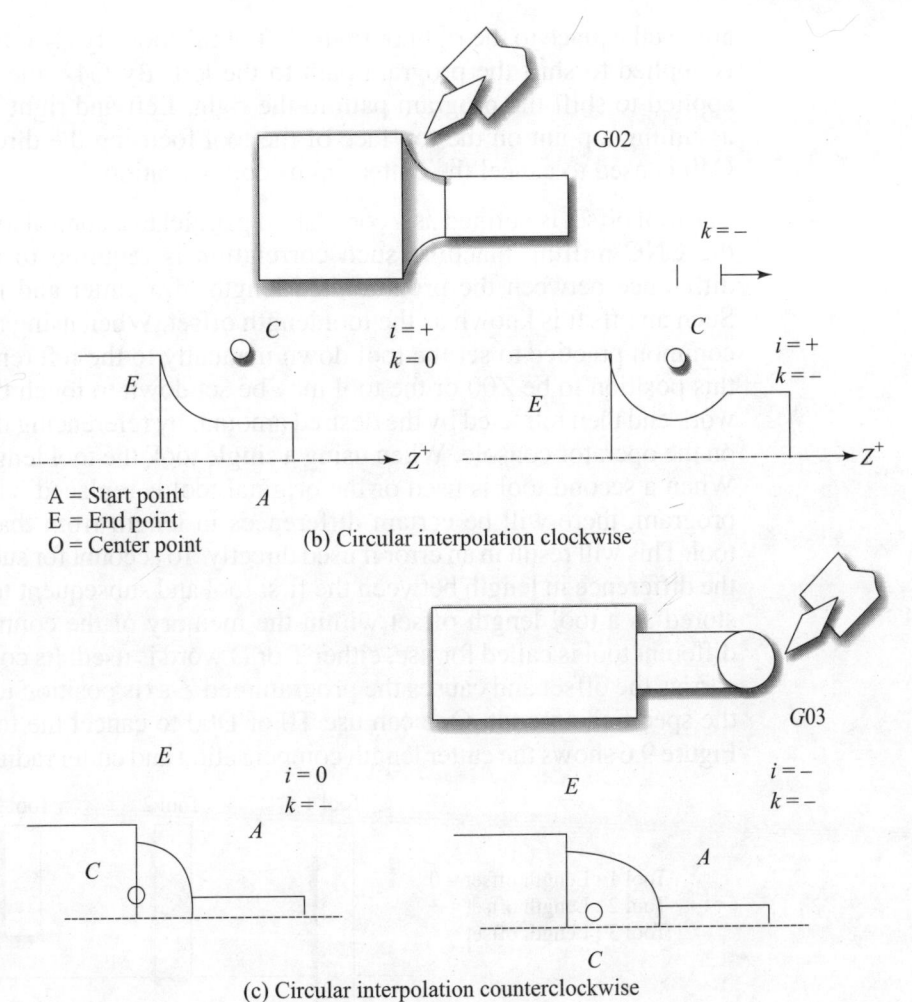

Fig. 9.5 Interpolation

9.2.5 Cutter Compensation

The important consideration in programming is to program the required path with respect to the centre of the tool considering a nominal radius value rather than at the point on its periphery where actual cutting takes place. The difference between the programmed radius and the actual radius of the cutter tools can be easily accommodated using the cutting radius compensation facility. The cutter radius compensation value is entered into the control unit adjacent to each tool number. Cutter compensation is defined as the correlation normal to a controlled axis. When cutter compensation is provided, the control unit generates a new tool path which is equidistant from the programmed cutter path by the difference between the programmed radius and the actual radius of the cutter. This is the compensator dimension. It is necessary to indicate whether these corrections are to take place to the right or to the left of the tool. By G41, the compensation is applied to shift the program path to the left. By G42, the compensation is applied to shift the program path to the right. Left and right are identified by assuming a point on the top face of the tool focusing the direction of the cut. G40 is used to cancel the cutter radius compensation.

The tool offset is defined as a correlation parallel to a control axis. In the case of the CNC milling machine such correlation is required to take care of the difference between the programmed length of a cutter and its actual length. Such an offset is known as the tool length offset. When using a single tool, it is common practice to set the tool down manually to the reference plane and set this position to be Z00 or the tool may be set down to touch the surface of the work and then retracted by the desired amount, by referencing the Z-axis display on the operator console. When using a single tool, the tool length offset is zero. When a second tool is used or the original tool is replaced with the some part program, there will be certain differences in length from that of the original tool. This will result in an error if used directly. To account for such discrepancies, the difference in length between the first tool and subsequent tools is noted and stored as a tool length offset within the memory of the control unit. When a different tool is called for use, either T or D word is used. Its corresponding tool carries the offset and causes the programmed Z-axis position to be modified by the specified amount. One can use T0 or D00 to cancel the tool length offset. Figure 9.6 shows the cutter length compensation and cutter radius compensation.

Tool 1 : Length offset = 0
Tool 2 : Length offset = +
Tool 3 : Length offset = −

(a) Tool length compensation

Tool 1 : G41—compensation to left
Tool 2 : G42—compensation to right

(b) Tool radius compensation

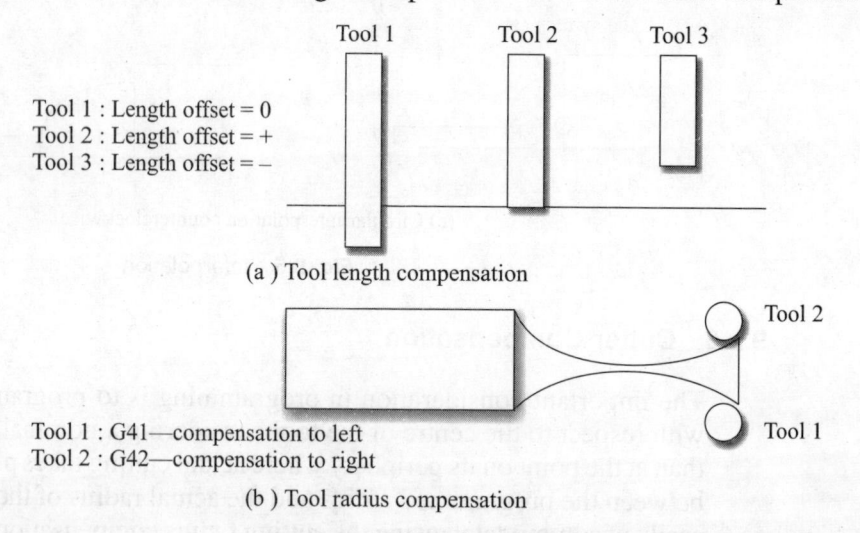

Fig. 9.6 Cutter compensation

An optional stop facility is available in some control systems in which some of the unwanted program blocks can be temporarily stopped while running a program by inserting 'l' or 'm' word before the start of the program block. The block in question will not be deleted from the program and if that block is to be executed again, it can be done just by erasing the character.

9.2.6 Part Program

A part program defines a sequence of NC machine operations. The information contained in the program can be dimensional or non-dimensional such as speed, feed, auxiliary functions, etc. The basic unit of a part program input to the control is called a block. Each block contains adequate information for the machine to perform a movement and/or a function. All blocks are terminated by the block end character,*. The maximum block length in each NC machine is fixed.

In the absolute dimensioning, all the dimensions are measured in the same direction from a fixed datum. In incremental dimensioning, adjacent dimensions are placed in series in a row or column. The disadvantage of incremental dimensioning method is that the tolerance on each dimension is accumulated. The following examples illustrate the part programming for absolute and incremental dimensional work pieces.

Figure 9.7 shows a part drawing. It requires X and Y axes to position first on point 1 and then to points 2, 3, and 4, with co-ordinates (100, 100), (150,125), (200,150), and (250, 175), respectively, from the datum.

```
N5    G71 *     ---- metric
N10   G90 *     ---- absdnle
N15   G00   X100   Y100 *
N20         X150   Y175 *
N25         X200   Y125 *
N30         X250   Y150 *
```

Fig. 9.7 Absolute program

The part drawing shown in Fig. 9.8 requires programmable X and Y axes to position from first point 1, point 2, and point 3, for which the incremental points in the X-direction are 20, 50, and 100 and in the Y-direction are 40, 60, and 80. The fixed datum is at (100, 100) from the edge of work piece.

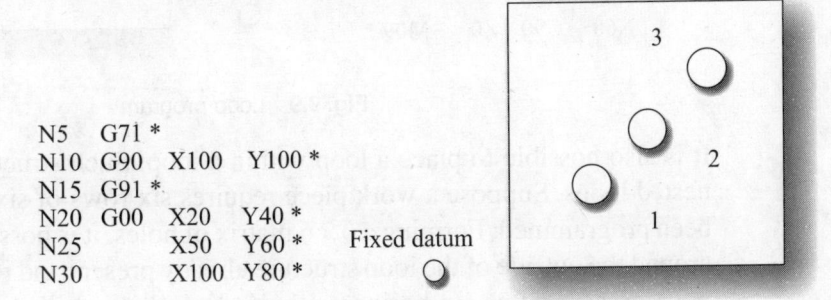

```
N5    G71 *
N10   G90   X100   Y100 *
N15   G91 *
N20   G00   X20    Y40 *
N25         X50    Y60 *
N30         X100   Y80 *
```

Fig. 9.8 Incremental program

When a certain set of operations is to be repeated within the same component, conventional programming methods take enormous time for program development. In order to minimize the length of the part program and its development time, repetitive programming techniques can be utilized. There are three types of repetitive programming techniques—looping, nested looping, and subroutine.

Looping provides the programmer with the ability to jump back to an earlier part of the program and execute the intervening program loop(s) back on itself. This is particularly useful when used in conjunction with incremental programming with a fixed or canned cycle. To repeat a sector of the part program a number of times, it is necessary to specify three things—the start of the loop, end of the loop, and the number of repeats. All the information is provided in a single block of information, e.g., = N100/3. In this the '=' character marks the start of the loop, 'N100' denotes the end block number, and '/3' specifies the number of repeats. When a start loop statement is encountered, the number of repeats is placed in a special type of memory called register. After every repeat, this register is decremented by 1. If the control of register is '0, the program will proceed to the block following the end of the loop. Figure 9.9 illustrates a loop program where an array of six holes at 30 mm needs to be drilled.

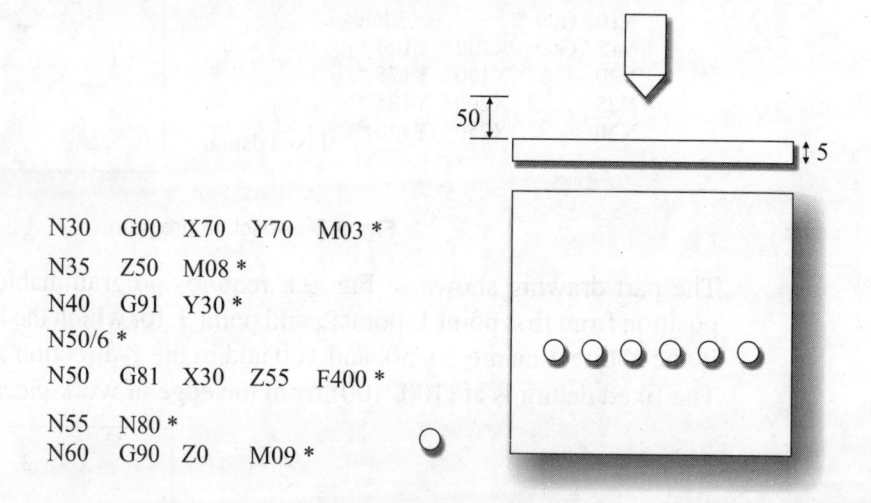

```
N30    G00   X70   Y70   M03 *
N35    Z50   M08 *
N40    G91   Y30 *
N50/6 *
N50    G81   X30   Z55   F400 *
N55    N80 *
N60    G90   Z0    M09 *
```

Fig. 9.9 Loop program

It is also possible to place a loop within a loop. Such structures are known as nested loops. Suppose a work piece requires six rows of six holes which have been programmed. Forming a 6×6 matrix of holes, it is possible to place a loop around the outside of the loop structure already present and repeated the spindle position at Y, after six horizontal holes have been drilled. The part program segment shown in Fig. 9.10 illustrates this nested loop.

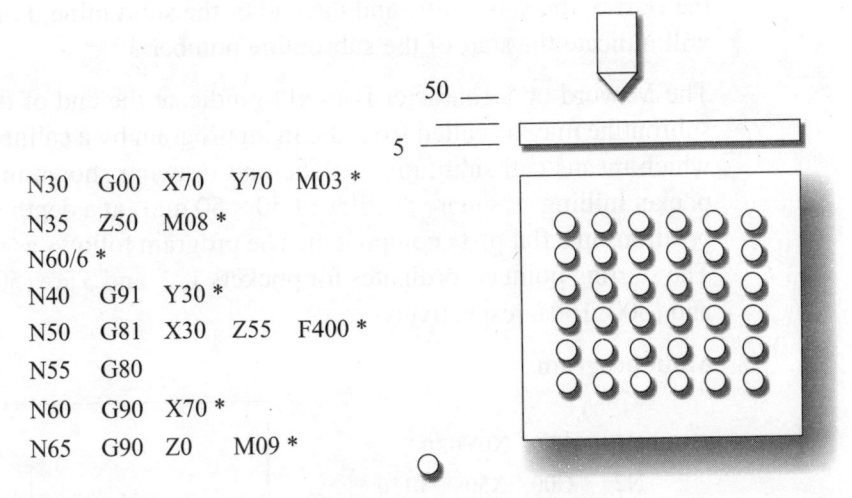

```
N30    G00   X70   Y70   M03 *
N35    Z50   M08 *
N60/6 *
N40    G91   Y30 *
N50    G81   X30   Z55   F400 *
N55    G80
N60    G90   X70 *
N65    G90   Z0    M09 *
```

Fig. 9.10 Nested loop

In the end milling, usually the cutter rotates on the axis perpendicular to the work piece. Flat surface as well as various profiles can be produced by end milling. End milling is ideally suitable for pocket milling. The part drawing shown in Fig. 9.11 requires pocket milling up to a predetermined depth of 2 mm with a 50-mm end mill cutter.

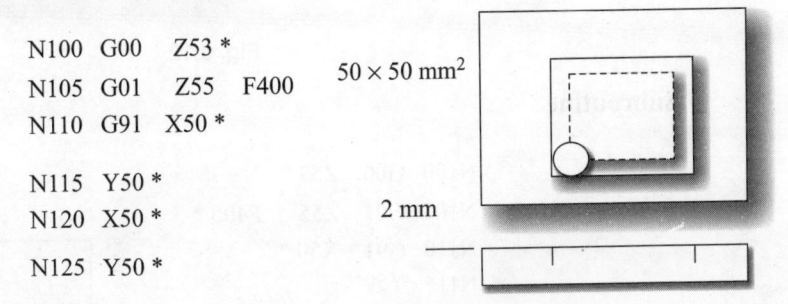

```
N100   G00   Z53 *
N105   G01   Z55   F400
N110   G91   X50 *

N115   Y50 *
N120   X50 *

N125   Y50 *
```

Fig. 9.11 Pocket milling

There may be certain repetitive features within a component which cannot be accommodated within a loop structure. In such cases, the repetitive element may be described in terms of a subprogram and is placed in the part program memory or at the end of the main program. Whenever such a feature is required within the part program, its associated subroutine is called for execution. When a subroutine is called, the flow of the program is transferred to the start of the subroutine to the start of subroutine to continue the execution. After the subroutine is completed, the flow of the program returns to the main program to the point immediately following the point at which the subroutine was called. Thus a subroutine may be called from many different positions within the main part program. To use and describe a subroutine, the following must be indicated—

the start of the subroutine and the end of the subroutine. For instance #3 or L3 will indicate the start of the subroutine number 3.

The M word or $ character is used to indicate the end of the subroutine. The subroutine may be called from the main program by a calling line such as = #3, which means call subroutine 3. The part drawing shown in Fig. 9.12 requires pocket milling of square profiles of 50×50 mm^2 at a depth of 2 mm at various positions in a flat plate component. The program follows as shown in Fig. 9.13. The starting point co-ordinates for pockets 1, 2, and 3 are (50, 150), (100, 100), and (300, 160), respectively.

Main program

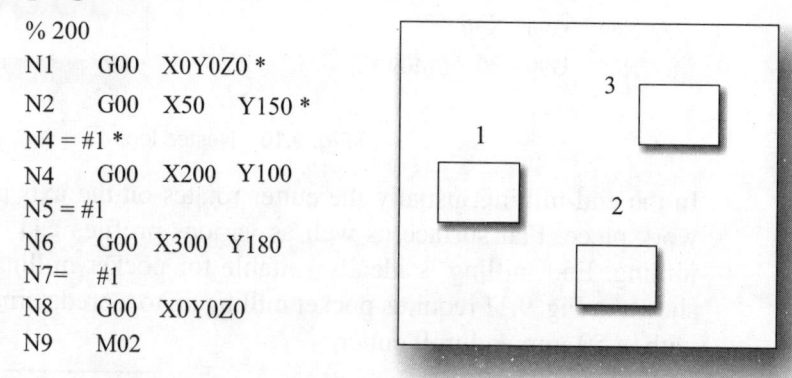

```
% 200
N1    G00   X0Y0Z0 *
N2    G00   X50   Y150 *
N4 = #1 *
N4    G00   X200  Y100
N5 = #1
N6    G00   X300  Y180
N7=   #1
N8    G00   X0Y0Z0
N9    M02
```

Fig. 9.12

Subroutine

```
#1
N100  G00   Z53 *
N105  G01   Z55   F400 *
N110  G91   X50 *
N115  Y50 *
N120  X50 *
N125  Y50 *
N130  G90 *
N135  G00   Z53 *
N140  $ *
```

50×50 mm^2

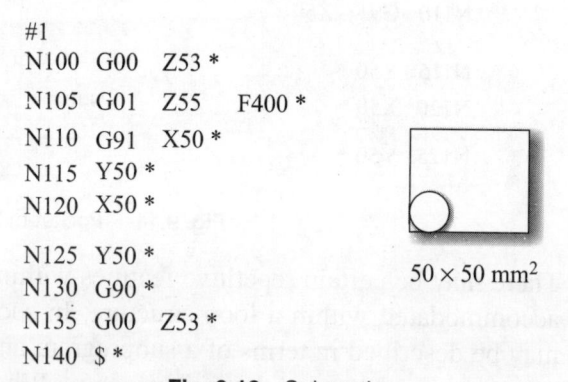

Fig. 9.13 Subroutine

A subroutine may be called after a subroutine. But subroutines must not be defined within a subroutine. A macro is a single command that generates a series of tool paths for the execution of a particular machining operation. The macro call within the part program must invariably be accompanied by additional information. Macros may be predefined as in-built control system facilities or they may be user-defined. In the latter case, the implementation of a macro is similar to that of a subroutine. In fact, the terms 'subroutine' and 'macro' are used without distinction by some manufacturers.

A macro can be thought of as a subroutine with the ability to pass values or parameters. A parameter is a value that acts as a constant value in the process, but may take different values from run to run. For example, the subroutine described earlier will not mill a single square projectile of 50×50 mm^2 only. It cannot mill a principle of any other dimension. With macro facilities, it will be possible to mill the same principle with different dimensions by simply passing different parameter values with the calling line. With macros, it is necessary to specify parameters that may change from call to call, a means of passing true values into these parameters. Within the macro definition the parameter likely to change is not given any numerical values, but is marked with an asterisk. True numerical values are supplied when the macro is called. Figure 9.14 shows a part drawing that requires pocket milling using macros. The starting point co-ordinates for pockets 1, 2, and 3 are (50, 150), (100, 100), and (300, 160), respectively. The starting point co-ordinates for points 1 and 2 are at (100, 100) and (200, 125), respectively.

Main program

```
#Z01 *
N1    G00    X0Y0Z0 *
N2    G00    X100  Y100 *
N3 = #1  X*  50  Y*  50   X*  50  Y*  50
N4    G00    X200  Y125
N5 = #1  X*  100  Y*  50   X*  100  Y*  50
N7    M02
```

Subroutine

```
#1 *
N100  G00  Z53 *
N105  G01  Z55   F400 *
N110  G91  X*
N115  Y*
N120  X*
N125  Y*
N130  G90*
N135  G00  Z53*
```

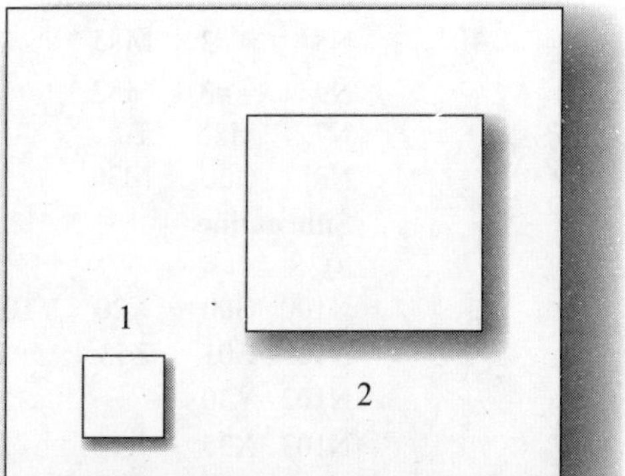

Fig. 9.14 Use of macros

Many control systems offer additional features that aid the programming of repetitive features. Most of these are invoked by G or M codes within the part program. Some common features available on a milling machine are:

- reflection or mirror imaging
- scaling
- rotation
- translation

Many engineering components need to be produced in handed pairs, right handed parts and left handed parts. The two components are invariably identical in dimensions but are geometrically opposite. For this purpose, the mirror imaging facility is employed. The line along which the component is mirrored is known as the axis symmetry. In the CNC control system, two axes' reflections are normally accommodated, reflections along X and Y axes. In this approach all X^+ dimensions become $X^- =$ dimensions for reflection along the Y-axis and all Y^+ dimensions become Y^- dimensions for reflection along the X-axis. The control system automatically arranges this on the receipt of the requisite G code within the program. Three possibilities are available, reflection along X, reflection along Y, and reflection along both X and Y simultaneously. Reflection normally occurs about the absolute programmed datum X0, Y0. Figure 9.15 shows the part drawing required for engraving the letter 'F' and its mirror images.

Main program

%	Z0			
N1	G00	X0Y0Z0 *		
*				
N2	G17	T1	M03 *	
*				
N3	#3			
N4	#3	M84 *		
N5	= #3	M83 *		
N9	= #3	M82 *		
N7	M80	T0		
N8	G00	M30		

Subroutine

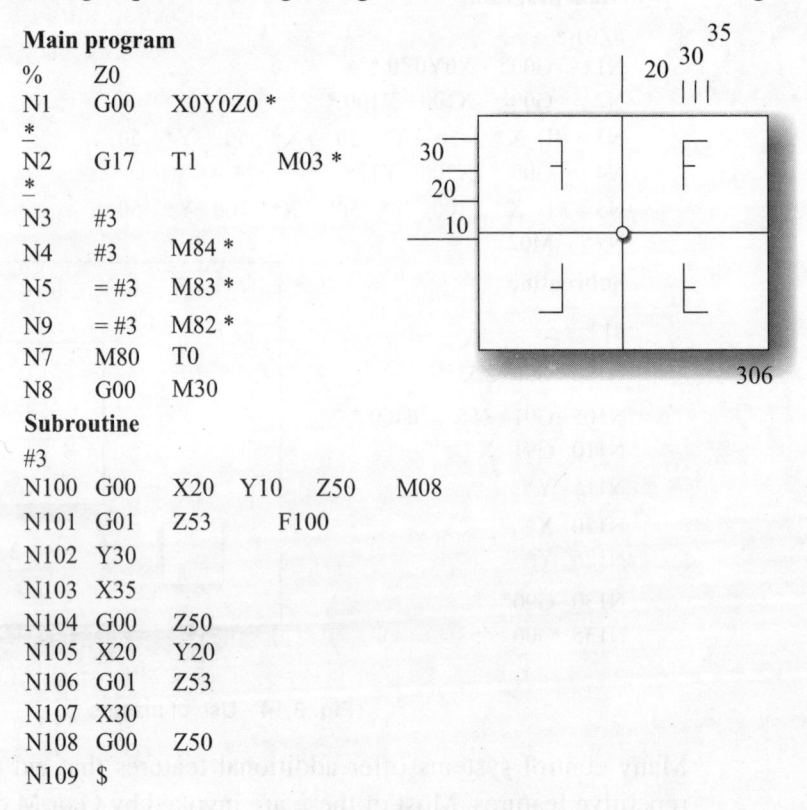

#3					
N100	G00	X20	Y10	Z50	M08
N101	G01	Z53	F100		
N102	Y30				
N103	X35				
N104	G00	Z50			
N105	X20	Y20			
N106	G01	Z53			
N107	X30				
N108	G00	Z50			
N109	$				

Fig. 9.15 Mirror imaging

Many components consist of a single element which rotates about a specified origin. The origin is normally the absolute X0, Y0. Common examples are holes or slots on a common pitch circle diameter or components comprising radial features such as arrows or spokes. If such a facility is specified using the preparatory function G code, it must be accompanied by the information specifying the angle of rotation and the pole of rotation.

Translation is used to produce a series of identical features shifted by a specified amount from the previous one. The shift may be in one or more axes. This can be accomplished by utilizing the appropriate preparatory G code within the part program and specifying the X and Y coordinates for the new starting point.

A scaling facility allows the cutter path to be transformed by increasing (scaling up) or decreasing (scaling down) it. Its magnitudes are relative to its programmed dimensions. The actual programmed dimensions are considered to have a scale value of 1. When a scaling facility is involved, extra information is supplied specifying the scaling required in the X and Y axes. A family of different-sized components may thus be machined using a single part program. The effects of rotation, translation, and scaling are illustrated in Fig. 9.16.

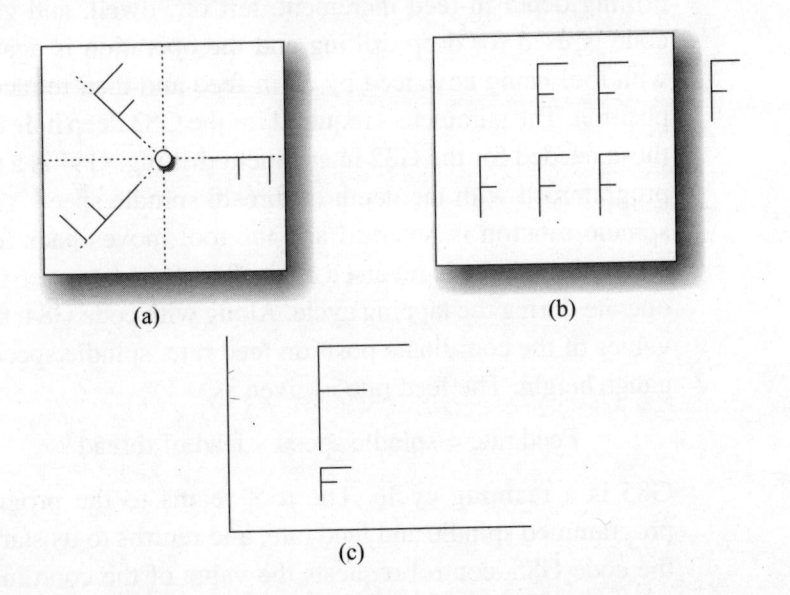

Fig. 9.16 (a) Rotation, (b) translation, and (c) scaling canned cycles or fixed cycles

A canned or fixed cycle is a fixed sequence of operations, in-built in the control system, which can be brought into action by a single command. Such cycles considerably reduce the programming time and effort in repetitive and commonly used machining operations. Preparatory function codes G81 to G89 are reserved for various canned cycles and G80 is used to cancel any canned cycle. G80, in addition to cancelling the cycle, usually positions the tool/work piece and returns the spindle to the gauge height. Canned cycles are modal functions and remain operative until superseded by a subsequent canned cycle. G81 is a drilling cycle. The program segment (using G81) N55 G81 X100 Y100 Z–55 F80 S1500 M08 drills a hole at the co-ordinate position $X = 100$, $Y = 100$, gauge height = Z–50, and required depth = 5 mm. Generally, the fixed cycle G-code must be accomplished by supplying some additional, required information in the same block. The information required is generally specified in the manufacturer's catalogue and the values are assigned by the operator or programmer as per the requirement.

G82 is a peck drilling cycle, i.e., drilling with intermittent feed. The operation is performed in steps, with the tool being advanced by an in-feed increment and then retracted by a small amount for breaking the chip. This procedure is repeated until the full depth has been reached. At the end, the tool returns to the start position at a rapid rate. The values to be given to the control along with the G82 code are coordinates of the hole, feed rate, spindle speed, depth of hole, first drilling depth in-feed increment, left off, dwell, and gauge height. The G83 code is used for deep drilling and the operation is again performed in steps, with tool being advanced by an in-feed and then retracted rapidly to the start position. The parameters required for the G83 deep hole drilling are the same as those needed for the G82 intermittent drilling. G84 is a tapping cycle which is programmed with the depth of thread, spindle speed, and feed rate. Then the spindle rotation is reversed and the tool moves back to the starting position where the spindle is reversed again. The feed rate over-ride control is does not operate during the tapping cycle. Along with code G84, the control requests the values of the coordinate position feed rate, spindle speed, threaded depth, and gauge height. The feed rate is given as

Feed rate = spindle speed × lead of thread

G85 is a reaming cycle. The tool reams to the programmed depth of the programmed spindle and feed rate, and returns to its start position. Along with the code G85, control requests the value of the coordinate position feed rate, spindle speed, reaming depth, dwell, and gauge height. G86 is a boring cycle. The tool bores to the programmed boring depth at the programmed spindle speed and feed rate. Then, the tool returns to its start position rapidly with the spindle standing still. Along with code G86, the control requires the values of the coordinate position, feed rate, spindle speed, boring depth, and gauge height. Figure 9.17 illustrates the canned cycles G81 to G86.

(a) G81: drill cycle

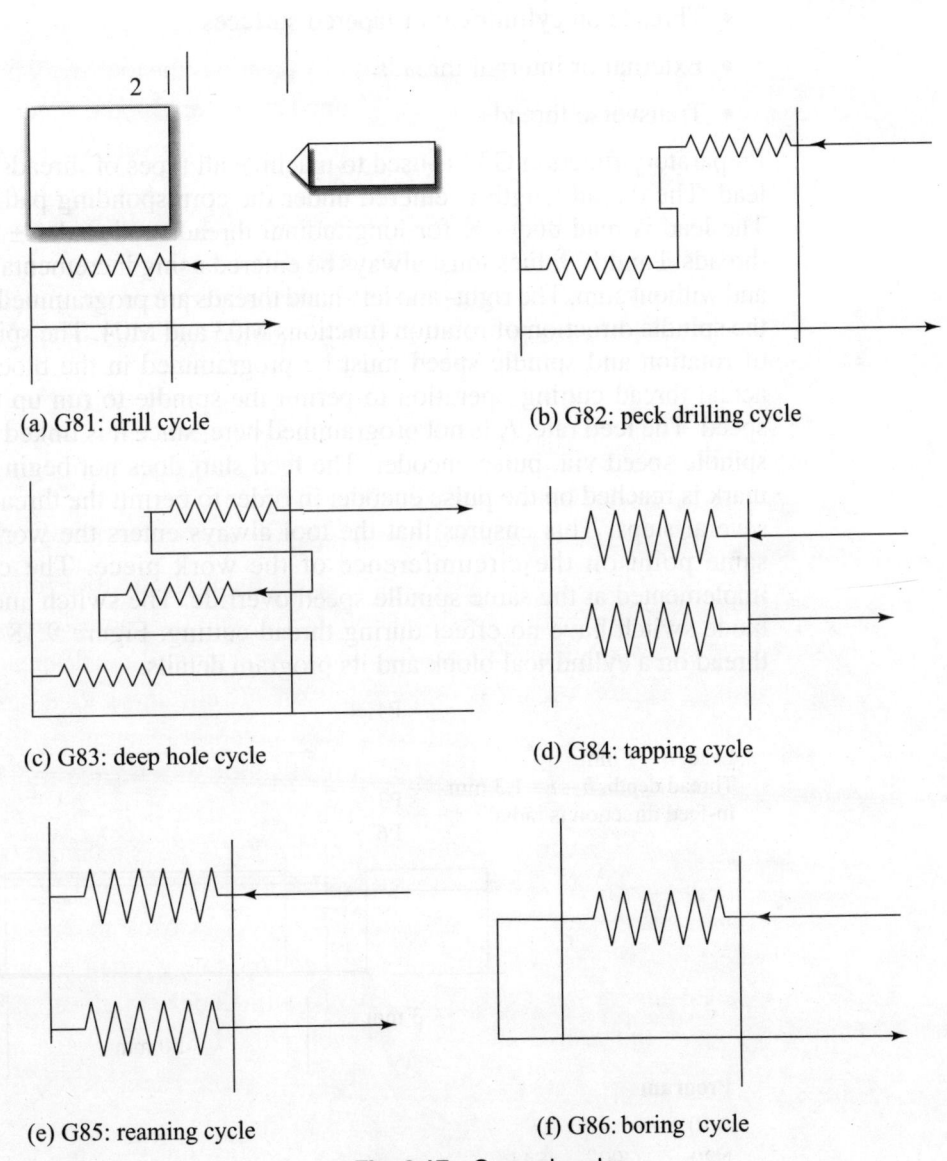

(b) G82: peck drilling cycle

(c) G83: deep hole cycle

(d) G84: tapping cycle

(e) G85: reaming cycle

(f) G86: boring cycle

Fig. 9.17 Canned cycles

Thread cutting

When threads are produced externally or internally by cutting with lathe tip of the tool, the process is called thread cutting or threading. There are various types of threads as follows:

- Threads with constant lead
- Threads with variable lead
- Single or multiple threads

- Threads on cylindrical or tapered surfaces
- External or internal threads
- Transverse threads

Preparatory function G33 is used to machine all types of threads with constant lead. The thread length is entered under the corresponding path address I, K. The lead is read under K for longitudinal threads and under ± for transverse threads. I and K values must always be entered using incremental position data and without sign. The right- and left-hand threads are programmed by specifying the spindle direction of rotation functions M03 and M04. The spindle direction of rotation and spindle speed must be programmed in the block prior to the actual thread cutting operation to permit the spindle to run up to its nominal speed. The feed rate, F, is not programmed here, since it is linked directly to the spindle speed via pulse encoder. The feed start does not begin until the zero mark is reached on the pulse encoder in order to permit the threads to be cut in several steps. This ensures that the tool always enters the work piece at the same point on the circumference of the work piece. The cut should be implemented at the same spindle speed override. The switch and single block mode switch have no effect during thread cutting. Figure 9.18 illustrates the thread on a cylindrical block and its program details.

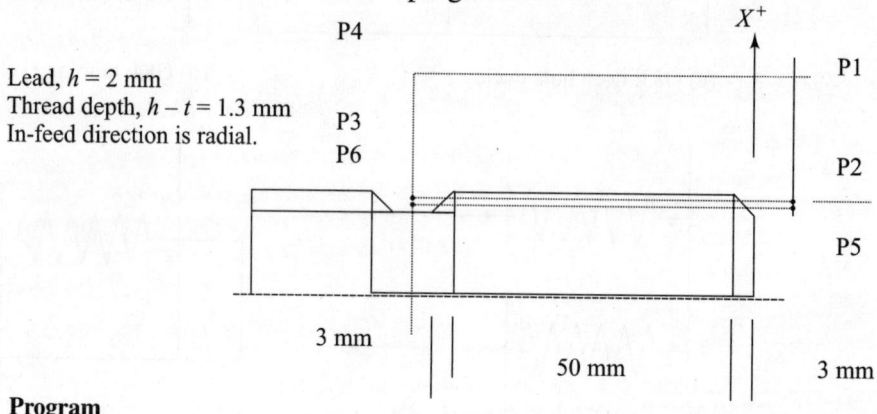

Lead, $h = 2$ mm
Thread depth, $h - t = 1.3$ mm
In-feed direction is radial.

Program
%10
N20	G90	S44*	
N21	G00	X46	Z3*
N22	X38.7*		
N23	G33	Z53	K2*
N24	G00	X46*	
N25	Z3	(P1)*	
N26	X37.4	(P5)*	
N27	G33	Z53	K2 (P6)*
N28	G00	X46	(P4)*
N29	M02*		

Fig. 9.18 Thread program: cylindrical thread

Taper thread

Taper thread fastening is generally used for the extension of rotating shafts. The advantage of taper thread is the quick removal of the assembly even if there is a cold welding between the bolt and the nut. Figure 9.19 shows a thread on a tapered block and its program details. In-feed direction is radial.

Lead $h = 5$ mm

Thread depth $t = 1.73$ mm

Angle $a = 15$ deg

Both the end position co-ordinates must be written.

The lead is entered under K.

Calculate thread start co-ordinates. A, B, C, etc. are diameters.

First cut P2 and P3: $t = 1$ mm

Second cut P5 and P6: $t = 1.73$ mm

$A = 70, B = A - 2t = 66.54$

$C = B - 2 (5 \tan 15°) = 63.86$

$D = C + Z(70 \tan 15°) = 101.366$

(a)

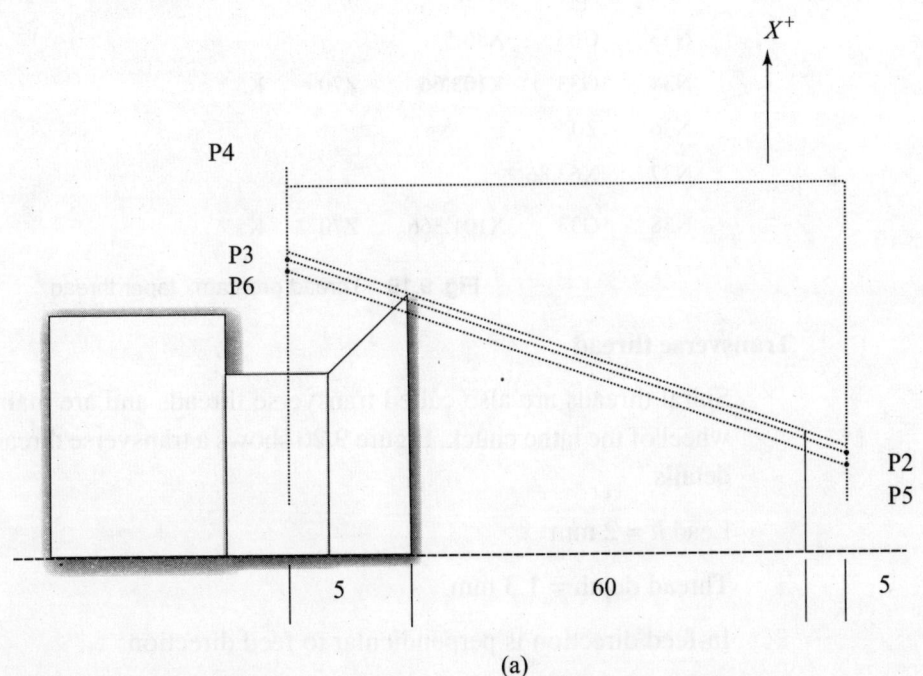

```
N39    G00    X110
N40    M02
```

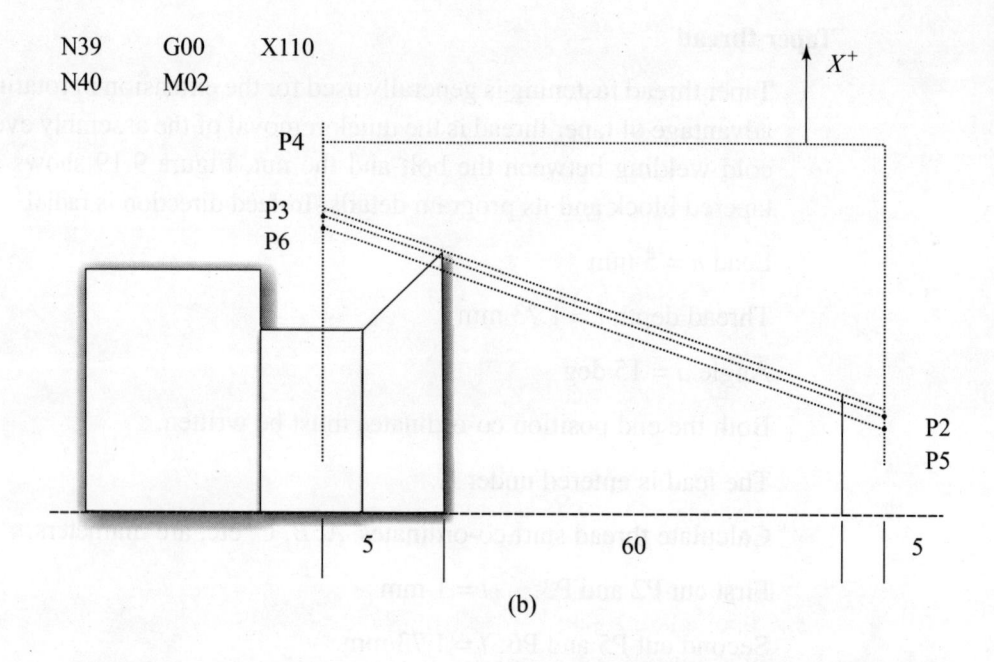

(b)

Program

```
%35 *
N31    G90       S49*
N32    G00       X110 *
N33    G65       X86 *
N34    G33       X103.66    Z70    K5 *
N36    Z0 *
N37    X63.86 *
N38    G33       X101.366   Z70    K5 *
```

Fig. 9.19 Thread program: taper thread

Transverse thread

Spiral threads are also called transverse threads and are mainly used in scroll wheel of the lathe chuck. Figure 9.20 shows a transverse thread and its program details.

Lead $h = 2$ mm

Thread depth = 1.3 mm

In-feed direction is perpendicular to feed direction.

Program

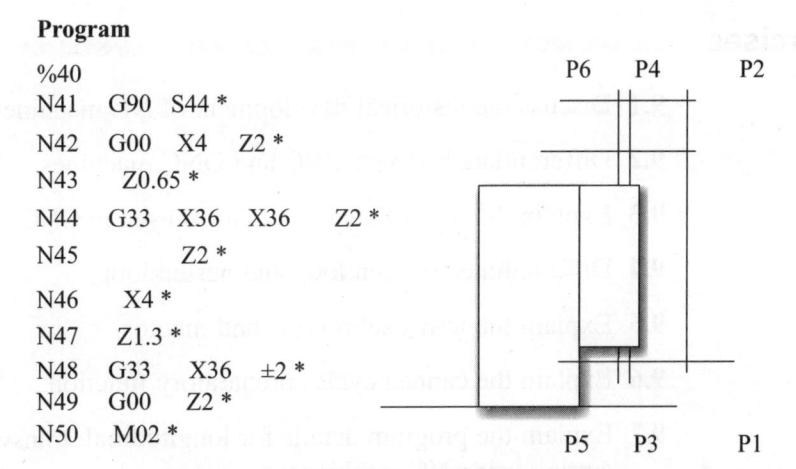

```
%40
N41   G90   S44 *
N42   G00   X4    Z2 *
N43         Z0.65 *
N44   G33   X36   X36   Z2 *
N45               Z2 *
N46         X4 *
N47         Z1.3 *
N48   G33   X36   ±2 *
N49   G00   Z2 *
N50   M02 *
```

Fig. 9.20 Thread program: transverse thread

Parametric and computer-assisted programming

Parameters are used in a program to represent the numeric value of an address. They are assigned values within the program and thus can be used to adopt a program to several similar applications such as different feed rates, different operating cycles, different sliding speeds for various materials, etc. All the necessary parameters for a parametric program are defined in the main program. A parametric program is also known as an R parametric program.

Computer support as an aid to part programming was not required during the early period of NC use. The parts to be unmachined were of two-dimensional configuration and required simple mathematical calculations. With the increased use of NC systems and growth in the complexity of parts to be machined, programs are no longer able to calculate efficiently, the required tool path. Hence the use of computers as an aid to part programming became necessary. The use of computers allows economical programming of the entire cutter paths of all parts that cannot be handled manually. Computers can perform required mathematical calculations quickly and accurately. The calculations required to program are performed by a software system contained in the computer. The programmer communicates with the system through programming languages. After defining the geometry of the part, the programmer defines the path of the cutter by writing appropriate instructions, which include tool parameters, tolerance, etc. The computer processes the information and performs the calculations and calculates the cutter centre path, the feed rate, codes, etc. The result of this processing is a file of data which is transmitted by a post-processor program to the code instructions necessary to generate the NC system. The post-processor is a computer program needed for every CPU/machine tool configuration. It can be directly applied to prepare the program using the same standard peripheral equipment.

Exercises

9.1 Discuss the historical development of NC machines.

9.2 Differentiate between CNC and DNC machines.

9.3 Explain the type of adaptive control used in CNC machines.

9.4 Differentiate between loop and nested loop.

9.5 Explain the terms subroutine and macros.

9.6 Explain the canned cycles preparatory function.

9.7 Explain the program details for longitudinal, transverse, and taper thread cutting using NC machines.

9.8 Write the part program to machine a component on a NC turning machine using the word-address format in the incremental mode assuming appropriate machining parameters for the component shown in Fig. 9.21. The material used is aluminium.

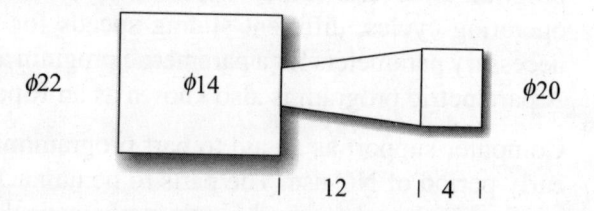

Fig. 9.21 Turning operation

9.9 The part outline of Fig. 9.22 is to be milled in two passes with the same tool. The tool is 20 mm diameter end mill. The first cut is to leave 0,05 mm of stock on the part outline. The second cut will take the part to size. Write the APT geometry and motion statement to perform the two passes. Assume appropriate machining parameters.

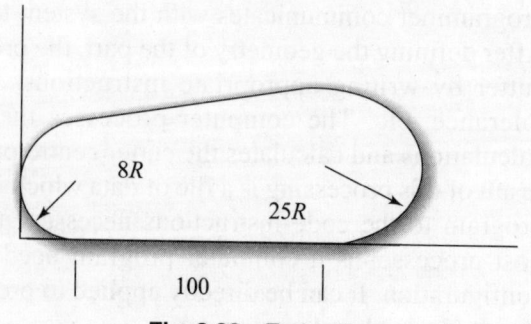

Fig. 9.22 End milling

9.10 Write the APT program for the component shown in Fig. 9.23. Assume the cutter diameter to be 20 mm, tolerance 0.005 mm, spindle speed 1740 RPM clockwise, rapid traverse 200 mm/minute and cutting speed of 500 mm/minute. Any missing data may be assumed.

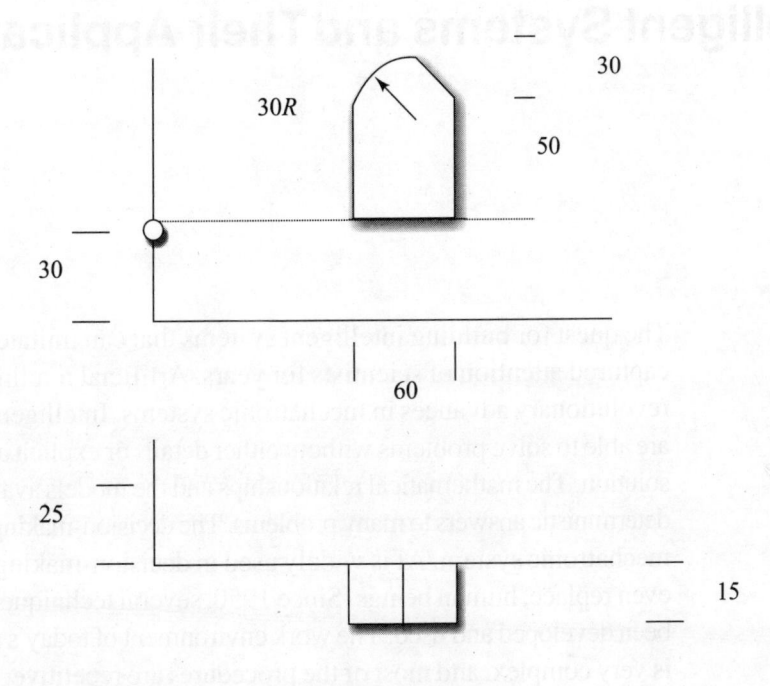

Fig. 9.23 End milling

Chapter 10

Intelligent Systems and Their Applications

The quest for building intelligent systems that can imitate human behaviour has captured attention of scientists for years. Artificial intelligence (AI) has yielded revolutionary advances in mechatronic systems. Intelligent mechatronic systems are able to solve problems without either details or explicit algorithms available for solution. The mathematical relationships and the models available provide complete deterministic answers to many problems. The decision-making process is an advanced mechatronic system. AI is widely used in decision-making processes to assist, or even replace, human beings. Since 1950, several techniques of AI modelling have been developed and used. The work environment of today's manufacturing systems is very complex, and most of the procedures are repetitive. A precise and accurate control and monitoring system is a must to ensure effective functioning of today's production systems. Such a control is not possible with a human operator. This has led to the development and incorporation of artificial intelligence based mechatronic systems to assist human beings in this field. Some of the tools used for the AI are as follows:

- Artificial neural network
- Genetic algorithm
- Heuristic decisions
- Fuzzy control systems
- Knowledge based systems

Conventional and modern control systems need precise knowledge of the model of the processes to be controlled and an exact measurement of the input and output parameters. However, due to inability and roughness of practical processes, it is not always possible/practical to ensure precise control on these process parameters. In many real processes, control depends heavily on human ability. Skilled human operators can control many processes successfully without using any quantities modelling method. The control strategy of the human operator is based mainly on linguistics and qualitative knowledge concerning the behaviour of an ill-defined process. Most modelling processes are based on the nature of the system. Non-linear and ill-defined processes can be easily controlled by fuzzy logic. The concepts of neural system and fuzzy control link the idea of linguistic control and deterministic decision making.

A recent technological advancement in mechatronic systems is the microsystem engineering. Microsystem engineering involves the designing, manufacturing, and packaging of micro-electro-mechanical systems (MEMS). The applications of microsystems are in aerospace, automotive, biotechnology, consumer products, environmental protection, safety health care, pharmaceuticals, and telecommunication. Many industries promoted drug microsystems and related products in the year 2000.

Nanoscience and nanotechnology are two hottest fields in science, business, and news today. MEMS, nanoscience, and nanotechnology are subdomains of mechatronics. Nanoscience, as the name suggests, is the study of fundamental principles of molecules and structures with at least one dimension roughly between one nanometre to hundreds of nanometres. Structures of this scale are known as nanostructures. Nanotechnology is the application of nanostructures in the useful nanoscale devices. Nanotechnology should not be confused with MEMS. MEMS are concerned with structures between 1000 and 1000000 nanometers, i.e., with structures much bigger than nanoscale structure. Nanoscience and nanotechnology are concerned with all properties of structures of the nanoscale, whether chemical, biological, or mechanical. Nanotechnology is more diverse and stretches into dozens of subfields.

To function effectively and optimally, all the smart technology needs artificial intelligence. Hence the tools used to incorporate artificial intelligence are equally important. Some of the tools used for artificial intelligence are discussed in the sections that follow.

10.1 Artificial Neural Network

An artificial neural network (ANN) is an information processing system acting as a black box device that accepts input and produces output. In earlier days, neural networks have been widely used for pattern recognition. In this case an input pattern is passed to a network and the network produces a representative class of output. The ANN has pattern matching, pattern completing, noise removal, and controlling abilities to create a desired response. A neural network consists of processing elements (PEs) and weighted connections. Figure 10.1 illustrates a typical neural network model. The first layer consists of a collection of processing elements. Each PE in the neural network collects the values from all its input connections to produce output values. The PEs are connected with weighted connections. The weighted connections are stored information. The value of a connection weight is often determined by the neural network procedure. Some important features of neural networks are as follows:

1. Each PE acts independently.

2. PEs rely only on local information.

3. A large number of connections provide a large amount of redundancy.

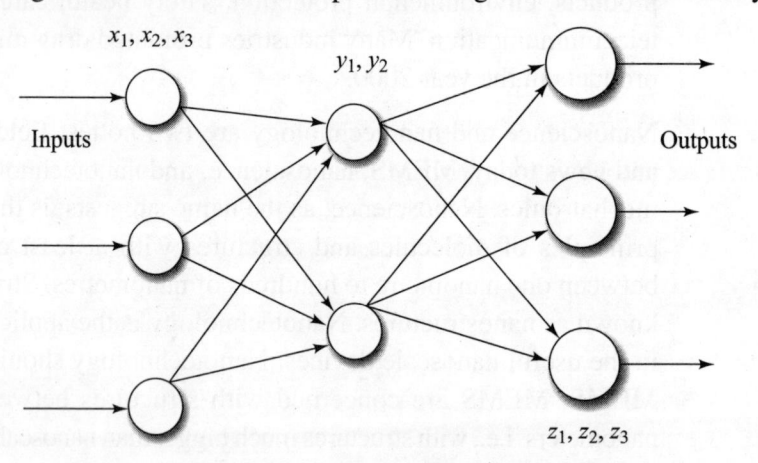

Fig. 10.1 Neural network model

The first two features allow the neural networks to operate efficiently in parallel. The third feature provides for generalization of the state of qualities, which is very difficult in a typical computing system. In addition to the introduction of non-linearity in the PE, an ANN arbitrarily learns non-linear mapping by appropriate learning rules. Major advantages of neural network are as follows:

1. A few decisions are required from among massive amount of data.

2. A non-linear mapping is required automatically.

3. An optimal solution to the problem is obtained very quickly.

A neural network consists of three principal elements, namely, topology, learning, and recall. Topology indicates how a neural network is organized into layers. Learning describes how the information is stored in the neural network. And recall means retrieving the stored information when required.

The following conventions, created in 1990, are used for ANNs. The input and output vectors are denoted by a capital letter from the beginning of the alphabet. This capital letter also has a subscript. For example,

Input: $A_k = (a_{k1}\ a_{k2}, \ldots, a_{kn})$, $k = 1, \ldots, m$

Output: $B_k = (b_{k1}\ b_{k2}, \ldots, b_{kn})$, $k = 1, \ldots, p$

where m is the number of inputs and p is the number of outputs. The processing elements in a layer are indicated by the same subscript variable as $F_x = (x_1, x_2, \ldots, x_n)$, where each x_i receives the input corresponding to the input pattern component a_{ki}. The next layer of PEs is denoted by F_y and then F_z, etc. The two layers of a neural network connection weights are stored as a weight matrix. Weight matrices are denoted by capital letters U, V, and W. In a two-layer network, the sets of connecting weights are between F_x and F_y and the weight w_i is the connecting weight from the ith F_x process element x_i to the jth F_y process element y_j. The basic function of a process element is to create an output for the input, referred to as the activation function or quality function of the process element. Five functions are regularly employed in major neural networks. These are as follows:

1. Linear process element (PE) function
2. Step function
3. Ramp function
4. Sigmoid function
5. Gaussian function

Figure 10.2 illustrates the different process elements and their characteristics.

Building blocks for a neural network are the neural network topologies involved in the pattern of the PEs and their connections. There are different types of network topologies that can be adopted.

The most important quality of neural networks is their ability to learn. Learning, in this sense, is defined as a change in the connection weight values that results in the capture of information that can be recalled later. Several procedures are available for changing the values of connection weights. The learning algorithms describe the pointwise notation with discretized time equations.

Learning networks can be classified into two categories, namely, supervised and unsupervised learning networks. Supervised learning network is a process that incorporates an external teacher or global information. The supervised learning algorithms include error correction as the learning reinforcement. Supervised learning information is classified into two subcategories, namely, structural learning and

$f(x) = \alpha x$, where α is a constant.

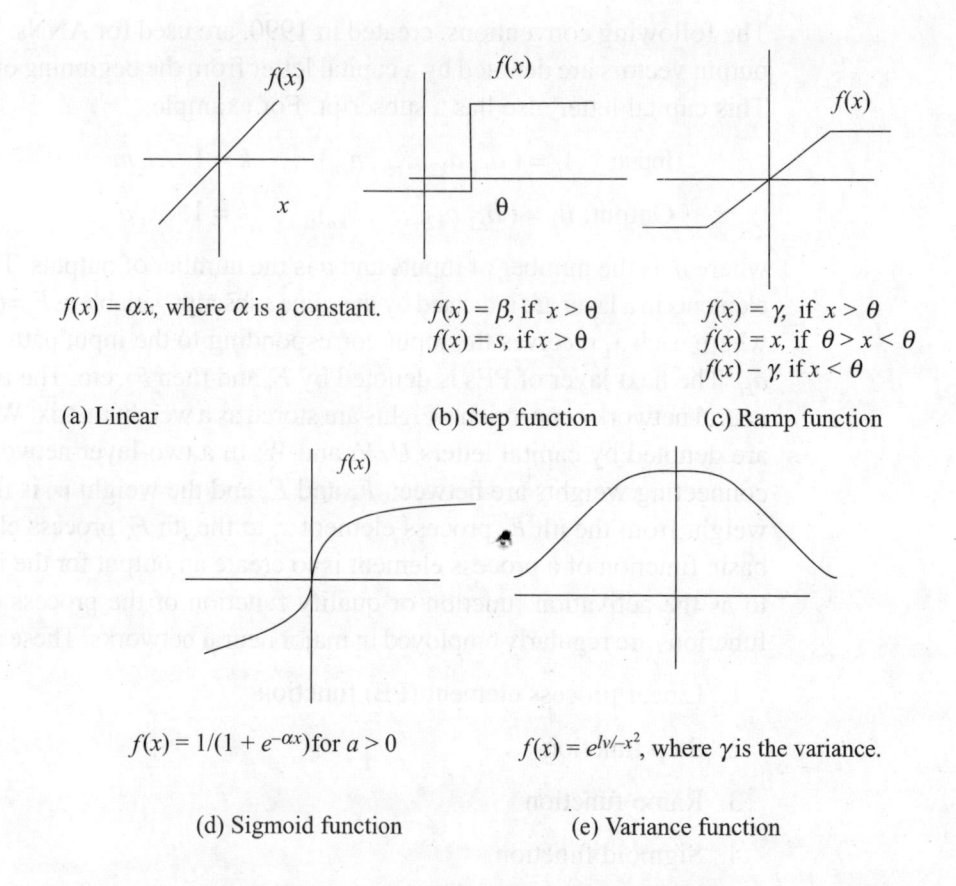

(a) Linear

$f(x) = \beta$, if $x > \theta$
$f(x) = s$, if $x > \theta$

(b) Step function

$f(x) = \gamma$, if $x > \theta$
$f(x) = x$, if $\theta > x < \theta$
$f(x) = -\gamma$, if $x < \theta$

(c) Ramp function

$f(x) = 1/(1 + e^{-\alpha x})$ for $a > 0$

(d) Sigmoid function

$f(x) = e^{|v/-x^2|}$, where γ is the variance.

(e) Variance function

Fig. 10.2 Basics of an artificial neural network

temporal learning. Structural learning is concerned with finding the possible input-output relationship for an individual pattern. Temporal learning is concerned with capturing a sequence of pattern necessary to achieve some final outcome. In temporal learning, the current response of the network depends on the previous input and output. In structural learning, there is no such dependence. Unsupervised learning is a self-arranging process in that it does not incorporate an external teacher. It relies upon only local information during the entire learning process. Examples of supervised learning include the Hebbian learning, principal component learning, differential Hebbian learning, in-max learning, and competitive learning.

In supervised learning, one adopts an ANN so that the actual output comes close to some target output. For a framing set, the goal is to adopt the parameters of the network so that it performs well for the pattern from outside of the framing set. Before adopting the parameters of a neural network, one must first obtain a framing database. Before specifying a learning ANN, one has to design how the operation of the neural network depends on its input and weights.

Back propagation (BP) is a very powerful tool for application to pattern recognition, dynamic modelling, sensitivity analysis, and control of system. In 1974, BP was used for estimating dynamic model to predict materialism and social communication. One of the serious constraints of BP is that the elementary subsystem must be represented by functions that are both continuous and differentiable. BP is an efficient and exact method for calculating all the derivatives of a single quantity with respect to a large set-up input quantities. It is currently the most popular method for performing supervised learning tasks. In basic BP, one assumes the following logic:

$$x_i = X_i, \quad 1 \leq i \leq m$$

$$\text{net} = \sum_{i=1}^{i-1} w_{ij}(x_i), \quad n < i \leq N + n$$

$$x_i = S(\text{net}_i), \quad n < i \leq N + n$$

$$y_i = X_{i+N}, \quad 1 \leq i \leq n$$

where w_{ij} are the weights, N is constant, and n is the number of neurons. The function net_i is usually the following sigmoid function: $S(z) = 1/(1 + e^{-z})$, where N is a constant, which can be an integer or can take any value as long as it is less than n. $(N + n)$ decides how many neurons are there in the network, which include the input and output neurons.

In the neural network terminology, a network is fully connected. In practice, the connections between neurons can be limited by simply fixing some of the weights w_{ij} to be 0. Most of the researchers prefer to use a layered network, in which all the connecting weights w_{ij} are 0, expect for those going from one layer to the next layer. In basic BP, one chooses the weight w_{ij} so as to minimize the squared error over the framing set:

$$E = \sum_{\delta=1}^{r} E(t) = \sum_{i=1}^{T} \sum_{i=1}^{n} \frac{1}{2} [Y_i(t) - y_i(t)]^2$$

where $Y_i(t)$ is the computed output using the model and $y_i(t)$ is the experimental value. This is a special case of the well known method of least squares used in statistics and engineering. The uniqueness of back propagation lies in how the expression is minimized. In back propagation, one starts with arbitrary values for w_{ij} ranging from -0.1 to $+0.1$. But it may be better to guess weights based on some prior information available. If increasing a given weight leads to a greater error, one must adjust the weight downwards. If increasing the weight leads to a lesser error, one must adjusts the weights upwards. After adjusting the weight, and when the error and weights are settled down, the studies are recommended for generalization. The researchers iterate the procedure till the errors are closed to zero. The uniqueness of back propagation also lies in the method used to calculate the derivative exactly for the weights. Back propagation suggests the need to use the conventional chain rule to calculate the derivatives of $E(t)$ with respect to all of the weights. Chain rules provide straightforward linear calculations of the derivatives $E(t)$ to all the input parameters as illustrated below:

$$F[y(t)] = \frac{\partial E(t)}{\partial y(t)} = [Y_i(t) - y_i(t)]$$

Here F indicates the order of derivatives of $E(t)$ for each pattern. Simply differentiating by the chain rule for order of derivatives, we get

$$F[x_i(t)] = F[y_{i-N}] \sum_{i=i+1}^{N+n} w_{ij} x F[\text{net}_i(t)]$$

$$F[\text{net}_i(t)] = S'(\text{net}_i) F[x_i(t)]$$

$$F(w_{ij}) = \sum_{i=1}^{T} F[\text{net}_i(t) x_j(t)]$$

where S' is the derivative of $S(z)$ and also $F[y(k)]$ is assigned to be zero for $k = 0$. From the calculus algorithm $S'(z) = S(z)[1 - S(z)]$, which can be used to compute the weight. Hence, new $w_{ij} = w_{ij} \eta F[x_{ij}]$, where the learning rate η is chosen arbitrarily. The usual procedure is to make it as large as possible up to 1, until the error starts to diverge. Once the topology of the network is determined from a partial derivative, the error function $E(t)$ is derived analytically. There are three ways for performing the back propagation through learning. One way is to update the parameter of the network wherever the input pattern differs from the target pattern. This method is called pattern learning. The other method updates the parameters of the network only after obtaining all the error terms of all the input pattern. This method is called batch learning. The third method is to update the parameter of the network once every preset data of a block of all the input examples. It is called block learning. Most of the manufacturing processes use the pattern learning approach. In this, the modelling procedure is repeated until the error is a minimum value. Pattern learning is the most commonly used method of learning. One major problem with the conventional back propagation algorithm is its slow convergence speed. Three methods can be used to improve the convergence speed. The first method uses a momentum factor as $w_{ij} = \gamma w_{ij} - \eta F(w_{ij})$, where γ is the momentum factor with value $0 = \gamma = 1$ and η is the learning rate. Even with the momentum factor, a large number of iterations are required for simple neural network mapping problems. The second method is to reduce the number of iterations by measuring the network complexity using Vapnik–Chervonenkis (VC) dimensions. In this approach, the performance of the back propagation depends on the type of the problem encountered. The third method to accelerate propagation convergence is by integrating the network with some traditional optimization technique such as the conjugate gradual method or variable matrix method. The momentum factor method is the most commonly used method for convergence of the back propagation.

The back propagation algorithm for an ANN can be applied to many different categories of dynamic systems. It is simply an efficient method for calculating all the derivatives of a single target quality with respect to a large set of input quantities. Back propagation is presently the most widely used tool in the field of ANN.

Illustration 10.1

Consider a logical OR gate with binary inputs x_1 and x_2 and the binary output Z. The summation block indicates that $Z = w_{11}x_1 + w_{12}x_2$ and the unit step threshold function block is $F(z)$. This can be represented in a single-layer neural network as shown in Fig. 10.3(a). The logical OR gate can be represented graphically using a linear function $x_2 = Ax_1 + B$, which is an (x_1, x_2) plane defining a straight line separating the point $Z = 0$ from point $Z = 1$.

The co-ordinates (x_1, x_2) of four points corresponding to the following value of Z as the defined logical 'OR' gate are: (0,0) for $Z = 0$, (0,1) for $Z = 1$, (1,0) for $Z = 1$, and (1,1) for $Z = 1$. The value of $Z = 0$ represents the origin and the other point $Z = 1$ is as shown in Fig. 10.3 (b). Assigning the straight line to intersect the points (0.5, 0) and (0, 0.5), the coefficients of A and B can be obtained as $B = 0.5$ and $A = -1$. From $x_2 = Ax_1 + B$,

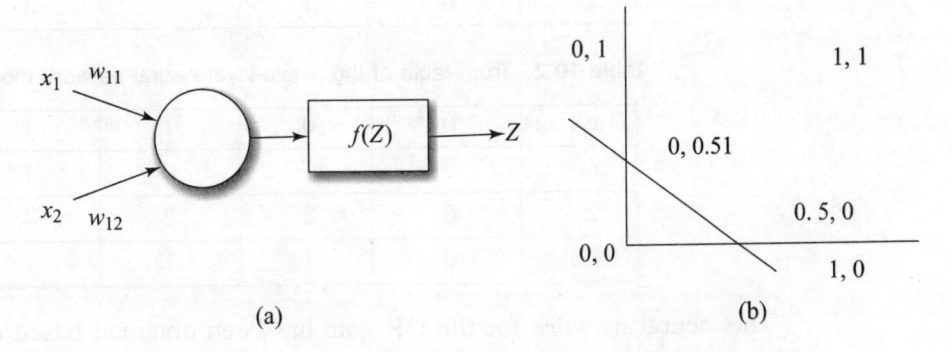

$$x_2 = -x_1 + 0.5, \text{ i.e., } 2x_1 + 2x_2 = 1$$

Fig. 10.3 (a) Single-layer neural network model and (b) graphical representation of the OR gate

This result permits one to interpret the geometrical single-layer neural network output as

$$Z = w_{11} x_1 + w_{12} x_2$$

with weights $w_{11} = 2$ and $w_{12} = 2$. The straight line $Z = 0$ separates the plane (x_1, x_2) as follows:

- For Z greater than 1, the plane above the line is defined.
- For Z less than 1, the plane below the line is defined.

The threshold function $f(z)$ corresponding to a step function can be defined as $Z = 0$ for $f(z) = 1$ and $Z = 1$ for $f(z) > 1$. This equation permits one to define the single-layer neural network with weights $w_{11} = 2$ and $w_{12} = 2$. The threshold function is

$$f(z) = \begin{cases} \text{for } Z = 2x_1 + 2x_2 \leq 1 \\ \text{for } Z = 2x_1 + 2x_2 > 1 \end{cases}$$

and the output is

$$Z = f(z)$$

The resulting neural network can be verified for the set of four values of the logical OR gate as shown in Table 10.1. The truth table for the single-layer neural network model is shown in Table 10.2.

Table 10.1 Truth table of the OR gate

x_1	0	0	1	1
x_2	0	1	0	1
Z	0	1	1	1

Table 10.2 Truth table of the single-layer neural network model

x_1	0	0	1	1
x_2	0	1	0	1
Z	0	2	2	4
$F(Z)$	0	1	1	1

This neural network for the OR gate has been obtained based on geometric consideration. A training algorithm is used to obtain weight functions w_{11} and w_{12} in a complex system. A single-layer neural network serves as a model for a system with much reduced complexity.

10.2 Genetic Algorithm

Genetic algorithms are search algorithms based on the mechanics of natural selection and natural genetics. These algorithms were first developed by John Holland in collaboration with his colleagues, and students at the University of Michigan. The goals of their research were two fold:

1. To abstract and rigorously explain the adaptive processes of natural systems.
2. To design artificial system software that retains the important mechanisms of natural systems.

This approach has led to important discoveries in both natural and artificial systems' science. The control theme of research on genetic algorithms has been robustness, the balance between efficiency and survival in many different environments.

Three main types of search methods are used conventionally. They are calculus-based, enumerative, and random methods. We will discuss the calculus-based methods in some detail here. These methods are subdivided into two main classes, direct and indirect. The direct method is a well-defined computation procedure that takes some value, or set of values as input and produces some value, or set of values, as output. In the direct method, a sequence of computational steps transforms the input into the output. For example, there may be a sequence of numbers which is to be sorted in a decreasing order. This problem arises frequently in practice. This is called the sorting problem. For the sequence (31, 41, 59, 26, 41, 58), a sorting algorithm will return the sequence (26, 31, 41, 41, 58, 59) as the output. The indirect method seeks local extremes by solving a non-linear set of equations resulting from setting the gradient of the objective function equal to zero. Depending upon the restrictive requirements, continuity, and derivative existence, calculus-based methods are unstructured for all types of the problems. These methods are insufficiently robust in unintended domains. Enumerative scales have been considered in many shapes and sizes. The enumerative method is a very human kind of search; such a scheme must ultimately discount in the robustness race for one single reason of lack of efficiency. Random search methods have achieved increasing popularity. The genetic algorithm is an example of the search procedure that uses random chance as the tool to guide a highly explanative search through a coding of parameters, for example, speed. Genetic algorithms differ from other types of algorithm in some very fundamental ways:

1. Genetic algorithms work with coding of the parameter set, not the parameter itself.
2. Such algorithms search from a population of points, not from a single point.
3. Genetic algorithms use pay-off information, not derivatives or other accumulating knowledge.

4. Genetic algorithms are themselves probabilistic; these do not use deterministic rules.
5. Genetic algorithms require a natural parameter set of the optimization problem to be coded as a finite length string over some finite alphabets.

The mechanics of a simple genetic algorithm is simple and involves nothing more complex than copying a string and swapping particle strings. The system of operation and power of effect are main attractions of the genetic algorithm approach. A simple genetic algorithm that yields good results in many practical problems is composed of three operations:

- reproduction
- crossover
- mutation

Reproduction is the process in which individual strings are copied according to their objective function values; biologically, this function is a fitness function. Copying of strings according to their fitness values means that strings with a higher value have higher probability of contributing one or more offspring in the next generation. In natural population, fitness is determined by a creative ability to survive predators, pestilence and other obstacles and subsequent reproduction. The reproduction operator may be implemented in the algorithm in a number of ways. The easiest way is to create a bias roulette wheel, where each current string in the population has a roulette wheel sized in proportion to its fitness. After reproduction, a single crossover may proceed in two steps. First, a number of newly reproduced strings mate at random. Secondly, each pair of strings undergoes a crossover. The mechanics of reproduction and crossover consists of random number of generation, string copies, and some potential string exchange. The combined emphasis of reproduction and the structured, through randomized, information exchange of crossover gives the genetic algorithm much of its power. Mutation is needed, because even though reproduction and crossover effectively search and recombine extent notaries occasionally, the strings become overzealous and loose some potentiality and useful genetic material. In artificial genetic systems, the mutation operator protects against such an unrecoverable loss.

The mutation operator plays a secondary role in the simple genetic algorithm. The frequency of mutation to obtain good results in empirical genetic algorithm studies is of the order of one mutation per thousand bit position transfers. Mutation rates are similarly small in natural populations, leading to the conclusion that mutation is an appropriate secondary mechanism of genetic algorithm adoption. Other genetic operator of reproductive plans has been abstracted from the biological examples. Reproduction, simple crossover, and mutation have proved to be both computationally simple and effective in attacking a number of optimization problems and new design techniques.

Illustration 10.2

Consider the black box switching problem with a bank of five input switches. For every setting of the five switches, there is an output signal γ. Mathematically, $f = f(s)$, where s is a particular setting of the five switches. The objective of the problem is to set the switches to obtain the maximum possible value using a genetic algorithm.

Let us first code the switches as a finite length string. A string over a finite set S is a sequence of elements of S. For example, there are 8 binary strings for length 3, i.e., 000, 001, 010, 011, 100, 101, 110, and 111. A switch is represented by 1 if it is on and by 0 if it is off. Other technique for solving this problem might start with one set of switch setting. Apply the same from citation rules and generate a new trial switch setting. Our genetic algorithm starts with a population of strings and, thereafter, generating successive populations of strings. For generating genetic algorithm studies, consider four strings at random, $A_1 = 01101$, $A_2 = 11000$, $A_3 = 01000$, and $A_4 = 10011$. After this, successive populations are generated using the genetic algorithm. The population used is chosen at random through 20 successive flips of an unbiased coin. The reproduction operator may be implemented in the algorithm form in a number of ways. The easiest way is to create a biased roulette wheel, where each current string has a roulette wheel slot size in proportion to its fitness. Fitness value of objective function value may be a measure of profit, utility, or goodness which one wants to maximize. Strings with higher values of fitness have a higher probability of contribution to one or more offspring. After a number of spins, let the fitness be 169, 516, 640, and 361 for A_1, A_2, A_3, and A_4, respectively. Summing the fitness over all four strings, we obtain the total fitness as 1170. Then, percentages of population of the total fitness are 14.4%, 49.2%, 5.5% and 30.9% for A_1, A_2, A_3, and A_4, respectively. The weighted roulette wheel for the generation reproduction is shown

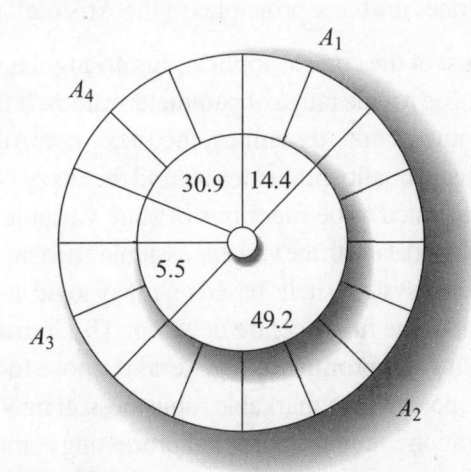

Fig. 10.4 Allocation of offspring string using roulette wheel

in Fig. 10.4. String A_1 has a fitness value of 169, which represents 14.4% of the total fitness. As a result, A_1 is given 14.4 of the biased roulette wheel and each spin turn of the string A_1 has probability of 0.144. Each time, one requires another offspring single spin of the weighted roulette wheel to yield the reproduction of candidate. In this way, more highly set string A_1 has higher number of string in the succeeding generation. Once a string has been selected for reproduction, an exact replica of the string is made. This string is then entered into a mating pool and a tentative new population for further genetic operative action.

10.3 Fuzzy Logic Control

Logic is the science that is used for formal principles of reasoning. It gives a precise reasoning view in limited cases. Fuzzy logic is unlike classical logic. It aims at modelling imprecise models of reasoning as does a human being to make rational decisions in an environment of uncertainty. Fuzzy logic depends on interfacing the approximate answer to a question based on the system with a store of knowledge. During the recent years, fuzzy logic has found numerous applications in fields ranging from finance to earthquake prediction. One of the important branches of fuzzy logic is called dispositional logic. This proposition deals with not always true or false statement; it is implicit in nature rather than explicit. The Aristotle logic deals with an explicit approach, as true or false. The dispositional logic normally refers to common sense knowledge. Thus, the main concern in the dispositional logic lies in the development of tools of interfacing from common sense knowledge. Fuzzy logic is viewed as the most valued logic. It includes truth in it, but it is a matter of degree. The greater expressive power of fuzzy logic is that it contains not only the classical two-valued or multi-valued logical systems but also probability logic. Figure 10.5 illustrates the basic principles of the Aristotelian logic and the fuzzy logic.

In most of the control applications, triangular type fuzzification is adopted since it works on a wide range of parameter values. If the system is non-linear, time varying with uncertainty dynamics, the fuzzy control strategies demonstrate better. The predicted quality brain idea behind the fuzzy control is that the controlled variables are assumed to be functions of state variables. The state variables are treated as linguistic data with the variable assuming a triangular possibility of distribution. Several expected system shells based on fuzzy logic are now commercially available. More recently, the fuzzy centre developed by Yamakawa of Kumanoto University has shown great promise as a general purpose tool for processing linguistic data at a high speed with remarkable robustness. It may be an important step towards a sixth generation computer capable of processing common sense knowledge. This capability is a prerequisite to solve many artificial intelligence problems.

Fig. 10.5 (a) Aristotelian logic and (b) fuzzy logic

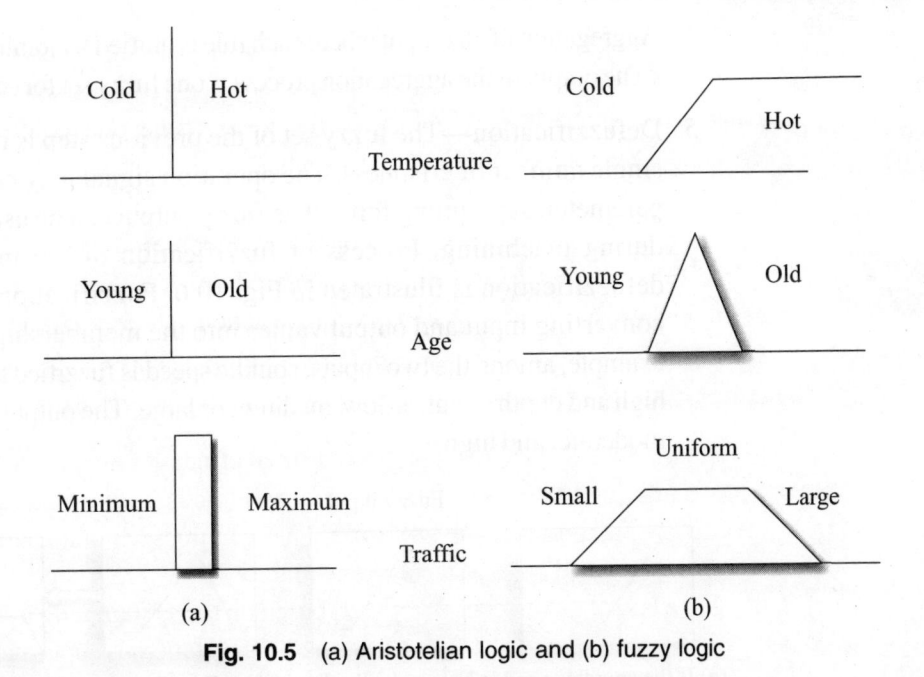

Illustration 10.3

During the manual operation of CNC machining, when the operator sends extra cutting force on the tool, he overrides the feed rate with the help of '<' or '>' button on the control panel. This performs an increase or decrease in the percentage of feed rate from the programmed feed rate in the process part program. The role of the operator can be easily substituted using an adaptive control package with a fuzzy logic tool box. The variation in the membership function for different speeds and depths of cut is obtained for a particular process parameter operational signal. The fuzzy logic tool box is used to create and edit the fuzzy inference system using a graphical tool or command line function. A fuzzy inference system is the actual process of mapping from a given input to output. The various steps involved in the inference system are as follows:

1. Fuzzifying input—All the fuzzy states are resolved in the antecedent to a degree of membership between 0 and 1.
2. Applying the fuzzy operator if there are multiple parts to be antecedent—The fuzzy logic operator is supplied and the antecedent of a single member between 0 and 1 is resolved.
3. Applying the implication method to resolve the degree of support for the entire rule is used to shape the output of fuzzy set—The consequent of the fuzzy rule assigns an entire fuzzy set to the output. If the antecedent is only partially true, then the output fuzzy set is processed according to the implication method.

4. **Aggregation of all output where each rule is unified by joining parallel threads**––The output of the aggregation process is one fuzzy set for each output variable.

5. **Defuzzification**—The fuzzy set of the previous step is input to generate a single number of crispness. The operation signal may be a process output parameter, say, cutting force. The fuzzy output can be used for fine control during machining. Process of fuzzification of the input and its final defuzzification is illustrated in Fig. 10.6. Fuzzification is the process of converting input and output values into the membership function. In this example, among the two inputs, spindle speed is fuzzified as low, medium, or high and depth of cut, as low, medium, or large. The output is fuzzfied as low, moderate, and high.

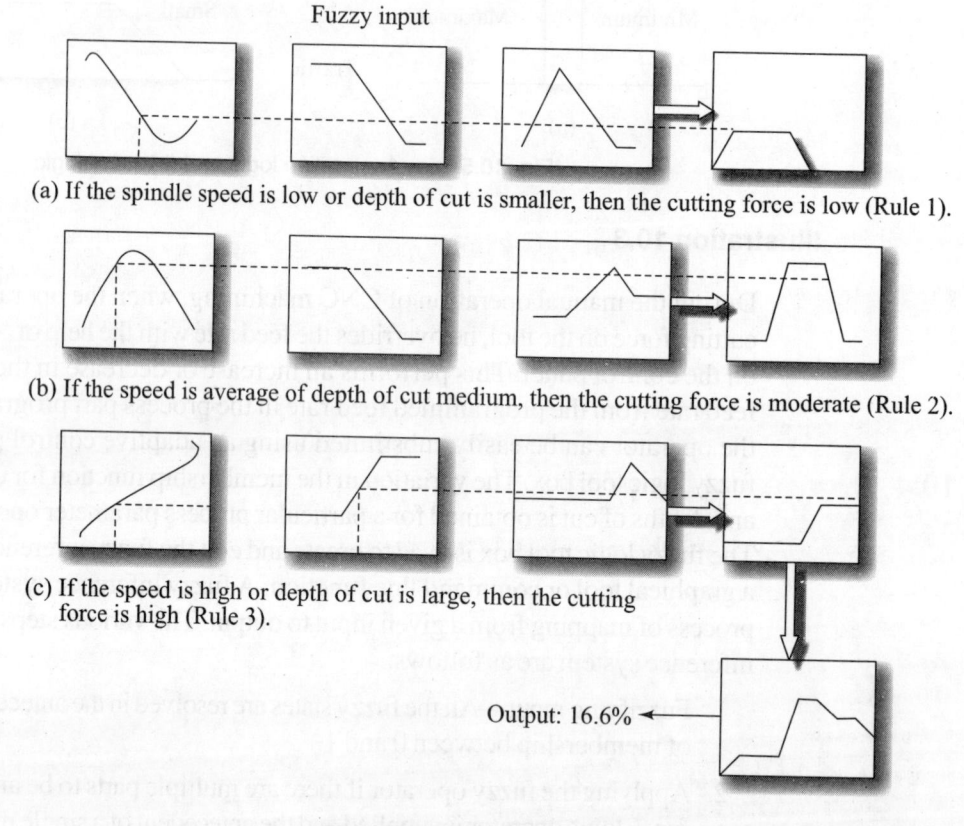

Fuzzy input

(a) If the spindle speed is low or depth of cut is smaller, then the cutting force is low (Rule 1).

(b) If the speed is average of depth of cut medium, then the cutting force is moderate (Rule 2).

(c) If the speed is high or depth of cut is large, then the cutting force is high (Rule 3).

Output: 16.6%

Fig. 10.6 Fuzzy inference system

Fuzzy inference rule base is the controller part of the system. The rule base is a collection of rules related to the fuzzy sets, the input variables, and the output variables and is meant to allow the system to decide what to do in each case. In this case, three fuzzy inference rules are used as indicated in Fig. 10.6.

Defuzzification is the conversion of the fuzzy output value into an equivalent crisp value for actual use. As fuzzy rules are evaluated and corresponding values are calculated for different output fuzzy sets, there are a number of different possibilities for defuzzification. Two common and useful techniques are the centre of gravity method and Mamdani's inference method. In the centre of gravity method, the membership value for each output variable is multiplied by the maximum singleton value of the output membership from the membership set in question. For example, suppose that the values obtained for the machining system membership sets are 0.4 for low and 0.6 for medium and further suppose that the singleton value for low is 10% and for medium is 21%, then the output value for the machining is $0.4 \times 10\% + 0.6 \times 21\% / (0.4 + 0.6) = 16.6\%$.

In Mamdani's inference method, the membership function of each set is truncated at the corresponding membership value as shown in Fig. 10.6. The resulting membership function is then added together as an OR function. This means that all the areas that are repeated are superimposed over each other as one layer only. The result will be a new area that is a representation of all areas. The centre of gravity of the resulting area will be the equivalent output. Most programs calculate the area by integrating the area under the curve. Hand calculation may be done. In this case, the value is 16.6%, which can be used to control the feed rate of the machine tool. A millinewton variation in cutting force is used to control the feed rate.

10.4 Nonverbal Teaching

The growth of information technology and electronics technology has made a great change in almost all walks of life. It has helped generate and disseminate knowledge across the globe. Today knowledge is the most valuable and most traded resource. Knowledge is different from all other kinds of resources. It constantly makes itself obsolete. Knowledge can be of two kinds, explicit and tacit. Explicit knowledge is easy to communicate through data, formulae, manuals, etc. Tacit knowledge is highly personal, hard to formulate, and difficult to communicate or store with a key. Welding and painting are examples of tacit knowledge. The knowledge of a certain creation is an interaction between explicit and tacit knowledge. The transformation of tacit knowledge into explicit knowledge creates new knowledge. New knowledge is created by an individual. Knowledge can create significant advantages. Conversion of tacit knowledge into explicit knowledge at a fast rate can be achieved by nonverbal teaching.

Painting is a skilled job that can be made by an experienced painter. If the painting gun is held very close to the body to be painted, paint drops will fall on the surface due to excess paint being projected on the surface. If the painting gun is held far away from the body, much paint will be carried away by the air without sticking to the body. An experienced painter will observe the body to be painted and will manipulate the distance between the body and the painting gun to get the correct position. If position sensors are used in the painting gun, when the painter performs the painting operation, the coordinate positions can be stored in a computer. For a trainee, the stored computer data acts as an instruction for the particular movement of hand. Any wrong movement, and the painting gun gives a warning or mild jerk to the hand. The trainee can adjust the painting gun as the master painter hand movement without any prior experience of painting. The tacit knowledge of the skilled painter can be easily transformed to help a trainee painter. This type of nonverbal teaching is used in advanced machining systems. Computer stored data can also be used for a robot to perform better work following a trial and error method of teaching to the robot.

10.5 Design of Mechatronic Systems

Engineering design is a complex process and involves interaction between many sketches and disciplines. The design process for any item involves a number of stages. The process begins with a need from a customer or client. This may be identified through market research, which is used to establish the needs of potential customer(s). The first stage in developing a design is to find out the true nature of the need/problem. This is an important stage in that defining the need/problem accurately makes the design serve the objectives very well. Following the analysis, a specification of the requirement can be preferred. This will state the problem, any constraints placed in the solution, and the criteria that may be used to judge the quality of the design. These might be a statement of mass, dimension, type, range of motion required, accuracy of input and output requirements of elements, interface power requirements, operating environment, relevant standards, and codes of practice. The next step is the generation of possible solutions. This is often termed as the conceptual stage. Outline solutions are preferred, which are worked out in sufficient detail to indicate the means of obtaining each of the required functions such as the approximate size, shape, material, and cost. The various solutions are explained and the most suitable ones are selected. The details of the selected design are then worked out. This might require the production of the prototypes in order to determine optimum details of the design. The selected design is then translated into a working drawing (circuit diagram) so that the item can be made. It should not be considered that each stage of the design process just flows on stage by stage. There will often be a need to return to an earlier stage and give it further consideration. Thus, when at the stage of generating a possible solution, there might be the need to go back and reconsider the analysis of the problem. The basis of the mechatronics approach is considered to lie in the inclusion of the disciplines of electronics, computer technology, and control engineering. The improved flexibility is the common characteristic of mechatronic systems. Microprocessors and microcontrollers are often embedded in a system so that control can be exercised. The term 'embedded system' is used for a microprocessor-based system designed to control a range of functions. The microprocessor programming is done by the manufacturer and is burnt into the memory system and control cannot be changed by the user.

10.5.1 Mechatronic System Design

Our standard of living and the wealth of our country are dependent upon strides in technology. The manufacturing of sophisticated products that we use in our day-to-day life requires well designed systems and products. Most of these products are mass produced, where least human intervention is required. To identify needs, today's world requires skills. A mechatronics engineer is expected to develop skills to identify the needs during the study of design and technology.

Today, there are numerous different materials and an enormous range of information and skills available. The process or system design should ensure efficiency, good looks, and safety. It is important to note that the manufactured product's use and disposal will have some effect upon people, wildlife, and environment. The design should make sure that the product appeals to a large number of people. A customer always tries to buy a product that is comfortable, stylish, and gives good value for the money spent. These features of a product form an important basis for the designers of the product.

10.5.2 Stages in the Designing Process

Designers and technologists help us solve practical problems that arise in our daily life situations. As an example, consider the situation in which old aged people need to climb up and down stairs. Before attempting to figure out a solution to this problem, it is important to analyse the situation to see the exact nature of the problem. Once the problem is fully analysed and understood, the next step is to write a design brief. A design brief is a short statement giving general outline of the problem to be solved. On the other hand, there are some problems that can be solved by using one's own knowledge and imagination. However, to obtain the best possible solution, one should keep oneself abreast of the advancements in the concerned knowledge domain. This requires research. To research the problem, one should have a good understanding of what is required and a clear understanding of the design limitations that will affect the ultimate product. Based on this combined knowledge, a specification can now be prepared. There must be an outline of specific details of the design that satisfy and identify the design limitations. The specifications of presently considered problem are as follows:

1. The final device must either carry or assist the person in climbing up and down the staircase.
2. It must be very easy to use and have a simple control.
3. It must be completely safe.
4. It must not obstruct the normal use of the staircase.
5. It must be neat and aesthetic.
6. It must be easy to operate.
7. The cost of the equipment should not exceed, say, Rs 30000.

All possible solutions to the design brief should now be visualized. The visualized ideas can be drawn on a paper. These ideas may not necessarily be the best. Using the ideas and information obtained through research, one should begin to move towards a refined solution. A design must be sketched now. This needs to be fine-tuned to incorporate all (or most) of the essential things listed in the specifications. At this stage, the working drawing of the design is prepared. The working drawing contains all the details of the design that are important to its construction. Planning for the work ahead is the next step. This stage ensures that work is completed in time.

Once the part drawings are ready, a prototype model can be prepared to verify its workability. A prototype is a miniature form of the ultimate product. This is the most interesting part of the work. It involves building, testing, and modifications, if any, of the design to satisfy the specifications. Testing and evaluation of the design involves checking the functionality, safety, and reliability of the designed product. At this stage, any fine-tuning of the product can be prescribed. All such adjustments will be documented and made part of the final design specifications. The areas of research include basically the following.

1. Practical function
    (i) Structural—support, protection, construction
    (ii) Mechanical—minimum parts and construction

(iii) Electrical—operation and construction

   (iv) Energy transfer—mode of energy transfer

2. Shape and form

   (i) Aesthetic quality, ergonomics

   (ii) Shape, stability, rigidity, safety

   (iii) Aerodynamics, surface texture, finish, and colour

3. Material properties

   (i) Strength, hardness, toughness, ductility

   (ii) Thermal conductivity, durability

   (iii) Material cost

4. Construction

   (i) Cutting—shaping, filing, drilling, etc.

   (ii) Fabrication—assembly of parts using screws, bolts, glass, solder, etc.

   (iii) Moulding—application of moulds

   (iv) Casting—use of mould to cast shapes, solidification of materials

5. Sales and environment effect

   (i) Manufacture, use, and dispose

   (ii) Health, safety factor, noise, and smell

   (iii) Aesthetic quality and pollution

Finally, a report must be prepared about the project, in which the designing process can be illustrated using a flow chart, as shown in Fig. 10.7. A flow chart is a simple, step-by-step, logical flow of actions oriented at a common, complex objective/product. The thick arrows in the flow chart show the general progression of activities from one step to the next. However, one must keep in mind that this progression is not so simple and more often needs some lateral inputs in the form of feedback/reinforcement at each stage. The long up-arrow on the side of the flow chart represents this feedback/reinforcement parameter. Documentation helps in achieving standardization of the product as per the specifications. It ensures a uniform product quality even when the same is being manufactured at more than one location.

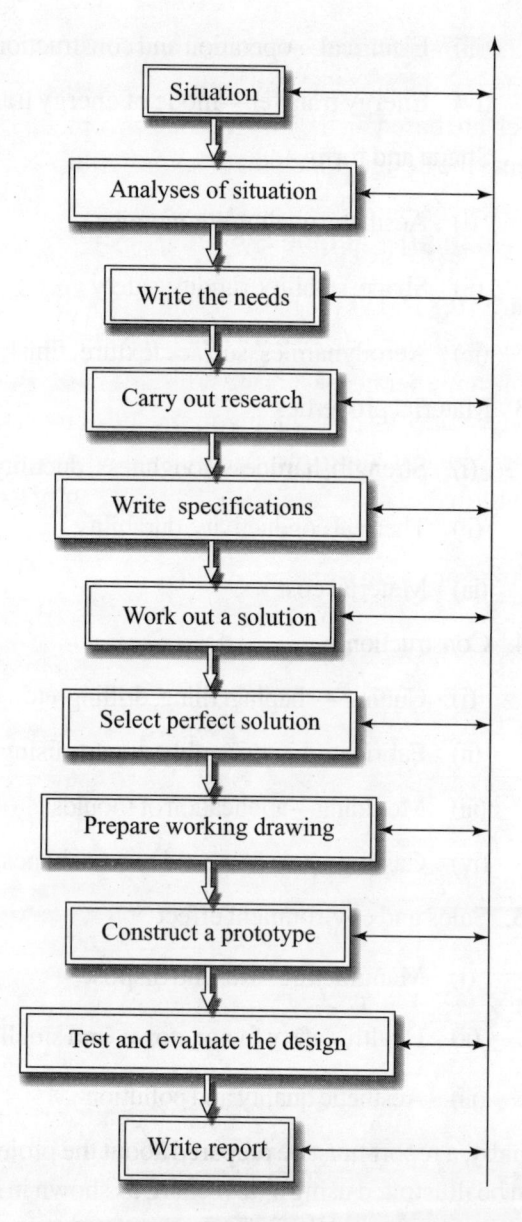

```
       ┌─────────────────┐
       │    Situation    │◄──────────┐
       └────────┬────────┘           │
                ▼                    │
       ┌─────────────────┐           │
       │Analyses of situation│◄──────┤
       └────────┬────────┘           │
                ▼                    │
       ┌─────────────────┐           │
       │ Write the needs │◄──────────┤
       └────────┬────────┘           │
                ▼                    │
       ┌─────────────────┐           │
       │Carry out research│◄─────────┤
       └────────┬────────┘           │
                ▼                    │
       ┌─────────────────┐           │
       │Write specifications│◄───────┤
       └────────┬────────┘           │
                ▼                    │
       ┌─────────────────┐           │
       │Work out a solution│◄────────┤
       └────────┬────────┘           │
                ▼                    │
       ┌─────────────────┐           │
       │Select perfect solution│◄────┤
       └────────┬────────┘           │
                ▼                    │
       ┌─────────────────┐           │
       │Prepare working drawing│◄────┤
       └────────┬────────┘           │
                ▼                    │
       ┌─────────────────┐           │
       │Construct a prototype│◄──────┤
       └────────┬────────┘           │
                ▼                    │
       ┌─────────────────┐           │
       │Test and evaluate the design│◄┤
       └────────┬────────┘           │
                ▼                    │
       ┌─────────────────┐           │
       │  Write report   │◄──────────┘
       └─────────────────┘
```

Fig. 10.7 Flow chart for the mechatronic system designing process

10.6 Integrated Systems

Integration is a process of incorporation of methods, tools, subsystems, and ideas in an optimized manner. A group of products based on mechatronic design incorporate intelligence through an application-specified integrated circuit (ASIC) chip. This has given rise to what is known as consumer mechatronic products in the market. Washing machines, automatic cameras, bathroom scales, inkjet printers, laser printers, alarm indicators, data loggers, toys, etc. are examples of consumer mechatronic products which are based on the integration concept. In the following subsections, we will discuss the design principles of some of these consumer mechatronic products.

10.6.1 Washing Machine System

Figure 10.8 shows the basic washing machine system. It gives a basic idea of the constituent elements of a washing machine. It is a mechanical system involving a set of cam-operated switches. When the machine is switched on, the electric motor slowly rotates its shafts, giving an amount of rotation proportional to time. Its rotation, in turn, controls the cams so that each cam operates electricals switches and switch circuits in the correct sequence. The contour of a cam determines the time at which it operates the switch. Thus, contours of the cams are the means by which the washing program is specified and stored in the machine. The instructions used in a particular washing program, and their sequence, are determined by a set of cam chosen. The modern washing machine controller is a microprocessor where the program is not guided by the mechanical arrangement of the cam.

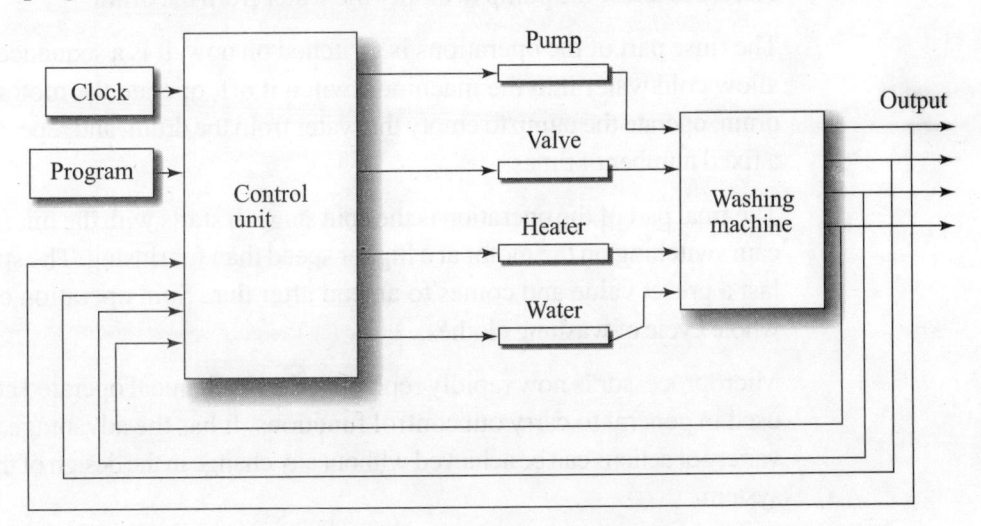

Fig. 10.8 Washing machine system

For a prewash cycle, an electrically operated valve is opened when current is supplied. This valve allows cold water into the drum for a period of time, which is determined by the profile of the cam. The output from the microprocessor is used to operate the switch. However, since a specific level of water in the washing machine drum is required, there need to be another mechanism that will stop the water going into the drum during the permitted time when it reaches the required level. A sensor is used to give a signal when the specified water level is reached. The preset level gives an output from the microprocessor, which is used to switch off the current to the valve. In the case of a cam-controlled valve, the sensor actuates a switch, which closes the valve admitting water to the washing machine drum. When the event is completed, the microprocessor gives out a signal to turn the prewash mode on for a fixed time. At the end of this time, the cam indicates the pump to empty the drum.

For the main wash cycle, the microprocessor gives output to start when the prewash part of the program is completed. In the case of a cam-operated system, the cam has a profile such that it starts operation when the prewash cycle is completed. It switches a current into a circuit to open a valve to allow cold water into the drum. The water level is continuously sensed and the water flow is shut off when the required level is reached. The microprocessor or cam then supplies a current to actuate a switch, which supplies a higher current to an electric heater to heat the water. A temperature sensor is used to switch off the current when the water temperature reaches the preset value. The microprocessor or cam switches on the drum motor to rotate the drum. This will turn off after some time, determined by the microprocessor or cam profile. Then the microprocessor or cam switches on the current to allow the pump to empty the water from the drum.

The rinse part of the operations is switched on now. It is a sequence of signals to allow cold water into the machine, switch it off, operate the motor to rotate the drum, operate the pump to empty the water from the drum, and repeat the sequence a fixed number of times.

The final part of the operation is the spin stage. It starts with the microprocessor or cam switching on the motor at a higher speed than for rinsing. The spin operations last a preset value and comes to an end after that. Spin operation completes the whole cycle of washing clothes.

Microprocessor is now rapidly replacing the mechanical operated controller. It is used in general to carry out control functions. It has the advantage that a greater variety of actions can be achieved without any change in the design of the mechanical system.

10.6.2 Automatic Camera System

Figure 10.9 shows the basic features of the Canon EOS model, automatic, autofocus, reflex camera. The camera has interchangeable lenses. There is a main microcontroller M-68HC11 in the camera body and another microcontroller in the lens housing. The two lenses communicate with each other, through the microcontrollers, when attached to the camera body.

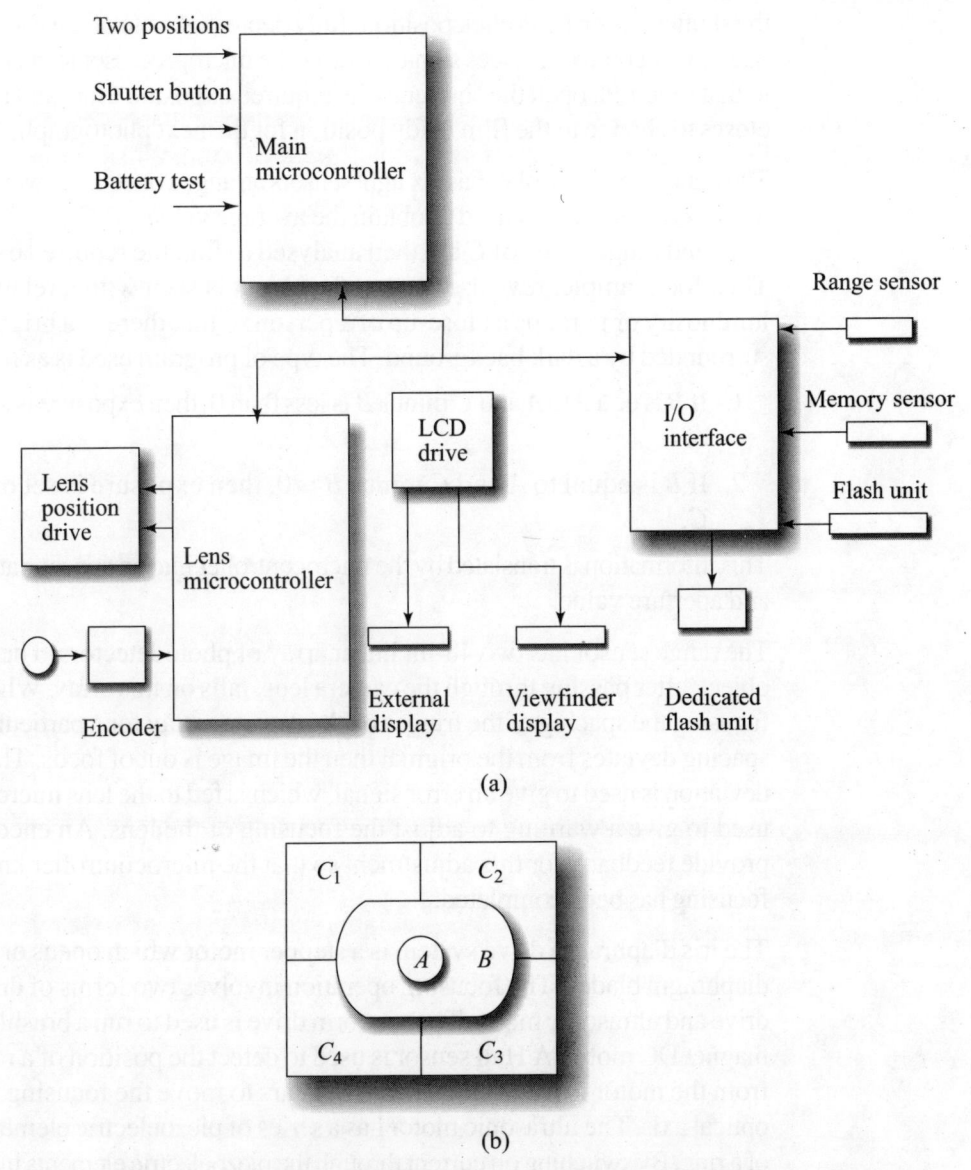

Fig. 10.9 (a) Block diagram of Canon EOS model, automatic, autofocus, reflex camera and (b) arrangement of light sensors

When the photographer presses the shutter button to its focus position (partially depressed state of the shutter button), the main microcontroller calculates the shutter speed, adjust the aperture setting from the input from the metering sensor, and displays the view focused in the viewfinder or an external LCD display. At the same time, the main microcontroller processes the input from the range sensor and sends a signal to the lens microcontroller. This issues signals to effect motion to adjust the focusing of the lenses. When the photographer presses the shutter button to its click position (fully depressed state of the shutter button), the main microcontroller issues signals to drive the microprocessor to change the aperture to that required, open the shutter to the required exposure time, and then the shutter closes to change to the film ready position for the next photograph.

The main microcontroller has six light sensors arranged on it, as shown in Fig. 10.9(b). Signal conditioning is used to obtain the average values of C_1, C_2, C_3, and C_4. The A, B, and range value of C are then analysed to find the required exposure value. This, for example, reveals whether the object is seen with a relatively constant luminosity or perhaps a close-up of a person so that there is a bright central zone surrounded by a dark background. The type of program used is as follows:

1. If B is equal to A and C minus B is less than 0, then exposure is set on the value of B.

2. If B is equal to A and C minus B is 0, then exposure is set on the value of C.

This information is translated by the microcontroller into an appropriate shutter speed and aperture value.

The range sensor has two 48-bit linear array of photodetectors. The light from the object, after passing through the camera lens, falls on this array. When the range is focusing the spacing of the fringes on the detector array for a particular value, if the spacing deviates from the original then the image is out of focus. The value of this deviation is used to give an error signal, which is fed to the lens microcontroller and used to give a warning to adjust the focusing of the lens. An encoder is used to provide feedback for this adjustment so that the microcontroller knows when the focusing has been completed.

The iris diaphragm drive system is a stepper motor which opens or closes a set of diaphragm blades. The focusing operation involves two forms of drives: arc-form drive and ultrasonic motor. The arc-form drive is used to run a brushless permanent magnet DC motor. A Hall sensor is used to detect the position of a rotor. The drive from the motor is transmitted through gears to move the focusing lens along the optical axis. The ultrasonic motor has a series of piezoelectric elements in the form of a ring. By switching on current through the piezoelectric elements in an appropriate sequence, a displacement wave can be made to travel along the ring of elements in either a clockwise or counterclockwise direction. This, consequently, rotates a rotor, which is in contact with the surface of the ring elements, hence driving the focusing element.

10.6.3 Inkjet Printer System

The inkjet printer uses a conductive ink, which is forced through a small nozzle to produce a jet of very small drops of ink of constant diameter at constant frequency.

In one form of inkjet printers, a constant stream of ink passes along a tube and is pulsed to form fine drops by a piezoelectric crystal, which vibrates at a frequency of about 100 kHz. Another form of inkjets uses a small heater in the point head with vapourized ink in a capillary tube, thus producing gas bubbles, which push out drops of ink. In the former version, each drop of ink is given a charge as a result of passing through a charging electrode. These charged drops are deflected when passing between the plates, between which an electric field is maintained. In the other version, a vertical stock of nozzles is used and each jet is just switched on or off on demand. The inkjet printer can give colour prints by using three different colour inkjet systems. The fineness of the drop is such that prints can be produced with more than 608 dots per inch (dpi). Figure 10.10 illustrates the principle of the inkjet printer.

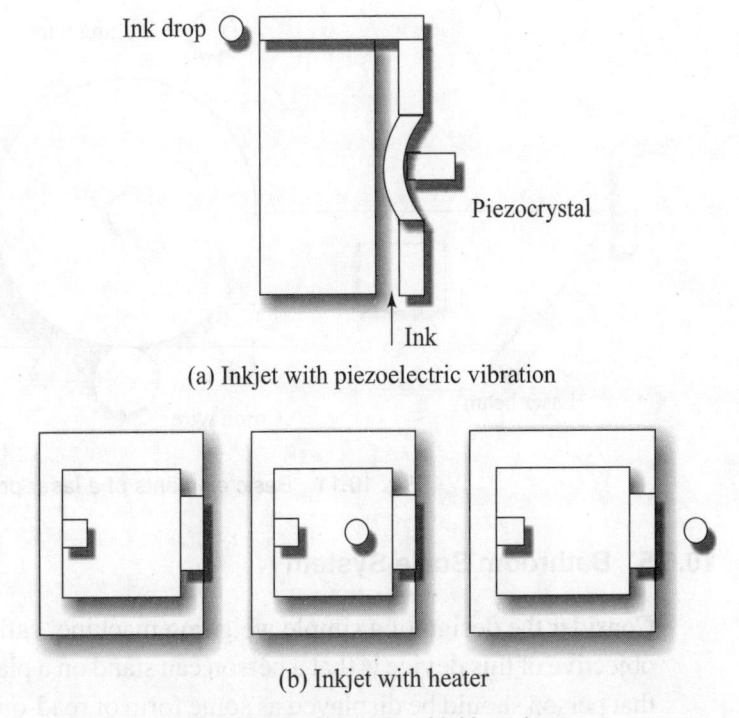

(a) Inkjet with piezoelectric vibration

(b) Inkjet with heater

Fig. 10.10 Principle of inkjet printer

10.6.4 Laser Printer System

The laser printer has a photosensitive drum coated with a selenium-based light-sensitive material. In the dark, selenium has a high resistance and consequently becomes charged as it passes close to a high-voltage charging wire, of which charge leaks. A light beam is made to pass along the length of the drum by a small rotating eight-sided mirror. When light strikes selenium, the resistance of selenium drops and it can no longer remain charged. By controlling the brightness of the beam of light, parts on the drum can be discharged or left charged. As the drum passes over the toner reservoir, the charged area attracts particles of the toner. The toner particles stick to the areas that have not been exposed to light, leaving out the areas that have been exposed to light. The sheet of paper is given a charge as it passes along another charging wire, the so-called corona wire. When this charged paper passes along the drum, it attracts the toner off the drum. A hot passing roller is then used to melt the toner particles so that, after passing between rollers, the finishings adhere to the sheet of paper. Laser printers currently in use are able to produce 600 dots per 25 mm. Figure 10.11 illustrates the working principle of the laser printer.

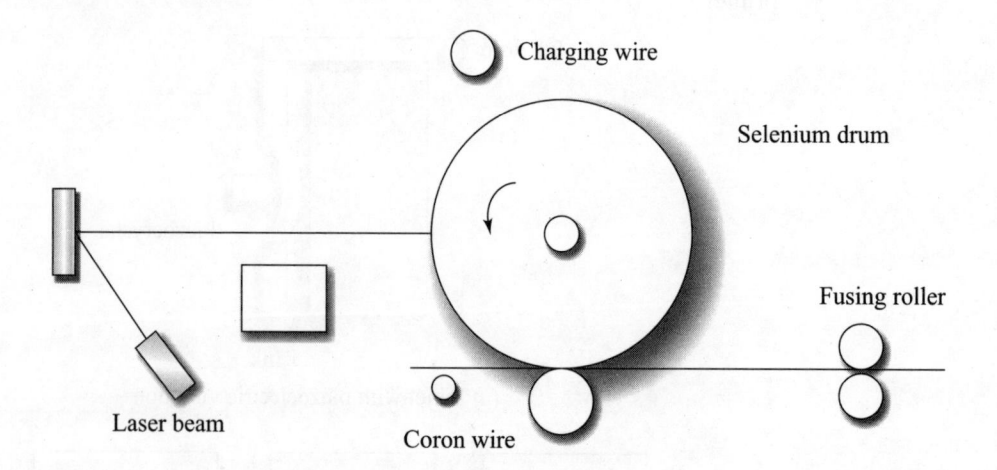

Fig. 10.11 Basic elements of a laser printer

10.6.5 Bathroom Scale System

Consider the design of a simple weighing machine, called bathroom scale. The objective of this device is that a person can stand on a platform and the weight of that person should be displayed as some form of read-out. The weight should be given with reasonable speed and accuracy and irrespective of the location of the person on the platform. One possible solution may be to use the weight of the person on the platform to deflect an arrangement of two parallel leaf springs. With such an arrangement, the deflection is virtually independent of where on the platform the person stands. The deflection can be transformed into a movement of a plate across the scale by using the arrangement shown in Fig. 10.12. A rack and pinion arrangement is used to transform the linear motion into a circular motion about the horizontal axis. This is then transformed into a rotation about the vertical axis, and hence into the movement of a pointer across a scale.

Intelligent Systems and Their Applications 275

(a) Platform detector

(b) Rack and pinion arrangment

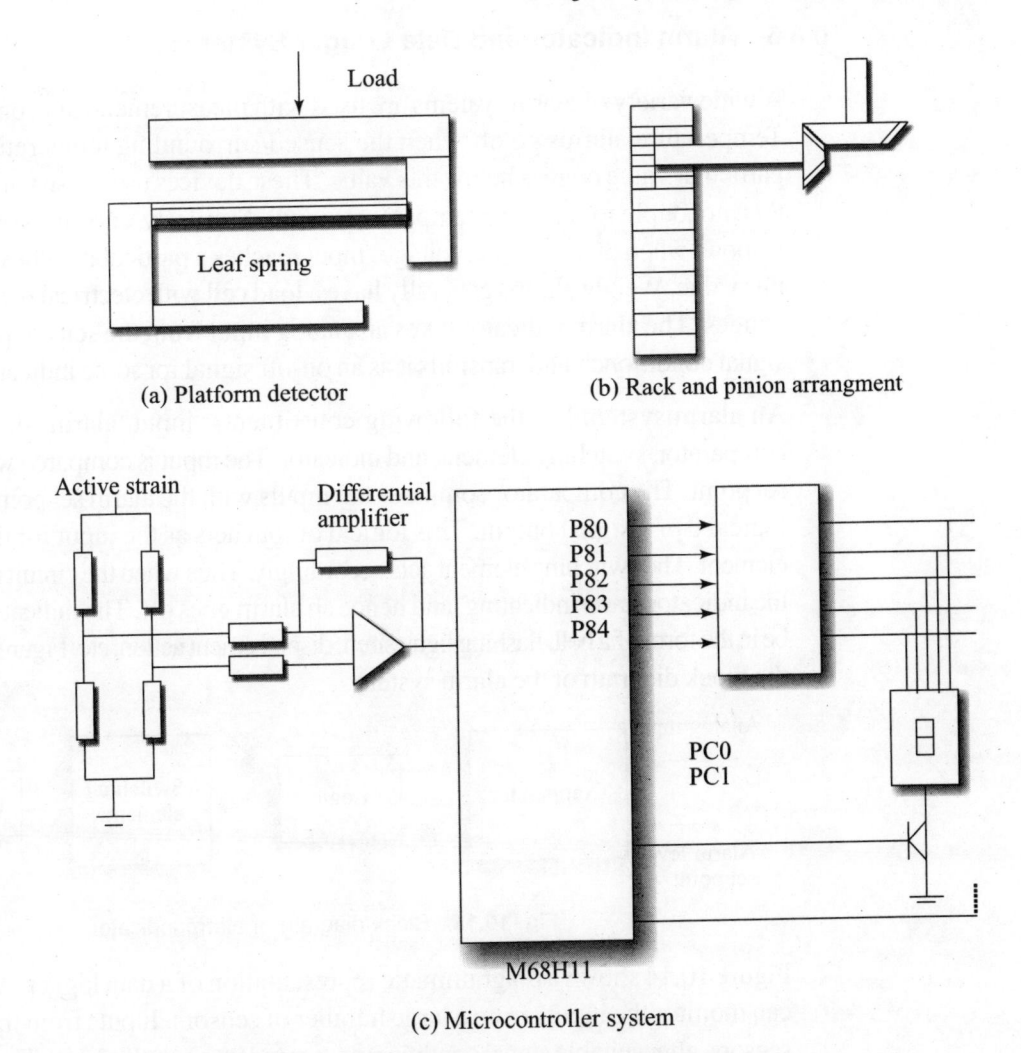

(c) Microcontroller system

Fig. 10.12 Bathroom scales

Another possible solution involves the use of a microprocessor. The platform can be mounted on load cells employing electrical resistance strain gauges. When the person stands on the platform, the gauges suffer strain and change resistance. If the gauges are mounted in a four-active-arm Wheatstone bridge, then when out of balance, the voltage output from the bridge is a measure of the weight of the person. The output voltage can be amplified by a differential operational amplifier. The resulting analog signal can then be fed through a latched analog to digital converter for inputting to the microprocessor. If a microcontroller is used, then, memory is present within a microprocessor chip, and by a suitable choice of microcontroller, the output can be calculated. For instance, M68HC11 can obtain analog to digital conversion for inputs. The output can be passing through a suitable decoder to give an LED display.

10.6.6 Alarm Indicator and Data Logger Systems

A wide variety of alarm systems are used with measurement and control systems. Temperature alarms go off when the sensed/surrounding temperature reaches a particular value or goes below this value. These devices use a resistance element or thermocouple to sense the temperature input. Similarly, current (voltage) alarms respond when their current (voltage) input reaches a particular value or falls below this value. Weight alarms generally have a load cell with electrical resistance strain gauges. The alarm indicator takes an analog input from the sensor, possibly via a signal conditioner, and transmits it as an on-off signal for some indicator.

An alarm system has the following constituents: input, alarm set point, logic, comparator, switching element, and indicator. The input is compared with the alarm set point. The comparator compares the inputs with the alarm set point and gives a logical 0 or 1 signal output. This logical output acts as the input for the switching element. The switching element acts accordingly. Thus when the circuit is completed, the indicator starts indicating, and hence an alarm goes off. The indicator device can be in the form of a bell, flashing light, siren, displacement action, etc. Figure 10.13 shows the block diagram of the alarm system.

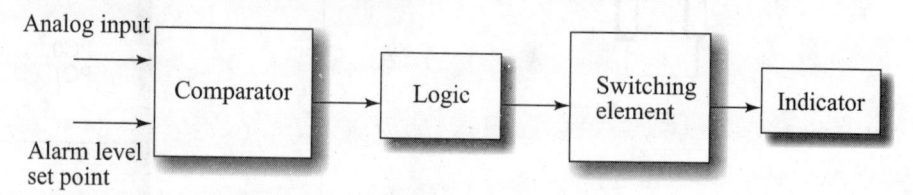

Fig. 10.13 Block diagram of alarm indicator

Figure 10.14 shows a diagrammatic representation of a data logger. A data logger can monitor the inputs from a large number of sensors. Inputs from the individual sensors, after suitable signal conditioning, are fed into a multiplexer. The multiplexer is used to select one signal, which is then fed, after amplification, to the analog to digital converter. The digital signal is then processed by a microprocessor. The microprocessor is able to carry out single arithmetic operation, that of taking the average of a number of measurements. The output from the system can be displayed on a digital meter that indicates the output and its channel number. The channel number is used to record the data for further printing, storing, or analysis by a computer.

Since data loggers are often used with thermocouples, there are often special inputs for the thermocouples, providing cold junction compensation or linearization. The multiplexer can be switched to each sensor one by one, so the output consists of a sequence of samples. Scanning of the inputs can be done by programming the microprocessor to switch the multiplexer to (a) just sample a single channel,

(b) carry out a single scan of all channels, (c) a continuous scan of all channels, or (d) carry out a periodic scan of all channels, say, every 1, 8, 15, 30, 60 min. Typically, a data logger may handle 20–100 inputs, though some may handle considerably more, about 1000. A blogger may have a sample and conversion time of 10 ns and be used to make 1000 readings/sec. The accuracy is typically about 0.01% and linearity is about ±0.005% of full scale input. Cross-talk is typically 0.01% of the full scale input or any one input. The term cross-talk is used to describe the interference that can occur as a result of signals from other sensors while a sensor is being sampled.

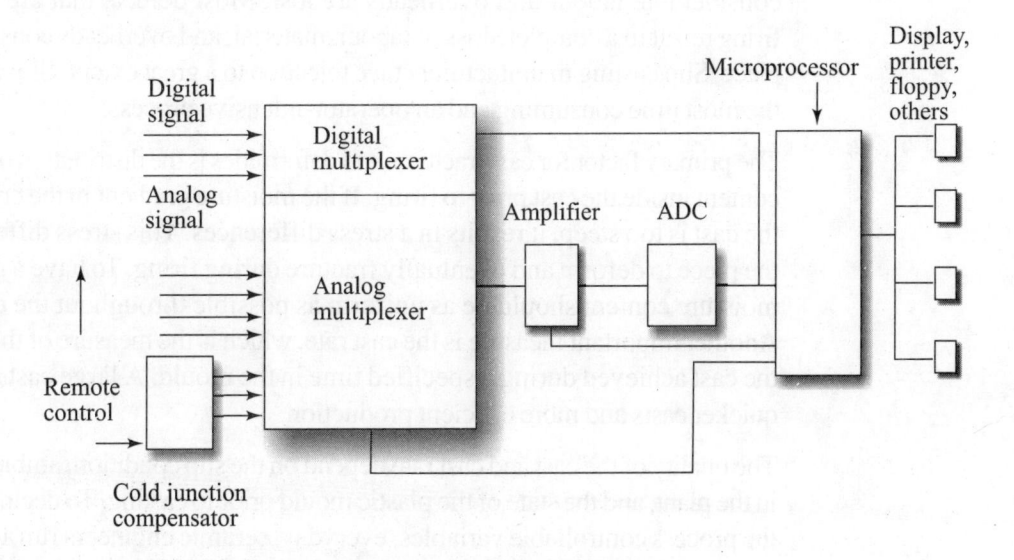

Fig. 10.14 Data logger system

Case Study 10.1: Slip Casting Process

Slip casting is used as the primary manufacturing process for many complex ceramic products when ceramic pressing is not practical, e.g., in the manufacturing of sanitaryware, giftware, tableware, etc. Slip casting is generally the second choice because it is considerably more difficult to achieve good results in both product quality and productivity. Pressing has the drawback in that it can only manufacture products of simple shapes. However, slip casting can be employed to manufacture items such as bowels, pitchers, statues, and sinks. The manufacturing of slip casting ceramicware involves the following steps:

1. Preparation of slip (clay)
2. Casting the slip in a plastic mould for a specified duration
3. Removing the mould
4. Air-drying the cast piece

5. Firing the glazed product in a kiln
6. Inspection of finished product

Step 2 of the slip casting process, wherein slip is poured into the plastic mould and allowed to cast through a specified time period to form a solid product, is of practical importance since it largely determines the quality of the final product. Cracks, slumps, and instability in the cast manifest in the quality of the product in the subsequent casting process. Some defects that are found before the ware is fired can be repaired. For the defects that cannot be repaired, the clay material can be recovered. However, considerable labour and overheads are lost. Most defects that are formed after firing result in a complete loss of labour, material, and overheads consumed for that piece. Slip casting manufacturers face rejection to a great extent. Slip casting is also the most time consuming and an operator-intensive process.

The primary factor for cast fractures and deformities is the distribution of the moisture content inside the cast prior to firing. If the moisture gradient in the cross section of the cast is too steep, it results in a stress differences. This stress difference causes the piece to deform and eventually fracture during firing. To have a good cast, the moisture content should be as uniform as possible throughout the cross section. Another important measure is the cast rate, which is the measure of the thickness of the cast achieved during a specified time in the mould. A large cast rate results in quicker casts and more efficient production.

The quality of the cast and cast rate depend on the slip condition, ambient conditions in the plant, and the state of the plastic mould prior to casting. To decide on values of the process controllable variables, everyday, ceramic engineers run a series of tests that emulate the behaviour of the slip during casting. These tests are time consuming, prone to human error, and use a trial and error approach to decide on the test cast quality. The process parameters required for a slip cast production batch are given in Table 10.3.

Table 10.3 Slip casting process parameters

Parameters	Definition
Plant temperature	Temperature of the plant
Relative humidity	Humidity level of the plant
Cast time	The time for which the slip is left in the mould before drainage
Sulphate content	Properties of solvable sulphate
Brook field 10 RPM	Viscosity of the slip at 10 RPM
Brook field 100 RPM	Viscosity of the slip at 100 RPM
Initial reading	Initial viscosity

Build up	Initial viscosity (taken at 3.5 min)
20 minute gelation	Thixotropy (viscosity versus time)
Filtrate rate	The rate at which the slip filter rates
Slip cake weight	Approximation of the cast rate without considering a mould
Cake weight water retention	Moisture content of the cake
Slip temperature	The temperature of the slip

A neural network and fuzzy logic technique can be used to improve the quality and the efficiency of the slip casting process. A hybrid approach is adopted because the moulds can be used individually or collectively depending on the decision to be made. The predictive module estimates the cast rate neural network (CRNN) and moisture gradient neural network (MGNN) to determine the cast quality and efficiency for the given ambient conditions, process settings, and slip characteristic as specified in Table 10.3. The fuzzy logic expert system recommends the processing time customized to a production line given the localized ambient state and the conditions of the plaster mould.

Many of the process variables are uncontrollable. For instance, ambient conditions, particle distribution of the raw material and slip properties cannot be changed on day-to-day basis. The interactive behaviour of the process variables is mostly unquantified. The neural network is an empirical universal approximation technique. That is, a neural network can theoretically model any relationship through examination of inputs and outputs. The neural network predictions can be used for the daily test casts. The predictions are also used in change with the process improvement algorithm for production control.

In this case, large data, both input specified in Table 10.3 and outputs, are considered, which contain thousands of observations in four to five years. The neural network architecture, training parameters, and stopping criteria were selected through experimentation and examination of preliminary networks. An ordinary back propagation learning algorithm was used because of its documented ability as a continuous approximator. With experience, it was observed that the neural network for CRNN was multi-layered, fully connected perception with 13 inputs, two hidden layers with 11 hidden neurons in each layer, and single output, which means the cast rate gives a convergence result. The MGNN model was developed in the same way, using the same input variables, two hidden layers and 27 hidden neurons in each layer, single output, and moisture gradient percentage change.

The more traditional random separation of the data into testing and training sets is in the ratio of 3 : 7, respectively. This system is for the industrial use. Fivefold group validation has been used to validate the neural network as this allows all data to be used for both validation and construction of the neural network model. The estimation of the neural network prediction error by the data splitting method is not accurate or reliable than with a single network. Finally, the neural network used in the system can be constructed using all the data rather than just the training subset to ensure the best possible prediction. The summation of the test set over five validation networks showed a mean absolute error of 4.8% and 68.4% for CRNN and MGNN, respectively. This may be due to the considerable human error in the measurement of the moisture gradient. This imprecision, however, did not preclude the MGNN use in the system. The MGNN does a fairly good job of prediction for a small moisture gradient and can discriminate between a good moisture gradient and a bad moisture gradient. Taking a categorical approach, the MGNN has a prediction ability of adequate precision for use in the plant.

The neural network prediction model estimates the cast rate and moisture gradient for each combination. The optimization algorithm selects the best three combinations that minimize the moisture gradient and maximize the cast rate. Providing the three best process settings allows the ceramic engineer to review the options prior to implementation, imparting some human expertise in the decision making. If the user wishes to further study the three recommended settings by considering other factors, they can use the analytic hierarchy process.

The fuzzy logic expert system model is incorporated to consider the important factor(s) affecting the cast quality, which is individual to a particular product line. These factors are the age of the plastic moulds (which ranges from 0 to 6 weeks) and the ambient conditions. It is necessary to use a knowledge-based approach rather than a pure numeric model because there is no data that relates the mould age and local ambient environment to the cast quality. It is impractical to begin to record such data in the plant. Instead, these two important factors are included in two bases that work as hierarchy, as shown in Fig. 10.15. The use of fuzzy logic is included to allow more precise handling of the variables, reduce the number of rules required, and result in a smoother decision surface.

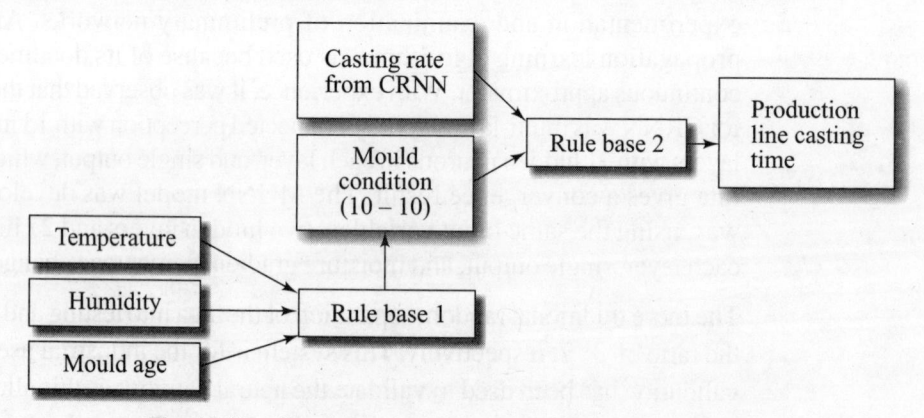

Fig. 10.15 Structure of the fuzzy logic expert system module

Referring to Fig. 10.15, the ambient temperature, relative humidity, and age of the moulds in weeks for the production line can be presented as crisp, continuous inputs to the first fuzzy rule base that predicts the mould conditions. The mould condition is rated on an ordinary scale of 0 to 10, with 0 being the dirtiest and 10 being the wettest. This prediction of the mould condition is then paired with a crisp, continuous prediction of the cast rate from CRNN. Together, these are used in the second fuzzy rule base in order to generate a recommended cast time in minutes for a particular production line. Defuzzification to a crisp output is done using the centroid method, by computing the centre of mass of the region of the output variable defined by the fuzzy output.

The hybrid modular system is a comprehensive decision support system and uses computational intelligence for the ceramic slip casting process. The integration of the neural network, dual optimization algorithm, and the fuzzy logic rule allows modelling of the highly non-linear relationship among the process variables and the capture of qualitative factors and expert knowledge from the plant personnel. Utilization of this system enables a casting decision to be made consistently and correctly without the need for judgemental speculation or expensive trial and error test casts. Figure 10.16 shows the block diagram that integrates the neural network, dual optimization algorithm, and fuzzy logic technique to improve the quality and efficiency of the slip casting step.

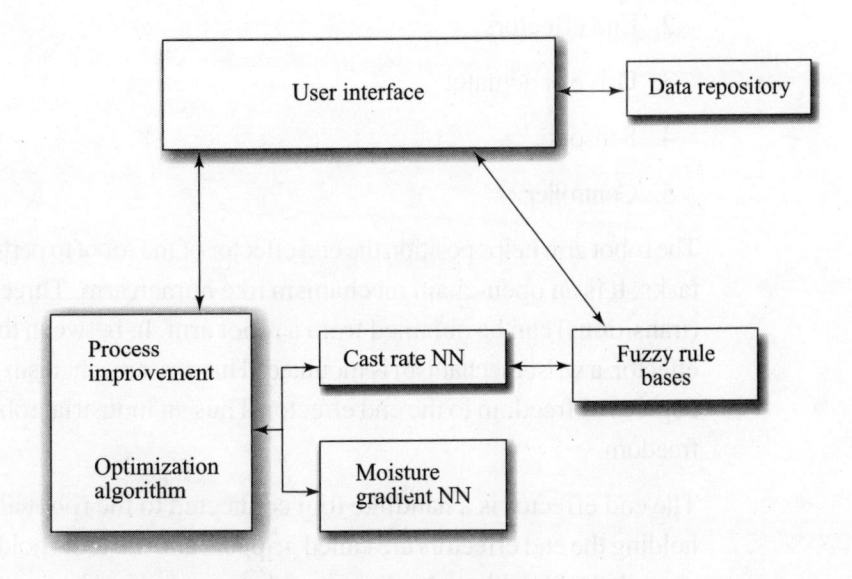

Fig. 10.16 Block diagram of the hierarchical system of slip casting

Case Study 10.2: Pick-and-Place Robot

Robots are widely used in automated industries. A robot is a mechatronic system capable to replace or assist the human operator in carrying out a variety of physical tasks. The Webster dictionary defines robot as 'a machine that looks like a human being and performs various complex acts (as walking or talking) of a human being'. An industrial robot is very difficult to define. The industrial robot continuously interacts with the surrounding environment. This continuous interaction is achieved through sensors and transducers. Computer-controlled interaction systems emulate human capabilities. The Robotic Industries Association has defined the manipulation robot used in industry as follows: 'A robot is a reprogrammable, multifunctional manipulator designed to move materials, parts, tools or special devices through variable programmed motion for performing a variety of tasks.'

There is a variety of robots used in industrial applications. Broadly, they are classified as fixed robots and mobile robots. Mobile robots are also called automated guided vehicles (AGV) or walking robots. Fixed robots have different configurations as per the requirement of their application in industry. Fixed robots are further classified as cartesian robots, cylindrical robots, polar robots, and articulated arm or spline robots. For the pick-and-place operation, usually the articulated arm robots are commonly used. An articulated arm robot contains basically five subsystems, namely,

1. Robot arm
2. End effectors
3. Drive or actuator
4. Sensors
5. Controller

The robot arm helps position the end effector of the robot to perform the programmed tasks. It is an open-chain mechanism like human arm. Three degrees of freedom (transitions) can be obtained from a robot arm. In between the robot arm and end effector, a wrist mechanism is mounted. This wrist mechanism gives three rotational degrees of freedom to the end effector. Thus an industrial robot has six degrees of freedom.

The end effector is a handlike tool connected to the robot arm. The work pieces holding the end effectors are called grippers and the tools holding the end effectors are called tool holders. A gripper may be a mechanical lever, adhesive, vacuum cup, magnetic hook, or a snooper.

Drives or actuators are parts of the controller element that act as per the command of the controller and translate the signal from the controller into an operation. Electrical, pneumatic, or hydraulic actuators can be used to perform the final actuation.

Sensors play an important role in the functioning of a robot. Both internal and external sensors are used in a robot arm. Robots using an external sensor to identify the external object are called smart robots.

Every robot is connected to a computer. The computer along with the necessary software is known as the controller. The controller also allows compatibility with other systems of the manufacturing environment. A controller with artificial intelligence makes a robot more self-reliant and independent.

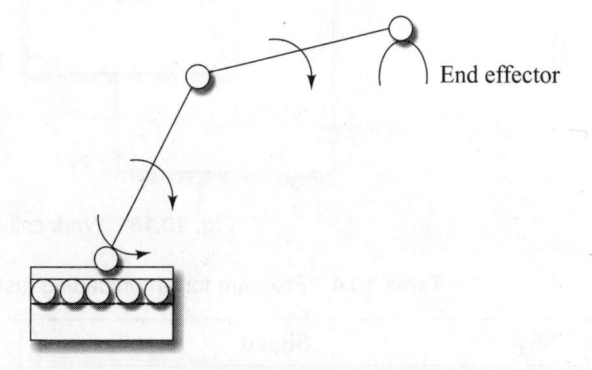

Fig. 10.17 Configuration of a jointed-arm robot

Figure 10.17 shows the configuration of a jointed-arm robot used for pick-and-place operations. Its configuration corresponds to the human forearm. Its upper arm is mounted on a vertical pedestrian. The components of this arm are connected by two rotary joints corresponding to the shoulder and elbow. A wrist mechanism attached to the end of the forearm provides three additional degrees of freedom.

A robot program is accomplished in several ways. Two basic methods to accomplish robot programming are

- Lead through method
- Textual robot language

The lead through method requires the program to move the manipulator through the desired motion path. It also requires that the path be committed to memory by the robot controller. Robot programming through a textual language is accomplished somewhat like computer programming. The programmer types in the program mode on a cathode ray monitor (CRM) using high level English language. Textual languages appeared around 1979 and are in use since then in the robot industry. To illustrate a textual program, consider a work cell as illustrated in Fig. 10.18. A press path is indicated by the pillar boundaries lines. The robot has to pick at point 8, 1 and drop a tote pan at 1, 8. The point 8,1. is the position to wait for the robot to operate before entering the press to remove the part. The robot must start from the point 1, 1, which is a safe place for the robot during the press work. The program given in Table 10.4 accomplishes the unloading task in the press work.

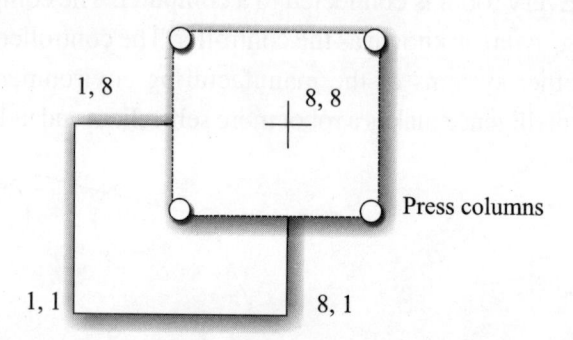

Fig. 10.18 Work cell

Table 10.4 Program for an unloading task in the press work

Step	Signal	Comments
0	1, 1	Start at home position
1	8,1	Move to wait position 1,1
2	WAIT 1,1	Wait for press to open
3	8, 8	Move to pick up point
4	SIGNAL 5	Signal gripper to close
5	8,1	Move to safe position
6	SIGNAL 4	Signal press to actuate
7	1, 1	Move around press column
8	1,8	Move to tote pan
9	SIGNAL 6	Signal gripper to open
10	1, 1	Move to safe position

Exercises

10.1 Why are advanced mechatronic systems required in manufacturing?

10.2 Explain the working principle of an artificial neural network.

10.3 Why is the sigmoid function considered an input signal for ANN?

10.4 Explain the back propagation algorithm with examples.

10.5 How can genetic algorithms be used for process control?

10.6 What is nonverbal teaching?

10.7 What are the advantages of fuzzy logic control over conventional control system?

10.8 Explain the application of a mechatronic system with the sequence of operation.

10.9 Explain the working of an automatic camera system.

10.10 Explain the working principles of inkjet and laser printers.

10.11 Write a short note on data logger systems.

Chapter 11

Autotronics, Bionics, and Avionics

Mechatronic products have become increasingly dominant in every aspect of commercial marketplace as technologies, electronics, and computers continue to be developed. Presently major commercial markets for mechatronic products are in the form of automobiles, aeronautical/astronautical systems, and biological systems. The automobile industry has been the major user of mechatronics technology during the recent decades. Another commercial application of mechatronics is in biomedical and biological sciences. Aircraft and rocket technologies are the other areas where the use of mechatronic systems has been growing rapidly. Advancements in mechatronics in the areas of automobile engineering, biotechnology, and aircraft and rocket engineering, have given rise to specialized disciplines of autotronics, bionics, and avionics, respectively. We will discuss these three major specialized study areas of mechatronics in this chapter.

11.1 Autotronics

The primary motivation for adopting mechatronic systems in automobiles is to make automobiles safer, more comfortable, fuel efficient, and less polluting systems. A widespread use of mechatronic products can indeed make an automobile smarter for consumer needs. Smart vehicles are based on an extensive use of mechatronic systems to detect the environment or road conditions. Microactuators are used to execute whatever actions are required to deal with these conditions. The modern car comes with many electronic control systems. As the mechatronics technology developed, a new discipline called autotronics has emerged. The design and production of mechatronic systems for automobiles that perform at extreme environmental conditions presents a major challenge to engineers. Application of mechatronic systems in automobiles is in the following major areas:

- Safety
- Engine and power train
- Comfort and convenience
- Vehicle diagnostics and health monitoring

The airbag deployment system to protect the driver and passenger from injury in the event of a serious vehicle collision is a mechatronic safety system. The antilock braking system, air suspension system, navigation system, and obstacle avoidance all give security to the driver and passenger.

Various mechatronic systems are used in engines of modern automobiles. Some of these are listed below:

- Manifold control pressure sensor
- Air flow control
- Exhaust gas analysis and control
- Crankshaft position control
- Fuel pump pressure and fuel injection control
- Transmission force and pressure control
- Engine knocking detection for high power output

The manifold absolute pressure (MAP) system was one of the first micromechatronic systems adapted by the automobile industry. In the early 1980s, MAP was used to measure the engine speed in revolutions per minute (RPM) to determine the ignition advance. The ignition timing with optimum air/fuel ratio can optimize the power performance of the engine with a low emission. Various mechatronic systems used in the modern automobile engines are:

- Seat control for passenger comfort
- Defogging of wind shield
- Automatic dim and bright control
- Satellite navigation

The comfort and convenience of passengers are considered while designing mechatronic systems. For vehicle diagnostic and health monitoring, mechatronic systems are used for monitoring the following conditions:

- Engine coolant temperature and quality
- Engine oil pressure, level, and quality
- Tyre pressure
- Transmission fluid
- Fuel pressure, etc.

288 Introduction to Mechatronics

Power and Nicastin in 1999 reported that 52 million vehicles were produced worldwide in 1996. There has been tremendous growth in the world economy and the number of vehicles is increasing at a tremendous rate. And so are the consumer expectations on safety, comfort, and performance of cars. There is every reason to believe that vehicles in the future will have even more microprocessors with many more microsensors and actuators to make good use of autotronic systems for true smart vehicles.

11.1.1 Car Engine Management

The aim of the engine control system is to ensure that the engine is operated at its optimum setting. Figure 11.1 shows a generalized block diagram of such a system. The system consists of sensors supplying, after suitable signal conditioning, the input

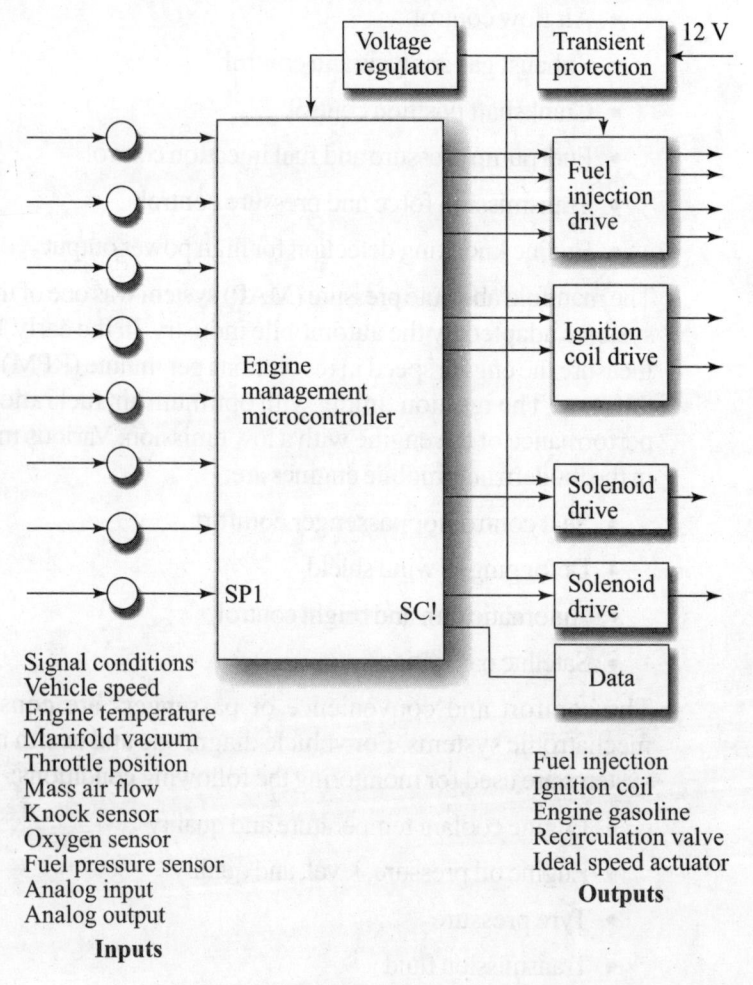

Inputs:
Signal conditions
Vehicle speed
Engine temperature
Manifold vacuum
Throttle position
Mass air flow
Knock sensor
Oxygen sensor
Fuel pressure sensor
Analog input
Analog output

Outputs:
Fuel injection
Ignition coil
Engine gasoline
Recirculation valve
Ideal speed actuator

Fig. 11.1 Engine management system

signals to the microcontroller and providing output signals via the driver to actuate actuators. The engine speed sensor is an inductance type that changes as the teeth of the sensor wheel pass it and so gives an oscillating voltage. The temperature sensor is usually a thermistor. The mass air flow sensor can be a hot wire anemometer and as air passes over the heated wire, it will be cooled. The amount of cooling depends upon the mass flow rate. The oxygen sensor is generally a closed-end tube made of zirconium oxide with porous platinum electrodes on the inner and outer surfaces. Above near 300°C, this sensor becomes permeable to oxygen ions and so measures the voltage flowing between the electrodes.

The engine management system of a car is responsible for managing the ignition and fuelling requirements of the engine. In a four-stroke internal combustion engine, there are several cylinders, each having a piston connected to a common crankshaft and each of which comes out of a four-stroke engine sequence of operation. When the piston moves, the valve opens, and the air-fuel mixture is drawn into the cylinder. When the piston moves down, up again, the valve closes and the air-fuel mixture is compressed. When the piston is near the top of the cylinder, the spark plug ignites the mixture with the resulting expansion of the hot gases. This expansion causes the piston to move down again and so the cycle is completed. The piston of each cylinder is connected to a common crankshaft and the power strokes occur at different times, so that there is a continuous power supply for rotating the crankshaft. The power and speed of the engine are controlled by varying the ignition timing and air-fuel mixture. In the modern car engine, this is done by a microprocessor and mechatronic systems. For ignition timing, the crankshaft drives a distributor, which makes electrical contacts with each spark plug and, in turn, with a timing wheel. The timing wheel generates pulses to indicate the crank position. The microprocessor then adjusts the timing at which high-voltage pulses are sent to the distributor, so they occur at the right time. To control the amount of air-fuel mixture entering into a cylinder during the intake strokes, the microprocessor varies the time for which a solenoid is activated to open the intake valve on the basis of inputs received in the form of the engine temperature and the throttle position. The amount of fuel to be injected into the air stream can be determined by the input from a sensor on the mass rate of air flow, or can be computed from other measurements. The microprocessor then gives an output to control the fuel injection valve.

11.1.2 Windscreen Wiper Mechanism

For cleaning/wiping the windscreen of the car, a device that will oscillate an arm back and forth in an arc is needed. The windscreen wiper is a mechanical device that can be operated by the use of stepper motor. Figure 11.2 shows how a programmable logic circuit (PLC) or microcontroller can be used with the stepper motor for operating a wiper device. The stepper motor needs an input to cause it to rotate a number of steps in one direction and then reverse the same number of steps with the reverse direction.

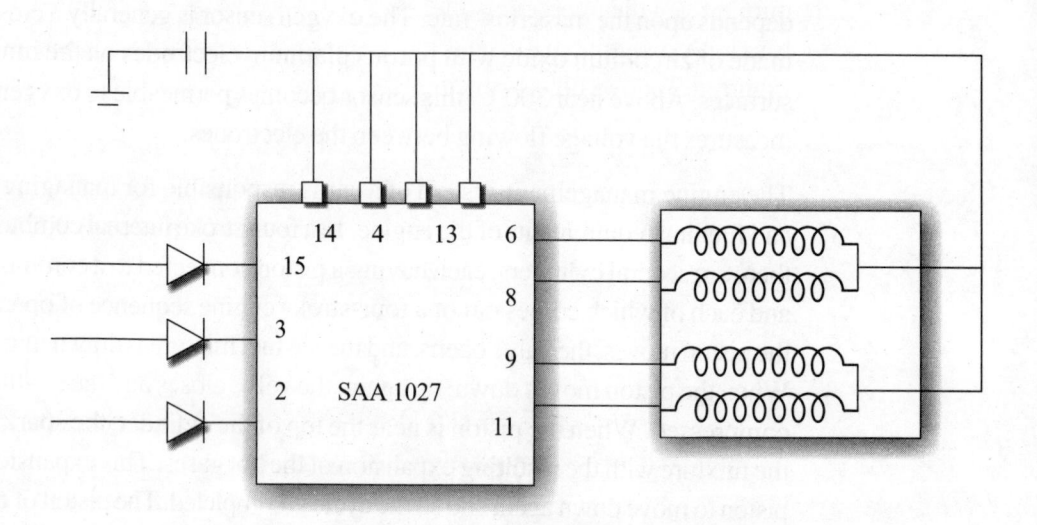

Fig. 11.2 Interfacing a stepper motor

If the stepper motor is to be in the full-step configuration, the output needs to be as shown in Table 11.1. Thus to start and rotate the motor in a forward direction, it involves the sequence A, 9, 5, 6 and then back to the start with 1 again. To reverse, one could use the sequence 6, 5, 9, A and then back to the start with 6 again. If the half-step configuration is used for forward motion, then it involves the sequence A, 8, 9, 1, 5, 4, 6, 2 and then back to A. To reverse, the sequence is 2, 6, 4, 5, 19, 8, A and back to 2. If many steps involved, a single program is to increment a counter with each step and loop control the counter value till it reaches the required number.

Table 11.1 Truth table for interfacing stepper motor

	Bits	Code
1	1 0 1 0	A
2	1 0 0 1	9
3	0 1 0 1	5
4	0 1 1 0	6
5	1 0 1 0	A

Integrated circuits are available for the stepper motor control, and their use can specify the interfacing with the software. All that is needed is the requisite number of steps, if input pulses to trigger the motor stepping on the low to high transition of a high-low, low-high pulse. A high on the rotation input causes the motor to step counterclockwise and the low gives clockwise rotation. Thus one just needs an output from the microcontroller for output pulses to the trigger and one output to rotation. An output to the set is used to reset the motor back to its original position. Figure 11.3 shows how an integrated circuit can be used for the stepper motor drive.

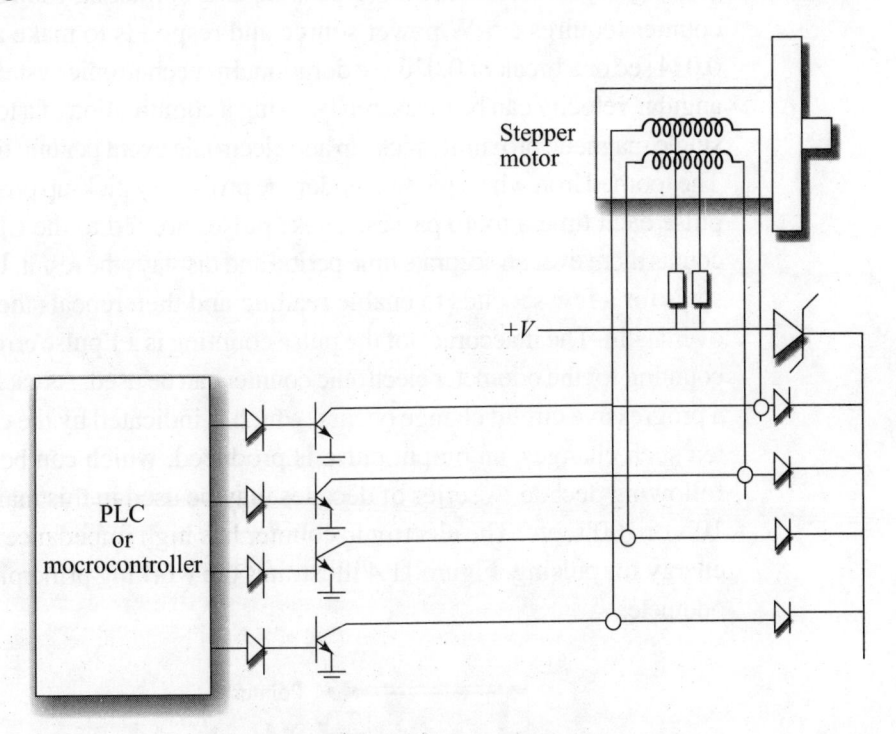

Fig. 11.3 Integrated circuit for stepper motor

11.1.3 Digital Speedometer and Odometer

In the conventional speedometer, an eddy current tachometer is used for speed measurement. The rotation of a magnet induces voltage into the aluminium cup, which thereby produces circulating eddy current in the cup material. These eddy currents interact with the magnetic field to produce a torque on the cup in proportion to the relative velocity of the magnet and cup. This causes the cup to turn through an angle until the linear spring torque just balances the magnetic torque. This steady state twist angle of the cup is directly proportional to the angular speed obtained on the wheel through the rotating cable. The conventional automobile odometer is a drum-type mechanical counter. The counter may be driven either mechanically or by directly coupling it to a rotating shaft or wheel, or they may be actuated by oscillating a reciprocating motion introduced through an appropriate linkage and ratchet mechanism. Mechanical counters that can be actuated by electrical impulses are also available. Such electrically actuated counters may be energized by a single switch or relay, or by photocells or by a pulse with a supply power of several watts. One high-speed counter of this type is capable of making 1000 counts/min. Such a counter requires a 5-W power source and responds to make a pulse as short as 0.024 sec or a break of 0.036 sec duration. In mechatronic system applications, the angular velocity can be measured by using a combination of a toothed wheel and a single magnetic proximity pick-up and electronic event per unit time (EPUT) meter. The toothed iron wheel passing under the proximity pick-up produces an electrical pulse each time a tooth passes. These pulses are fed to the EPUT meter, which counts them over an accurate time period and displays the result. Usually, the display stays for a few seconds to enable reading and then repeats the process over and over again. The inaccuracy of the pulse counting is ±1 pulse error. The high-speed counting for the odometer electronic counter can be used. As each pulse is received, a progressive circuit change occurs, which is indicated by the counter tube. After ten such changes, an output pulse is produced, which can be used to drive the following decade. A series of decades may be used in this manner to record 1's, 10's or 100's, etc. The electronic counter has high impedance and requires little energy for pulsing. Figure 11.4 illustrates the working principle of speedometer/odometer.

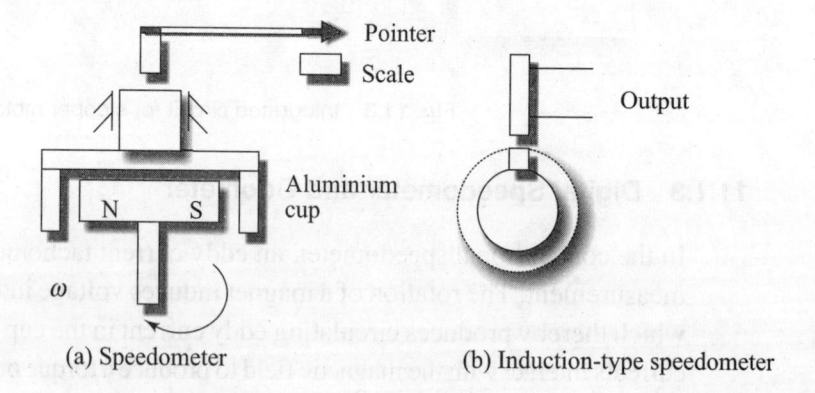

Fig. 11.4 Speedometer

11.1.4 Automatic Dim and Bright Control

Photosensors can be used to control the dimness and brightness of headlights. The circuit diagram for such an automatic dim and bright control actuation is shown in Fig. 11.5. An operation amplifier is used to compare the voltage generated from the photocell to the reference voltage. The operation amplifier is actuated with a volt battery, with typical values of resistance $R_a = 8.15\,k\Omega$, $R_1 = 1\,k\Omega$, and $R_g = 100\,k\Omega$. When a vehicle from the opposite direction is within the range of the source vehicle, the light from the opposite vehicle actuates a detector and the analog input actuates a relay to adjust to the dim light.

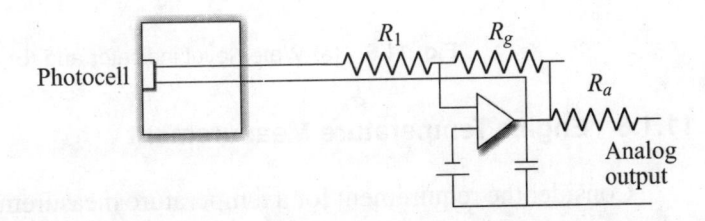

Fig. 11.5 Automatic dim and bright control system

11.1.5 Radiator Water Level Indicator and Control of Radiator Fan

The level of liquid in a vessel can be measured by monitoring the position of the liquid surface indirectly by measuring some variables related to the height. Direct methods involve floats and the indirect method includes the monitoring of weight of the vessel. The weight of the liquid is proportional to its height. A direct method of monitoring the level of liquid in a radiator is made easier by monitoring the movement of a float. Figure 11.6(a) shows a water level indicator. The displacement of the float lever causes it to make contact with a microswitch. When the water level goes down, the microswitch makes contact and a circuit is completed, which, in turn, gives or sends off a warning signal in the panel indicator.

A bimetallic device can be used to actuate the cooling fan to ON when the temperature of the engine body is high. A bimetallic device is a strip of two different metal strips bonded together [Fig. 11.6(b)]. The metals have different coefficients of expansion. When the temperature changes, the composite strip bends into a curved strip. This deformation can be used as a temperature control switch, as in a simple thermostat commonly used in domestic heating systems. The actuation makes the cooling fan to turn ON or OFF.

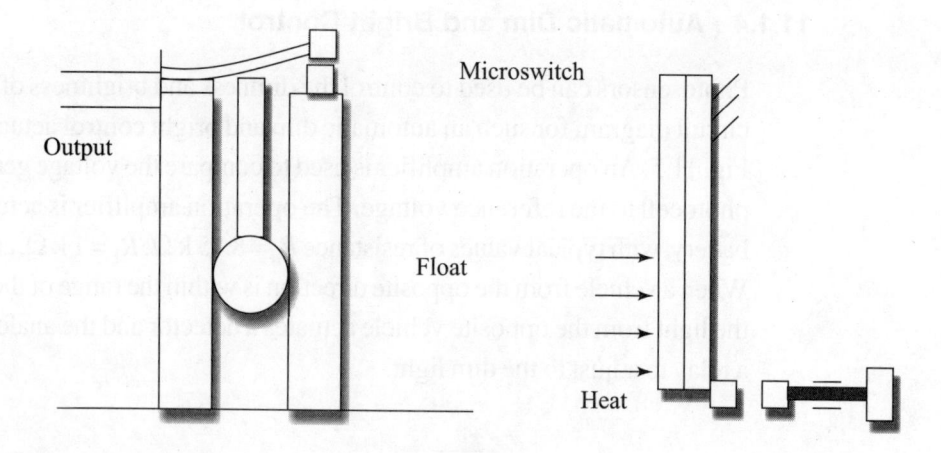

Fig. 11.6 (a) Water level indicator and (b) autoradiator

11.1.6 Engine Temperature Measurement

Consider the requirement for a temperature measurement system for measuring temperature in the range 0–100°C, which is the case of the body temperature of the engine of an automobile. The system gives an 8-bit binary output with a change in 1 bit corresponding to the temperature change of 1°C. The output is intended for inputting to a microprocessor as part of a temperature indicating system. Thermistor LM35 can be used since a linear temperature sensor is required. LM35 gives an output of 10 mV/°C when fed with a supply voltage of 5 V. If one supplies from LM35 with an 8-bit analog to digital converter (ADC), then a digital output can be obtained. The resolution of the ADC should be 10 mV so that each strip of 10 mV can generate a change in the output of 1 bit. If one uses a successive approximation ADC, ADC0801, then it requires an input of the response voltage, which when subdivided into 256 bits gives 10 mV per bit.

This reference voltage input to the ADC0801 has to be $V_{ref}/2$ and so an accurate input voltage of 1.28 V is required. Such a voltage can be obtained by using a potentiometer circuit across the 5-V supply with a voltage follower to avoid loading problems. Because the voltage has to be steady at 1.25 V even if the 5-V supply voltage fluctuates, a voltage regulator is likely to be used for a 2.54-V supply, ZN458/B. Such a circuit is shown in Fig. 11.7.

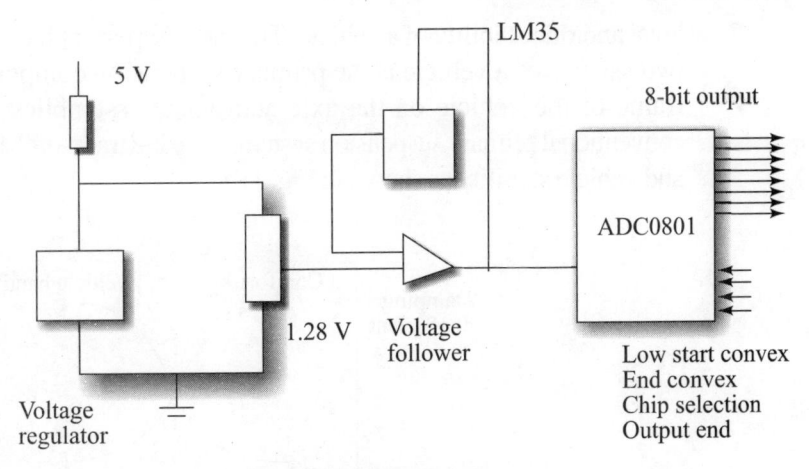

Fig. 11.7 Temperature sensor

11.1.7 Vehicle Suspension System

Vehicles are used for industrial, agricultural and other transport purposes. The increased number of vehicles, along with the extended driving hours and exposure to the severe working environments such as rough road conditions and complex ride behaviours, inevitably results in an increased demand for improved ride comfort. Suspension is the tool that is used to provide some buffer for the vibrations caused by uneven and bumpy road surfaces so that the rider is not exposed to the actual levels of discomfort that can be caused by such vibrations. Suspension is observed based on the measured ride vibration at the driver's seat as well as the passenger's seat. The vibration level in commercial vehicles is 9–15 times higher than that in cars. In most of the passenger vehicles, drivers and passengers are exposed to ride vibration for 10–20 hr/day. The passenger's and the driver's perception of ride comfort is based upon the road shock, impact, and vibration transmitted though the seats. Ride vibration has a significant influence on the passenger's and driver's fatigue and safety.

Much improvement in the ride comfort can be achieved by making use of appropriate tyres and primary and secondary suspension. The suspension system used between the tyre and chassis of the vehicle is known as the primary suspension and that used between chassis and seats is known as the secondary suspension. The suspension for the seat has proved to be a simple and effective option to attenuate vertical as well as sidewise vibrations. In the majority of the commercial vehicles, the seat suspension system is the only mechanism to control the transmission of low-frequency and high-amplitude ride vibrations.

The perceived comfort level and ride stability are two of the most important factors in a vehicle's subjective evaluation. There are many aspects that influence the comfort level and ride stability of a vehicle. The most important factor that influences these two aspects of a vehicle is the primary suspension component that supports the frame of the vehicle on the axle and wheel assemblies. In the design of the conventional primary suspension system, there is a trade-off between the ride comfort and vehicle stability as shown in Fig. 11.8.

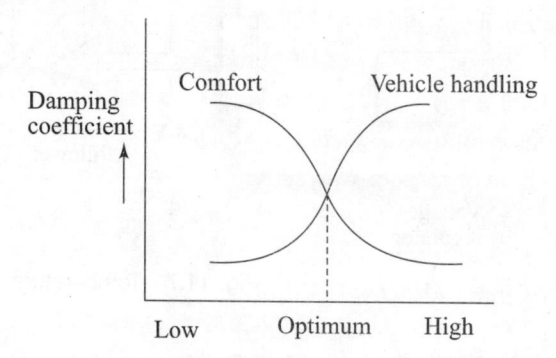

Fig. 11.8 Ride comfort and vehicle stability versus the damping coefficient

If the primary suspension is designed to optimize the handling and the stability of the vehicle, the operator is often subjected to a large amount of vibration and perceives the ride to be rough and uncomfortable. On the other hand, if the primary suspension is designed to be a soft cushion, the vibrations in the vehicle are reduced; but the vehicle may not be too stable during maneouvres such as cornering and line changing. As such the performance of the primary suspension is always limited by the compromise between ride and handling. A good design of passive or semi-active suspension system for the passenger and driver seats eliminates the compromising character of the two opposing goals of comfort and handling.

Isolating and/or protecting a piece of delicate equipment from the vibrations of its base structure is of practical importance in a number of engineering fields. Seats in moving automobiles are one of the examples that require three-dimensional vibration isolation. In many of the cases, the base structure vibrates with an unpredictable waveform of a very broad band spectrum. The active isolation of vibration of the passenger seat from a vibrating base should meet the challenges of high and low frequency isolation.

As the need for reduced noise and vibration increases, suspension systems of machines and structures are becoming even more complex and more important than ever. Passive anti-vibration mounts are widely used to support the passenger or driver seat to protect it from the severe base vibrations. However, conventional passive mounts suffer from inherent trade-off between the poor high frequency isolation and the amplification of vibration at the fundamental mounted resonance frequency. Generally, the best isolation performance is achieved by using an active system in combination with a passive mount where the fundamental resonance can be effectively controlled without the high frequency performance. Such a system is called the semi-active damping vibration isolation system.

The semi-active suspension system combines the advantages of both passive and active suspension systems. It provides for better performance than the passive suspension system. It does not require high power actuators or high supplies. Its performance may not be better, but it costs far less than the active suspension system. Although the control implementation remains almost identical, semi-active actuator control mechanisms, which are often developed by modifying active control mechanisms, require an active and robust mathematical model of the structure and control of devices. Owing to its robustness to change in the system parameters, the clipped optimal control is one of the most commonly used semi-active control algorithms. However, semi-active control devices possess inherent non-linearity, which makes the development of optimal control laws challenging. As the mechatronic suspension technology advanced, many passive suspension systems have been replaced by semi-active suspension systems. The stringent performance requirement for the semi-active suspension system is the simple implementation, not the best isolation. Thus numerous semi-active ON-and-OFF control algorithms have been developed and adopted for semi-active control systems. Such algorithms are robust to modelling and certainties. The skyhook damper control algorithm has been widely adopted as a vehicle suspension system and has demonstrated its improved performance over the passive system when applied to the single degree of freedom system.

Air (or pneumatic) suspension is used in some heavy-duty passenger vehicles. In the air suspension system, suspension springs are replaced by air bags. The air bag is an elastic element, as shown in Fig. 11.9, and consists of a housing, piston group, and a diaphragm with a suitable air inlet. The housing is linked with the spring mass. The housing contains the piston group, which is linked with the spring mass. It also contains a piston groove, which is linked with an unsprung mass. The housing and the piston are connected by the diaphragm. For better sealing and to lessen friction between the members of the elastic element, each air bag is filled with compressed air. A compressor mounted on the vehicle supplies the pressurized air into a reservoir. The pressure in the reservoir is maintained at about 20 Pa. The compressed air in the air bag supports the weight of the vehicle. Whenever the wheel encounters a bump of the road surface, the air in the air bag gets further compressed to absorb the shock caused by vibrations. Air is pumped into the air bag through a series of circuits. In one of the circuits, the pressure of the air is reduced to 2 Pa with the help of a regulator. The pressurized air is then admitted into the air bag through a levelling valve. Whenever there is sufficient air in the air bag, the car will ride low. This state causes the levelling arm to move two linkages. The air inlet valve of the air bag thereby admits more air into it. The other circuit has air at 40 Pa. This air is used to care for the additional loading of the car. A special regulator in this circuit ensures a uniform distance between the wheel and the vehicle body under various loadings. This circuit maintains the car level, irrespective of the load conditions, i.e., irrespective of whether there are passengers in the vehicle or not. The main advantages offered by the pneumatic suspension are the high smoothness of run due to the low rigidity and favourable nature of change in the elastic characteristics, regulation of the suspension rigidity within wide limits, keeping the body at a permanent height under all static loads taken by the elastic elements, high response of the suspension to roughness due to the complete absence of friction in the elastic element, and a larger service life of the vehicle due to improved smoothness of run. However, the pneumatic suspension needs an intricate design and is costly.

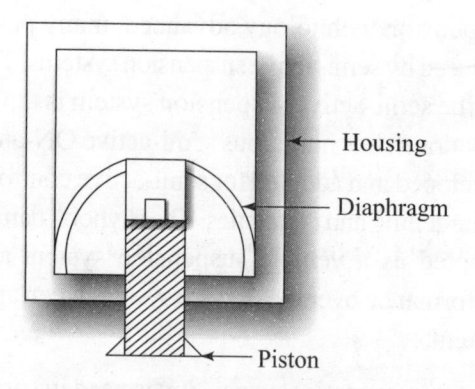

Fig. 11.9 Air suspension system

11.1.8 Car Park Barrier

Consider the cam-operated barrier for a car park. The barrier opens and allows a car in when the correct money is inserted into the collection box. The barrier opens again to allow the car out on its detection on the park side of the barrier. Figure 11.10 shows the type of the wall system that can be used to lift and lower the pivoted barrier. When a current flows through the solenoid of valve A, the piston in the cylinder moves upwards and causes the barrier to rotate about its pivot and raise to let a car through. When the current through the solenoid of valve A ceases, the return spring of the valve results in the valve position changing back to its original position. When the current flows through the solenoid of valve B, pressure is applied to the lower barrier. Limit switches are used to detect whether the barrier is in down or up position.

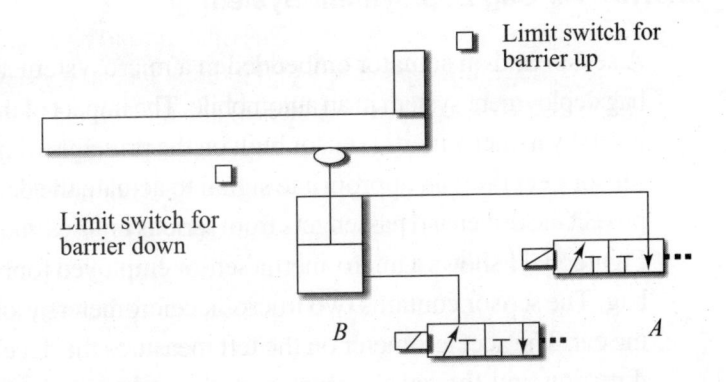

Fig. 11.10 System for raising and lowering a car park barrier

11.1.9 Antilock or Antiskid Device

A vehicle stops more quickly when the brakes are applied just hard enough to get maximum static friction between the tyres and the road. If the brakes are applied harder than this, the tyres will skid or slide on the road and lesser kinetic friction will result. In this situation, applying brakes is less effective. Several devices have been developed to prevent a vehicle from skidding and thus provide maximum effective braking. Skid control is employed generally for the rear wheel only. As long as the wheels are turning/rotating, the antiskid device permits normal application of the brakes. But if the brakes are applied so hard that wheels stop turning, skid starts to develop. At this point, the antiskid device starts operating and partially releases the brakes so that the wheels continue to turn/rotate. However, intermittent braking continues, but it is held to just below the point where a skid would start. The result is maximum braking effect. The distance in which a vehicle can be brought to rest from a steady speed depends upon the following factors:

- Braking efficiency
- Condition and inflation pressure of tyres
- Nature of road surface
- Air resistance encountered by the vehicle

Braking causes a retarding force on the vehicle, which in turn gives rise to deceleration. Braking efficiency is measured in terms of the rate at which it will bring the vehicle to a stationary position from a given speed. It is expressed in terms of the ratio of the deceleration rate to the acceleration rate due to gravity.

11.1.10 Air Bag Deployment System

A sensor and an actuator embedded in a microsystem are used to operate the air bag deployment system in an automobile. The impact of the car in a serious collision is felt by a micro-inertia sensor built on the principle of micro-accelerometer. The sensor generates an appropriate signal to actuate the deployment of an air bag to protect the driver and passengers from serious injuries due to the impact of collision. Figure 11.11 shows a micro-inertia sensor employed for rapid deployment of an air bag. The sensor contains two micro-accelerometers mounted onto the chassis of the car. The accelerometer on the left measures the deceleration in the horizontal direction and the accelerometer on the right measures the deceleration in the transverse direction. Both these accelerometers are mounted on the same integrated circuit chip along with a signal transducer and processing unit. The entire chip has an approximate size of 3 mm × 2 mm, with the micro-accelerometer occupying about 10% of the overall chip area. Accelerometers that can sense in the range of ±2 g are used in car suspension systems and antilock braking systems. Accelerometers in the sensing range of ±50 g are used to actuate the air bag deployment. In the case of micro-accelerometer, significant different arrangements are necessary because very limited space is available in miniature devices. A miniature silicon beam with attached mass constitutes a spring-mass system and the air in the surrounding space is used to produce the damping effect. The structure that supports the mass acts as a spring.

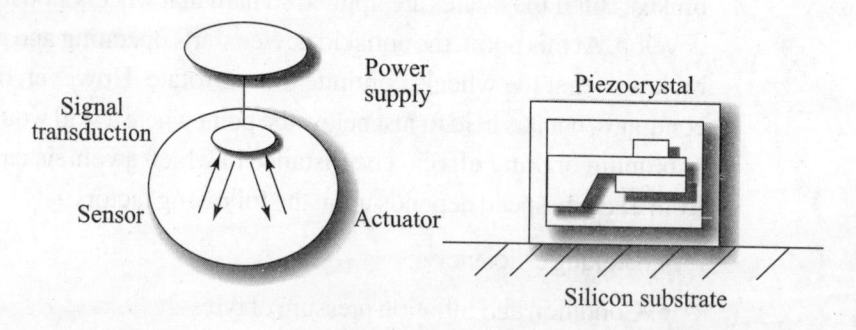

Fig. 11.11 Micro-inertia sensor air bag deployment

11.2 Bionics

Bionics is poised to have significant stake in mechatronic sensors market in the near future. Biomedical sensors are mainly used for diagnostic analyses. Because of its miniature size, a biomedical sensor requires less amount of sample and can produce results significantly faster. These sensors can be produced in batches, thus resulting low unit cost of the sensor. Another cost cutting factor is that most of these sensors are disposable, thus manual labour involving cleaning and proper treatment for reuse is saved. Biosensors are extensively used in analytical chemistry and biomedical care as well as genetic engineering. These sensors usually involve biological molecules such as antibodies or enzymes, which interact with analytes that are to be detected. Major advantages of the use of mechatronic systems in biomedicine are as follows:

- Functionality for biomedical operators
- Adaptability to existing instruments and equipment
- Compatibility with biological systems
- Controllability, mobility, and easy navigation facilities for operators
- Possibility of the fabrication of mechatronic structures with a high aspect ratio (the ratio of the depth dimension to the surface dimension of the structure)

11.2.1 Biological Systems

Of the 91 naturally occurring elements, many are found in biological systems. Traces of metals such as zinc, iron, vanadium, manganese, selenium, and copper are used for specific biological functions in human body. Well over 95% of the mass of plants and animals is made of four elements: hydrogen, oxygen, nitrogen, and carbon. These are also the elements that dominate in most of the synthetic polymers. These elements can combine (bond) in various ways to give rise to a number of simple and/ or complex biological structures. The nature has used this feature of these elements to build various kinds of biological systems to accomplish the various life processes. Scientists, on similar lines, have used them to make new materials.

The numerous molecules in our own body are responsible for respiration, digestion, temperature regulation, protection, and all other functions that our body carries out. The molecules found in nature are complex. For these molecules to perform useful functions, they must be easy to assemble, recognize, and bind to other molecules. They must also be products of biological processes and have variable properties.

There are four broad classes of biological molecules. The first three are nucleic acids, proteins, and carbohydrates, which are all polymer structures. The fourth is the catalytical category (enzymes) composed of particularly small molecules that perform special tasks.

Proteins make up much of the bulk of biological mass. Our nails are mostly made up of the protein keratin. Oxygen is carried in our blood by the protein haemoglobin and the protein nitrogen is responsible for taking nitrogen out of the air. There are thousands of proteins, most of which are very well understood in terms of their structures and functions. However, some proteins are still a mystery as far as their structures and functions are concerned. Proteins are machines of biology, the functional agents that make things happen.

Nucleic acids are of two types: ribose nucleic acid (RNA) and de-oxyribose nucleic acid (DNA). Though both RNA and DNA are needed to make proteins, the former (i.e., RNA) has not yet been of major interest in developing nanostructures.

The DNA consists of a sugar and side chain containing a negative charge due to the presence of phosphorus and oxygen atoms in it. DNA molecules are stacked planar molecules, which are on top of one another like a pile of poker chips. Each of the stacked chips consists of two separate planar molecules held together weakly by bridges between oxygen and nitrogen or hydrogen. Because each chip is held at both its right and left ends and because the structure is helical, the DNA has the structure of a double helix or double spiral staircase. It also looks like a spring when it is tightly wound.

The DNA is an almost unique molecule because each chip can have one of the four compositions, namely, AT, TA, CG, or GC. For each position on the strand, it is possible to control which of the base pairs is present because the two planer molecules that compose them can only be chosen from a set of four molecules adenine, thiamine, guanine, and cytosine (which are abbreviated A, T, G, and C). A and T will only bond to each other and not to G and C. Similarly, G and C will only bond to each other and not to A and T. Because of this limitation, the only possible base pairs are AT and GC and their opposites TA and CG. These are placed on a double helix in a particular order and it is the code for all the functions of an organism. The genetic code is simply an arrangement of base pairs in the DNA double helix. This code is read in a very sophisticated way by RNA and by proteins, which use the information to make protein-based biological systems necessary for carrying out biological process of life.

Another class of macromolecules is called polysaccharides. Polysaccharides are simply sugars made of very large molecules. They are crucial to the functioning of a cell. Some of these are formed in ligaments and in other biological structural materials. However, these are not yet of major use in the synthetic material technology.

The fourth class of biological molecules consists of very small molecules. These molecules include water and oxygen as major energy sources. Carbon dioxide is the raw material for making plants. Nitric oxide is a very small molecule consisting of nitrogen and oxygen bonded together. It plays many roles in biology from acting as a secondary messenger for communication or relay messenger for communication within in a cell to causing erectile function.

There are some other molecules that are infrequent and small but still crucial in biological systems. These include simple sugar and all drug molecules. A drug generally works by binding itself either to a protein or to a DNA and causing changes in the functions of the host structure/cell. Often the binding of these small molecules is very specific and very important. The ability of one molecule to attract and bind onto another molecule is often referred to as molecular recognition. Molecular recognition is generally very specific. Thus, molecular recognition can be used for sensory applications involving biological matter. Molecular recognition is one of the key features of nanotechnology. Much of nanotechnology depends on building from bottom-up, making molecules that can organize themselves on their own or with a supporting surface such as a metal or plastic.

11.2.2 Biostructure

Sam Stupp, Head of the Institute of Northwestern, conducted research on human repair, which involved utilization of self-assembly and nanostructures to repair damaged body parts or organs. His research was focused on nanoscale biostructures that can mimic or effect a biological process. One example of nanoscale biostructures is the self-assembling artificial bone. The molecules that make up the bone are held together by chemical bonds. These bonds hold the molecules together with each other in a particular shape. The molecules in the bone are designed to occupy space in a particular way so that they may assemble spontaneously to form a desired shape. Once assembled, the molecules are packed closely enough for the bone to be very strong. The structure of packed molecules can be made compatible with the human immune system by properly choosing the head groups of the molecules. A group of molecules ultimately forms the outer shell of the artificial bone template. The outer shell is designed so that natural bones begin to form around it like corals on a reef or gold on a plated jewellery. The key to human repair is by allowing the body to fix damaged tissue naturally rather than replacing with steel or ceramic implants.

11.2.3 Glucose Detection and DNA Sensing

Detection of glucose levels in human body is a classic case of biosensing. Diabetic patients cannot control their insulin level if the level of blood glucose fluctuates tremendously. If the level gets either too high or too low, their condition can be life threatening. Currently such patients must actually draw blood on a daily basis or even more often to monitor the blood glucose level. Sensing the blood glucose level can be done in many ways, using optical, conduction, or molecular recognition methods. None of these have yet been shown to be compatible with an implantable simple device that could automatically show or continuously sense the glucose level in the blood. This remains one of the major challenges in chemical sensing and nanoscale structures.

The DNA sensing is potentially an enormous area in which the application of nanoscience can prove to be path breaking. One can sense the structure with the sequence GEGEAAG by using a strand GCGCAAG. This means that a single strand of, say, six bases can contain 4096 different combinations. Consequently, a particular biological target such as botulism or strap or scarlet always has a unique DNA sequence. It is possible to target a short sector of the DNA sequence, say, a section of 10–15 bases. It can be uniquely sensed without any error.

The most important application of DNA sensing will probably come in the generalization of a laboratory on a chip concept. By using the powerful analytic capability of such dense micro-laboratories, it will be possible to include several screening sensors on a chip. This chip can be used to recognize a viral or bacterial DNA associated with several different diseases found in the body. This chip could also be used to sense the presence of toxic species, either natural or artificial. Since the entire human genome is known, a biochip can be used to sense either a particular DNA signature or a particular protein signature known to be characteristic of a disease. It is also possible to create a sensor that takes advantage of the DNA recognition technique. The simplest DNA recognition sensor works by introducing a strand of DNA complimentary to the analyte into a solution to be tested. If the analyte is present, it will hybridize with the test DNA and form a double strand. Hybridization confirms that the analyte is present, or finding out that hybridization has occurred is trivial. One cannot see the double strands without very sophisticated instruments. Therefore, one of the great challenges in DNA sensing is to amplify the effect of hybridization so that it is easy for measurement. One way to provide amplification is to change the optical properties of gold or silver nanodots that are attached to the DNA. The change in the colour of gold upon changing the size of the gold cluster and the molecular recognition is called quantum optical effect. The colour change is measured by a device called calorimetric sensor, which can be read by simply looking at it. Nanosphere lithography is used to prepare the tiny gold dots on a surface. A sensor is designed to recognize a particular portion of the analyte appearing in the solution. If one wants to construct an explosive detection sensor, the problem is much more complex. Nitrates, which are common to most explosives, are common in household items including fertilizers. If one detects them to an accuracy of a single molecule, then even fertilizers are carrying a bomb. A great deal of research is underway in this direction.

11.2.4 Drug Delivery

The size of the human body is very large compared to the size of a molecule. It is important for the thermofusion effectiveness that drug molecules find/reach the place in the body where they are needed/effective. Bio-availability refers to the presence of drug molecules where they are needed in the body and where they will do the most good. The issue of drug delivery aims at maximizing bio-availability both over a period of time and at the specific place in the body. Increasing the bio-availability is seldom as simple as increasing the amount of drug used. The drugs used in chemotherapy are actually somewhat toxic and need to be target-specific to avoid damage to the normal/healthy tissue. It is necessary to keep the drug doses to a minimum, otherwise the amount used can adversely affect or even kill a patient. Taking these issues into account, drug delivery assumes a lot of significance. Nanotechnology and nanoscience are very useful in developing entirely new ways for increasing bio-availability and improving the drug delivery.

Magnetic nanoparticles used for computer memory can be used for drug delivery also. For drug delivery, the molecular recognition method is used to bind a nanomagnet to the drug to be delivered. External control is exercised over the magnetic field created by magnetic nanoparticles to improve local bio-availability of the drug. Effectively, a doctor can drag drug molecules through the body in the same way as you drag an iron filing across a table with a hand magnet. One interesting combination of smart materials and drug delivery is the triggered response. This consists of placing drug molecules within the body in an inactive form that works upon encountering a particular signal. A simple example is antacid implored in a coating of a polymer that dissolves in a highly acidic spot. The antacid is released only when the outer polymer coat encounters a highly acidic spot in the digestive track.

11.2.5 Photodynamic Therapy

In photodynamic therapy, a particle is placed within the patient's body. This particle is illuminated with a light source from outside of the body. The light may come from outside from a laser or light bulb. The light is absorbed by the particle, after which several things might happen. If the particle is simply a metal nanodot, the energy from the light will heat the dot, which, in turn, will heat any tissue within its neighbourhood. With the same particular molecular dot, light can also be used to produce highly energetic oxygen molecules. Such oxygen molecules are very reactive and will chemically react with (and, therefore, destroy) many organic molecules that are next to them.

The photodynamic therapy is attractive for many reasons. One reason is that, unlike the traditional chemotherapy, it is directed at the damaged/diseased cell. The chemically reactive excited oxygen or quantum data is released only where such cells are present and where the light is illuminated. This ensures that, unlike the traditional chemotherapy, the photodynamic therapy does not leave a fixed trail of highly aggressive and reactive molecules throughout the body.

11.2.6 Molecular Motors

Molecular motors are proteins or protein complexes that transform chemical energy into mechanical energy at a molecular level scale. These are responsible for producing and transducing the energy stored in ATP, which is the common energy currency of the body for powering every process and action. The sodium/potassium, called ATP molecule, actually acts as a rotary motor. A critical amount of the nanostructure rotates around a pivot and the outside part of the nanostructure reacts differently with the chemical group around the periphery. This rotary motor is one of the several molecular motor mechanisms that are now understood to play important roles in the functional biology of a cell. Molecular motors also allow us to manage the availability of the different components of the cell as they move about within the cellular structure. Kinesin is a nanoscale molecular motor that carries molecular cargo through the cell by moving along nanoscale tracks, called microtubes, within the cell. In many respects, it is the "world's smallest type of train". Molecular motion is also responsible for signal transduction in the human ear. Molecular motion may provide acceleration and motional energy for both artificial nanostructures within the body and in more complex nanostructure assemblies.

11.2.7 Neuro-electronic Interface

The neuro-electronic interface involves the idea of constructing nanodevices that can permit computers to be joined and linked to the neurosystem. The construction of a neuro-electronic interface system requires the building of a molecular structure that will permit control and detection of nerve impulses by an external computer. The real challenge is to combine computational technology and bio-nanotechnology. The nerves in the human body convey messages by permitting electrical current to flow between the brain and nerve centre throughout the body. The most important ions for signals are sodium and potassium ions. These ions move along sheets and channels that have evolved specially to permit special, controllable, rapid ion motion. This is the mechanism that allows you to feel sensation. For example, when you put your foot in hot water, a signal is transmitted by the local nerve through the nervous system to the brain. The brain interprets the transmitted signal and processes it for a suitable reaction. Often this process results in a response being filtered into the muscular system. The aim of the neuro-electronic interface technology is to permit the registration and interpretation of these signals as well as response to them to be handled by a computer. The sensor must be able to sense ionic currents and cause current to flow backward so that the muscular system can be instructed to perform a desired motion. The most obvious structure will be a molecular conductor or molecule whose own conduction process, ions, or electrons can link with the ionic motion in a nerve fibre.

11.2.8 Electrochemical Sensor

The electrochemical sensor works on the principle that certain biological substances, e.g., glucose in human blood, can release certain element(s) by chemical reactions. These substances can alter the pulse flow pattern in a sensor, which can be readily detected. Figure 11.12 illustrates the working principle of a biomedical glucose detector. A small sample of blood is introduced to a sensor with a polyvinyl alcohol solution. Two electrodes are present in the sensor. A thin sheet of platinum and a thin Ag/AgCl film are used as the reference electrodes. The following chemical reaction takes place between the glucose in the blood sample and the oxygen in the polyvinyl alcohol solution:

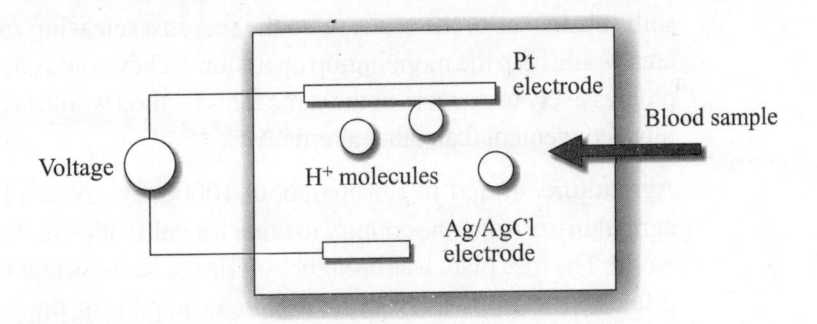

Fig. 11.12 Biomedical glucose detector

$$Glucose + O_2 = gluconolacton + H_2O_2$$

The H_2O_2 produced in the chemical reaction is electrolysed by applying a potential to the platinum electrode. This results in the production of positive hydrogen ions, which flow towards the electrodes. The glucose concentration in the blood sample is thus measured by measuring the current flow between the electrodes.

11.2.9 Biotechnology

Mechatronics plays an important role in biotechnology even though it is a small subdomain of biotechnology. Biotechnology includes all techniques that use living organisms or substances obtained from them to make or modify a product. It involves improvement of microbe, plant, and animal species. Genes and gene products are the basic tools in biotechnology. Biotechnology aims at harnessing the genetic diversity in the living organisms for the benefit of the humankind. Understanding of genes and the possibility to manipulate them are the very bases of modern biotechnology. Classical genes and manipulation of the genes at cellular level have played a major role in enhancing the productivity of crops, plants, and animals. Molecular manipulation of genes to obtain better products requires input from specialists in many different areas of biology, besides from specialist in other branches of science, especially mechatronics.

The splendors of biotechnology are so much that it almost seems to be the science of wish fulfillment. Bigger fruits, bright flowers, higher yields, super cattle, exotic colours and flavours, cheaper medicines, and more efficient vaccines are the products of biotechnology.

The first studies of genetic engineering have been reported on microbes. Bacteria in the soil that digest animal waste or even burial carcasses are an example of genetic engineering involving microbes. Imagine bacteria that would be able to digest petrol or crude oil. Such bacteria could be used for clean-up operations when oil tankers spill substances of the crude oil in the sea. Just releasing such bugs at the spillage area would help the mopping-up operations. They could eat off the oil and multiply progressively to do more of the same. This method would be much cheaper, fair, and more convenient than other alternatives.

Agriculture started in Europe about 10000 BC. New plants were brought by agriculturists from one country to other for cultivation and consumption across the world. The rose plant was brought by a British ambassador to the Moghul court and gifted to the empress Noorjahan. She arranged for the propagation of the rose flower in India some 400 years ago. Many of the vegetables that we eat in India were imported from other countries. For example, potato is a native of South America. It was exported to other countries of the world from there. Potato was brought from South America to Europe in 1570 and to India around the eighteenth century. People started cultivating potatoes in India from early eighteenth century onwards.

The choice of potato seeds for planting is largely influenced by its size, shape, colour, and tastes, skin appearances, storing capabilities, and processing properties. A potato has to be high in solid or dry matter in order to be considered suitable for making good potato chips. Researchers at Monsanto Company in the USA have inserted a starch-producing gene from the common intestinal bacterium that can make potato 20% more starch rich than the best potato in the market. Genetic engineering has yielded superpotatoes. Calgene Pacific, a Melbourne based biotechnology company, has announced that it has isolated a gene that is capable of providing twice the normal number of tubers. This new variety of potato is termed superpotato. This has helped growers to double potato yield per hectare. The company also claims that cultivation of superpotato provides more nutrition from less land and in less time than rice, wheat, or corn. Potatoes are cultivated in 130 countries in the world. An increase in the yield of potato has considerable impact on the potato production the world over. Biotechnology facilitates incorporating borrow genes from other organisms into a target species. The greater wax moth has a protein that stores large amounts of tyrosine. This protein is the product of a moth gene. If this gene is incorporated into potato, the gene might change tyrosine location and form with tuber cells. That might make tyrosine less accessible to the blacking enzyme in the potato.

11.3 Avionics

Considerable effort and progress have been made in recent years in the development of mechatronic systems in the aerospace industry. Numerous and complex mechatronic systems are used in advanced commercial and military aircrafts. With the ever-increasing emphasis on robustness and safety, there is a trend towards using more mechatronic systems in aerospace industry. This has given rise to a new area of mechatronics in the form of avionics. The major applications of mechatronic systems in aerospace industry can be classified as follows:

- Cockpit instrumentation
- Safety devices
- Wind tunnel instrumentation
- Sensors for fuel efficiency and safety
- Microgyroscope for navigation and stability
- Microsatellites

11.3.1 Cockpit Instrumentation

Air-data systems vary in complexity from a light airplane to advanced commercial or military aircrafts. However, all air-data computations are based upon four sensed parameters, namely, static pressure, total pressure, temperature, and the angle of attack. Pressure and temperature of the atmosphere are functions of the altitude above the sea level, latitude, season, and time of day. Static pressure can be easily and accurately measured because the actual altitude for a given pressure varies only slowly with respect to time and distance.

A Pitot tube is a pressure measuring instrument used to measure fluid flow velocity and, more specifically, to determine the airspeed of an aircraft. The Pitot tube is named after its inventor, Italian born French engineer Henri Pitot. The opening on the smooth side of the Pitot-static tube provides a source of the atmospheric pressure, termed static pressure. The open end of the Pitot-static tube is headed into the stream and provides a source of the total pressure resulting from the impact of a body travelling through the atmosphere. The difference between the total pressure and the static pressure gives the dynamic force, which contains the velocity term. The speed of the aircraft can thus be determined. Most of the Pitot-static tubes are electrically heated to melt off any ice that might form. Otherwise, ice might partly or completely seal off the opening and the instrument will give erroneous results.

Most of the aircraft altimeters operate like the aneroid barometer, except that altimeters are calibrated in height instead of pressure. Static pressure surrounding one or more evacuated diaphragm capsules operates a pointer through a system of linkages and gears. Corrections must be made for changes in the barometer pressure. A sensitive altimeter shows changes in pressure for an altitude of 2 m or less. Since atmospheric pressure reduces to half for approximately every 6000 m increase in altitude, non-linear diaphragm linkage systems are used to provide a complete altitude range. Pressure altitude information may be supplied by air-data computers to various forms of indicator and display systems.

Air-speed indicators work on the principle that the difference between the total pressure and the static pressure is the measure of the indicated air speed. The indicated air speed at which a given aircraft with a given load stalls is a constant over a wide range of pressure and temperature. The knowledge of this is of utmost importance during take-off and landing as also during flight. The true air speed is primarily useful in navigation. This factor is a function of not only the difference between the total and static pressures but also temperature. A temperature indicator is in effect a computer that operates on a complex relationship of two pressure measurements and the air temperature measurement to give the correct indication.

A rate climb indicator comprises an enclosed volume of air connected to the atmospheric pressure through constriction. As the altitude changes, the enclosed pressure lags the outside pressure. The pressure difference is measured in terms of the rate of change of altitude or the rate of climb. Since any change in temperature causes a proportional change in pressure of an enclosed volume of the air, the instrument enclosure must be a good thermal insulator to prevent all but very gradual temperature changes of the enclosed air. Otherwise, correction must be introduced for both static pressure and temperature to provide the correct rate of climb information.

The angle of attack transducer measures the acute angle between the velocity vector of the aircraft through the surrounding air and some reference such as the force-opt axis of the aircraft or the chord of an aircraft. The lift developed by an aircraft wing increases as a function of the angle of attack and the airplane velocity.

Velocity information is readily available on all aircrafts, but the angle of attack information is not. A knowledge of both the factors enables the pilot to fly the airplane more economically and more safely. The output of an angle of attack transducer may supply information to the ground system, speed control system, and the throttle system to optimize airplane performance during the maneouvres. In addition, the angle of attack transducer provides essential information to the stall warning system, which in some cases, will actually push strictly forward if the sensor indicates an angle of attack approaching the stall. The pressure difference over a cylinder to align the cylinder with the air stream is used as for the angle of attack measurement. Two rows of static pressure ports, at $45°$ on either side of the stagnation line at null, detect the pressure difference due to nonalignment. The pressure difference due to nonalignment is converted into an electrical error signal. The error is amplified. It then allows a motor to drive, which serves as the cylindrical probe into the wind. The amount of probe rotating is a measure of the angle of attack.

The ice detector utilizes the absorption properties of ice when subjected to beta-particle's radiation to indicate the presence of ice. An airfoil-shaped probe containing a radioactive source, a Geiger Mueller tube, and a heater are positioned in the air stream. An amplifier detector, which is packaged with the probe or placed in a remote location dependent upon specific application, receives ice information from the probe and causes the probe heater to turn on when ice acceleration is greater than 0.375 mm. The rate at which the heater turns on and off can be used as a measure of the severity of ice conditions. Beta-particles emitted by strontium-90 are directed in a collimated beam along the leading edge of the sensing portion of the airfoil toward the Geiger Mueller tube. As the ice accumulates, the beta-particle pulse rate falls. When the rate falls below 10 pulses/sec, the amplifier detector energizes a relay that turns the heater on. The heater is turned off when the pulse rate exceeds 40 pulses/sec. There is a visual indicator to indicate presence of ice.

The air-data computers centralize the computation of air data from a number of inputs, e.g., static pressure, total pressure, air stagnation temperature, and angle of attack. This information is used to ascertain the true air speed, temperature, Mach number, air density, and rate of climb. The information is then transmitted to indicators and displays and is used by various aircraft subsystems such as the flight control system, fire control system, and navigation system.

11.3.2 Alarms and Safety Devices

An alarm is a protective device for maintaining critical points in a system under constant surveillance. In basic principles, alarm differs little from other forms of indication and control instruments. Strong emphasis is placed on reliable performance of alarms. Such devices must remain operable over long periods of inactivity and need periodic, scheduled check, because the equipment does not function except in case of emergency. Most alarm installations are of audiovisual type, with illuminated name plates or bull-eye lights. There is a means to silence the alarm and switch the light to steady state. Another push button usually is provided for testing other components of the system.

Micro-accelerometers or micro-inertial sensors are used to eject the pilot seat from the aircraft, especially in military airplanes. The use of the seat eject system protects the pilot in the event of an emergency. The sensor opens the latch where the seat is loaded with a spring, which throws the seat along with the pilot a distance away from the aircraft. When it is detached from the aircraft, a parachute automatically opens to ensure safe landing of the pilot.

11.3.3 Aircraft Guidance and Control

To guide and control an aerospace vehicle successfully is a matter of measuring position, determining path errors, and controlling to correct the path. These three functions usually are called navigation, guidance, steering, and control. Navigation determines the position, guidance determines the error from the indicated path, steering is used to select a proper series of changes in the path, and control changes the forces on the vehicle to adjust them with the direct path changes. Initial guidance predicts the ballistic path. The aircraft guiding system measures accelerator and recalls the reference angle, computes velocity and position, predicts the destination, determines a preferred path to correct the error and controls the forces to change the path from steering instructions, all can be automated. Aerospace vehicle systems have many degrees of automation. Almost all aerospace vehicle systems use combinations of all or many of automatic, manual, remote, self-contained, and preprogrammed corrections.

Aerospace vehicles may be piloted by a human being or an automatic flight control system. Gyroscope instruments are among the most important elements of the flight instrumentation and control systems whether for assisting a human pilot or for providing input to a fully automatic flight control system. A wide variety of gyroscope instruments are used in aircrafts. The rate gyroscope is a device designed to measure the instantaneous angular velocity component of a body with respect to the inertial space. Its typical applications include autopilot damping, rate of turn indicator, limiting antenna stabilization, and telemetry instrumentation. The rate integrating gyroscope or floated gyroscope can be designed as either a single- or two-axis device. It is designed so that the fluid damping between gimbals and housing is the predominant torque, which balances the input rate precessional torque. Figure 11.13 illustrates the working principle of a gyroscope.

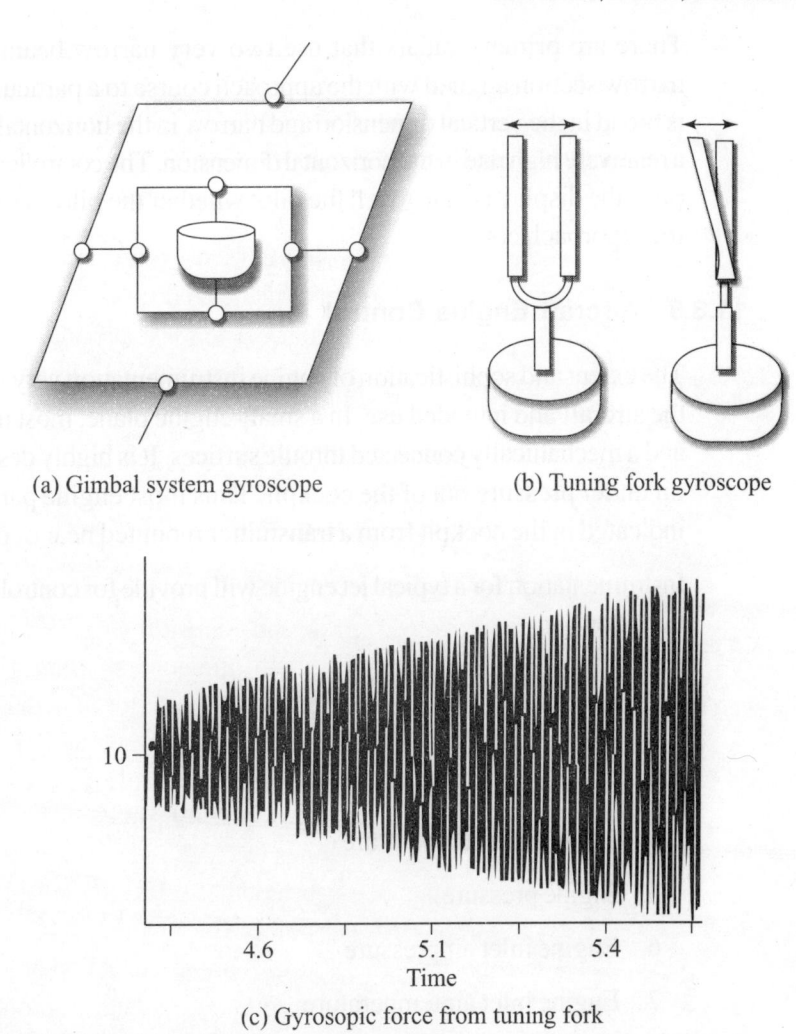

(a) Gimbal system gyroscope (b) Tuning fork gyroscope

(c) Gyrosopic force from tuning fork

Fig. 11.13 Working principle of gyroscope

11.3.4 Air Traffic Control

Position reporting by the pilot to the air traffic controller over a voice radio link is the basic source of air traffic control position data. The ground controller can ascertain the aircraft location independently using the following methods.

Method 1 It can use the primary radar, which operates on the reflection by the aircraft of the pulse signals that the radar transmits.

Method 2 It can use the secondary radar, which operates on replies from pulsed radio and is verified by the secondary radar pulses. Some airports are equipped for precision approach radius.

There are primary radars that use two very narrow beams to scan a relatively narrow section aligned with the approach course to a particular runway. One beam is broad in the vertical dimension and narrow in the horizontal dimension. It scans at a relatively high rate in the horizontal dimension. The controller watching the cathode-ray-tube display is able to tell the pilot whether the pilot is to the right or left of the true approach course.

11.3.5 Aircraft Engine Control

The extent and sophistication of engine instrumentation vary widely with the type of the aircraft and intended use. In a small-engine plane, most instruments are simple and a mechanically connected throttle suffices. It is highly desirable to keep fuel and oil under pressure out of the cockpit. Thus most engine parameters are remotely indicated in the cockpit from a transmitter mounted near or on the engine.

Instrumentation for a typical jet engine will provide for controlling and monitoring of the following:

1. Low-pressure rotor speed
2. High-compressor rotor speed
3. Fuel flow
4. Exhaust gas temperature
5. Engine pressure
6. Engine inlet air pressure
7. Engine inlet air temperature
8. Fuel pump inlet temperature
9. Fuel decreasing air shut off valve position
10. Fuel pump inlet pressure
11. Fuel filter pressure difference warning
12. Engine oil pressure
13. Engine oil and inlet temperature
14. Engine radial vibration

A modern two-spool turbojet engine uses an advanced control system to achieve robustness and avoid the deterioration of the output due to variation in input. Robustness of a controller gives good performance irrespective of the flight parameters such as flying at different speeds and altitudes. A jet engine receives three inputs, namely, fuel flow, variable nozzle area, and inlet compressor vane angle variation. Five outputs of the jet engine are required for controlling the jet. These are

- low-pressure turbine speed,
- high-pressure turbine speed,
- combustion chamber pressure,
- combustion chamber temperature, and
- exit input temperature of high-speed turbine.

It is noted that the low-pressure turbine work is used to run the low-pressure compressor and the high-pressure turbine is used to run the high-pressure compressor. The primary aim of the control system is to control the engine through regulating the compressor surge margin. Since aero-engines are highly non-linear system, it is normal for several controllers to be designed at different operating points and scheduled across the flight envelop. Also in an aero-engine, there are a number of parameters, apart from the ones being primarily controlled, that are to be kept within specified safety limits, for example, the turbine blade temperature.

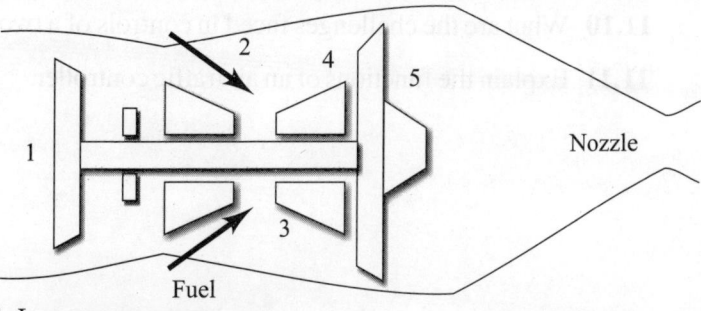

1: Low-pressure compressor
2: High-pressure compressor
3: Combustion chamber
4: High-pressure turbine
5: Low-pressure turbine

Fig. 11.14 Schematic diagram of a high-performance gas turbine engine

The number of parameters to be controlled and/or limited exceeds the number of controllable inputs. All the parameters cannot be controlled independently at the same time. This problem can be tackled by designing a number of scheduled controllers, each for a different set of output variables. For making aircraft, a robust multivariable controller is used to provide satisfactory performance over the entire operation range of the engine. Figure 11.14 shows the schematic diagram of a high-performance gas turbine engine used for modern military aircraft. The engine has two compressors and two turbines. The efficiency of the engine and the thrust produced depend on the pressure ratio generated by the two compressors.

Exercises

11.1 Identify the different mechatronic systems used in automobiles.

11.2 How are automobile engines managed using mechatronic systems?

11.3 Differentiate between passive, semi-active, and active vibration-suppressing systems used in vehicles.

11.4 What is a nanostructure? Give examples.

11.5 What is a molecular motor?

11.6 What is a neuro-electronic interface? What is its future potential?

11.7 Explain the principle of photodynamics.

11.8 What is hybridization?

11.9 Enumerate different mechatronic system instruments used in aircrafts.

11.10 What are the challenges faced in controls of a two-shaft turbojet engine?

11.11 Explain the functions of an air traffic controller.

Appendix A

Laplace Transform

	Time domain function	Laplace transform
1	$d(t)$, unit pulse input	1
2	$x(t)$, unit step input	$\dfrac{1}{s}$
3	$tx(t)$, unit ramp input	$\dfrac{1}{s^2}$
4	$t^2 x(t)$, unit parabolic input	$\dfrac{1}{s^3}$
5	$t^n x(t)$, $n = 1, 2, 3, 4, \ldots$	$s^{n!/(n+1)}$
6	$\dfrac{1}{\sqrt{\pi t}}$	$\dfrac{1}{\sqrt{s}}$
7	$2\sqrt{\dfrac{t}{x(t)^2}}$	$\dfrac{1}{\sqrt{s^3}}$
8	e^{-at}	$\dfrac{1}{s+a}$
9	be^{-at}	$\dfrac{b}{s+a}$
10	te^{-at}	$\dfrac{1}{(s+a)^2}$
11	$t^w e^{-at}$	$\dfrac{n}{(s+a)^{n+1}}$
12	$\dfrac{1}{a}(1 - e^{-at})$	$\dfrac{1}{s(s+a)}$

13	$e^{-at} - e^{-bt}$	$\dfrac{b-a}{(s+a)(s+b)}$
14	$\dfrac{1}{ab} - \dfrac{e^{-at}}{a(b-a)} - \dfrac{e^{-bt}}{a(a-b)}$	$\dfrac{1}{s(s+a)(s+b)}$
15	$(1-at)e^{-at}$	$\dfrac{s}{(s+a)^2}$
16	$\sin(w_n t)$	$\dfrac{\omega_0}{s^2 + \omega_n^2}$
17	$\sin(w_n t + q)$	$\dfrac{s\sin\theta + \omega_n \cos\theta}{s^2 + \omega_n^2}$
18	$\cos(w_n t)$	$\dfrac{s}{s^2 + \omega_n^2}$
19	$e^{-at}\sin(w_n t)$	$\dfrac{\omega_n}{(s+a)^2 + \omega_n^2}$
20	$e^{-at}\cos(w_n t)$	$\dfrac{s+a}{(s+a)^2 + \omega_n^2}$
21	$1 - \dfrac{1}{\sqrt{1-\zeta^2}} e^{-\zeta\omega_n t} \sin\left[\omega_n\sqrt{1-\zeta^2}\,t + \cos^{-1}\zeta\right]$	$\dfrac{\omega_n^2}{s(s^2 + 2\zeta\omega_n s + \omega_n^2)}$
22	$\dfrac{\omega_n}{\sqrt{1-\zeta^2}} e^{-\zeta\omega_n t} \sin\left[\omega_n\sqrt{1-\zeta^2}\,t\right]$	$\dfrac{\omega_n^2}{s^2 + 2\zeta\omega_n s + \omega_n^2}$
23	$1 - \dfrac{\omega_n^2}{\sqrt{1-\zeta^2}} e^{-\zeta\omega_n t} \sin\left[\omega_n\sqrt{1-\zeta^2}\,t + \cos^{-1}\zeta\right]$	$\dfrac{s\omega_n^2}{s^2 + 2\zeta\omega_n s + \omega_n^2}$

- a, b are constants.
- ω_n is the natural frequency of the system.
- ζ is the damping ratio.

Appendix B

Laplace and Z Transforms

$f(t)$	$F(s)$	$f(kT)$	$F(z)$
$\delta(t)$, unit pulse input	1	$\delta(kT)$, unit pulse train input	1
$x(t)$, unit step input	$\dfrac{1}{s}$	$u(kT)$, unit step pulse train input	$\dfrac{z}{z-1}$
$tx(t)$, ramp input	$\dfrac{1}{s^2}$	$kT\,x(kT)$,	$\dfrac{Tz}{(z-1)^2}$
$e^{-at}x(t)$	$\dfrac{1}{s+a}$	$(e^{-a})^{kT} x(kT) = c^k x(kT)$, where $c = e^{-at}$	$\dfrac{z}{z-e^{aT}} = \dfrac{z}{z-c}$
$te^{-at}x(t)$	$\dfrac{1}{(s+a)^2}$	$kT(e^{-a})^{kT} x(kT) = kTc^k x(kT)$	$\dfrac{Tze^{-aT}}{(z-e^{-aT})^2} = \dfrac{Tcz}{(z-c)^2}$
$\sin(bt)x(t)$	$\dfrac{b}{s^2+b^2}$	$\sin(bkT)\,x(kT)$	$\dfrac{z\sin(bT)}{z^2 - 2z\cos(bT) + 1}$
$\cos(bt)x(t)$	$\dfrac{s}{s^2+b^2}$	$\cos(bkT)\,x(kT)$	$\dfrac{z\{z-\cos(bt)\}}{z^2 - 2z\cos(bT) + 1}$
$e^{-at}\sin(bt)x(t)$	$\dfrac{b}{(s+a)^2+b^2}$	$(e^{-a})^{kT}\sin(bkT)\,x(kT)$ $= c^k \sin(bkT)\,x(kT)$	$\dfrac{z\{e^{-aT}\sin(bT)\}}{\{z - e^{(-a+jb)T}\}\{z - e^{(-a-jb)T}\}}$ $= \dfrac{zc\sin(bT)}{z^2 - \{2c\cos(bT)\}z + c^2}$

- a, b are constants.
- T is the sampling period.
- $k = 0, 1, 2, 3, 4, \ldots$

Bibliography

Textbooks

Aditya, P. Mathur (1987), *Introduction to Microprocessor*, Tata McGraw Hill, New Delhi.

Bayle, H.P. (1997), *Transducer Handbook*, Newman Publication, USA.

Bolton, W. (1996), *Instrumentation and Measurement*, Newman Publication, USA.

Bolton, W. (1997), *Pneumatic and Hydraulic Systems*, Butterworth Heinemann, London.

Bolton, W. (2003), *Mechatronics*, Pearson Education, India.

Bolton, W. (2003), *Programmable Logic Controller*, Newman Publication, USA.

Brodley, B.A., et al. (1991), *Mechatronics: Electronics in Products and Processes*, Chapman and Hall, London.

Buerk, D.G. (1993), *Biosensor: Theory and Applications*, Technomic Publication, USA.

Chiltan, J.A. and M.T. Goosy (1995), *Special Polymers for Electronics and Optoelectronics*, Chapman and Hall, London

Chong, E.K.P. and S.H. Zak (1996), *An Introduction to Optimization*, John Wiley & Sons, USA.

Colleib, I.M. (1974), *Electronics Motors and Control Techniques*, TAB Books, McGraw Hill, New York.

Croser, P. (1989), *Pneumatics: Basic Level*, Festo Didactic, Germany.

Dauglas, M. Considine (1991), *Encyclopedia of Instrumental Control*, McGraw Hill, New York.

David, D. Gold, Berg (2000), *Genetic Algorithms*, Pearson Education Asia, India.

De Silva, C.W. (1989), *Control Sensors and Actuators*, Prentice Hall Inc., Englewood Cliffs, New Jersey.

Del Toro, V. (1986), *Electrical Engineering Fundamentals*, Prentice Hall Inc., Englewood Cliffs, New Jersey.

Denny, K. Mia (1993), *Mechatronics*, Springer Verlag, New York.

Denton, T. (1995), *Automobile Electrical and Electronics Systems*, Arnold Publication, London.

Doebelin, E. (1998), *System Dynamics*, Marcell Dekkar Inc., New York.

Douglas, V. Hall (1987), *Microprocessor and Interfacing, Programming and Hardware*, Tata McGraw Hill, New Delhi.

Dubey, G.K. (1995), *Fundamentals of Electrical Drives*, Narosa Publishing House, New Delhi.

Groover, M.P. et al. (1998), *Industrial Robotics Technology: Programming and Applications*, McGraw Hill, New York.

Gaonkar, (1991), *Microprocessor Architecture: Programming and Application*, Wiley Eastern, USA.

Gene, H. Hsteller, Clemand J. Savant, and Raymond J. Stefa (1982), *Design of Feedback Control Systems*, Holt Saunders International Education, USA.

Gibbs, D. (1982), *CNC Part Programming: A Practical Guide*, Casse Publication, London.

Harry, L. Stiltz (1969), *Aerospace Telemetry*, Prentice Hall Inc., Englewood Cliffs, New Jersey.

Hartley, J. (1984), *FMS: A Tutorial Work*, IFS Publication, London.

Karunakarans, et. al. (1995), HMT Handbook, *Mechatronics*, Tata McGraw Hill, New Delhi.

Jain, R.K. (1997), *Engineering Metrology*, Khanna Publication, New Delhi.

James Reghg, A. and Henry W. Kraebber (2001), *Computer Integrated Manufacturing*, Pearson Education Asia, India.

James Garratt, (1991), *Design and Technology*, Cambridge University Press, Cambridge.

Hancock, John C. (1991), *Introduction to Principles of Communication Theory*, Tata McGraw Hill, New Delhi.

Kalisi, H.S. (1995), *Electronics Instrumentation*, Technical Education Series, Tata McGraw Hill, New Delhi.

Kamerichy, I. (1974), *Heat Treatment Handbook*, MIS Publications, London.

Kasop, S.O. (1997), *Principles of Electrical and Engineering Materials and Devices*, Irwin Publication, USA.

Ogata, Katsahiko (2002), *Modern Control Engineering*, Pearson Education Asia, India.

Krishnaswamy, K. (2003), *Industrial Instrumentation*, New Age International, India.

Kurk, H.L. (1997), *Electronic Materials*, PWS Publication, Boston.

Lenz, J.E. (2002), *Flexible Manufacturing*, Marcel Dekkar Inc., New York.

Mahalik, N.P. (2005), *Mechatronics: Principles, Concept and Applications*, Tata McGraw Hill, New Delhi.

Ratner, Mark and Daneil Ratner (2003), *Nanotechnology*, Pearson Education Asia, India.

Mattershead, A. (1971), *Electronic Devices and Circuits*, Prentice-Hall of India, New Delhi.

Meanie, Mitchel (1998), *An Introduction to Genetic Algorithm*, Prentice-Hall of India, New Delhi.

Merkle, B. and M. Thomas Schrador (1990), *Hydraulic Basic Level*, Festo Didactic, Germany.

Groover, Mikell P. (2001), *Automation, Production Systems and Computer Integrated Manufacturing*, Pearson Education Asia, India.

Groover, Mikell P. and Emarry W. Zimmer (1981), *CAD/CAM Computer Aided Design and Manufacturing*, Prentice-Hall of India, New Delhi.

Groover, Mikell P., Mckehell Weiss, Roger N. Nagal, and Nichoas G. Groover (1988), *Industrial Robotics: Technology, Programming and Applications*, McGraw Hill, New York.

Millmann, J. and C.C. Halekis (1982), *Electronic Devices and Circuits*, McGraw Hill, New York.

Morrise, W. (1974), *Advanced Industrial Electronics*, McGraw Hill, New York.

Neelm Kavl, F. (1987), *Computer Simulation and Modeling*, John Wiley & Sons, USA.

Nessulescu, Dan (2002), *Mechatronics*, Pearson Education Asia, India.

Nils, J. Nilson (1990), *Principle of Artificial Intelligence*, Narosa Publishing House, New Delhi.

Ogaa, K. (1997), *System Dynamics*, Prentice-Hall Inc., Englewood Cliffs, New Jersey.

Pittman, J.B. (1993), *Design with PLC Micro-controller*, Prentice-Hall Inc., Englewood Cliffs, New Jersey.

Pinchar, M.J. and B.J. Callear (1996), *Power Pneumatics*, Prentice-Hall Inc., Englewood Cliffs, New Jersey.

Rafiquzaman, M. (1992), *Microprocessor*, Prentice-Hall Inc., Englewood Cliffs, New Jersey.

Rahne, P. (1996), *Automation with Programmable Logic Controller*, Macmillan, USA.

Rao, V. and H. Rao (1995), *C++ Neural Network and Fuzzy Logic*, MIS Press, India.

Reeves, C.R. (1993), *Modern Heuristic Techniques for Combinatorial Optimization*, John Wiley & Sons, USA.

Rohner, P. (1996), *Automation with Programmable Logic Controller*, Macmillan, USA.

Rohner, P. and G. Smith (1990), *Pneumatics Control for Industrial Automation*, Wiley Publication, USA.

Ross, Timothy J. (1995), *Fuzzy Logic with Engineering Applications*, McGraw Hill, New York.

Saatyg, T. (1980), *The Analytic Hierarchy Process*, McGraw Hill, New York

Stadler, W. (1995), *Analytical Robotics and Mechatronics*, McGraw Hill, New York.

Tai Ran Hsu (2002), *MEMS and Microsystems Design and Manufacturing*, Tata McGraw Hill, New Delhi.

Tomasi, W. (1996), *Advanced Electronic Communication Systems*, Prentice Hall Inc., Englewood Cliffs, New Jersey.

Warnock, I.G. (1988), *Programmable Controllers: Operation and Application*, Prentice Hall Inc., Englewood Cliffs, New Jersey.

Journal papers

Angell, J.B., S.C. Terry, and P.W. Barth (1983), 'Silicon micro-mechanical devices', *Journal of Scientific American*, vol. 248, pp. 44–53.

Askin, R.G., H.M. Selin, and A.J. Vakhari (1997), 'A methodology for designing flexible cellular manufacturing systems', *IEEE Transactions*, vol. 29, pp. 599–895.

Bart, S.F., L.S. Tavrow, Mehregang, and M.L. Lang, 'Micro-fabricated electro-hydrodynamic pumps', *Journal of Sensors and Actuators*, vol. 21, pp. 193–97.

Berardinis, L.A. (1990), 'Mechatronics: A new design strategy', *IEEE Transactions on Mechatronics*, vol. 62, pp. 1–58.

Boothroyd, G.P. and Dewuhurst (1998), 'Product design and manufacture and assembly', *Journal of Manufacturing Engineering*, vol. 41, pp. 42–46.

Buzacott, J.A. and D.D. Yao (1997), 'Flexible manufacturing systems: A review of analytical models', *Journal of Management Science*, vol. 36, pp. 1239–57.

Cellier, F. and H. Elonquest (1993), 'Automated formula manipulator supports object oriented continuous system modeling', *IEEE transactions on Control Systems*, vol. 27, pp. 28–38.

Chen, C.D., J. Shi-Shang, and A.Y. Tsen (1997), 'Semi-continuous copolymer composition distribution predictive control using a double ANN model structure', *Journal of Chinese Institute of Chemical Engineers*, vol. 28, pp. 49–50.

Choi, I.H. and K.D. Wise (1986), 'A silicon thermopile based infrared sensing array for use in automated manufacturing', *IEEE Transactions, Electronic Devices*, vol. 31, pp. 77–79.

Fan, L.S, Y.C. Jai, and R.S. Muller (1998), 'IC processes for electrostatic micro-motors', *IEEE Transactions, Electronic Devices*, vol. 35, pp. 666–69.

Geman, S., E. Bienestock, and R. Doursat (1992), 'Neural networks and the bias/variance dilemma', *Journal of Neural Computation*, vol. 4, pp. 1–58.

Harrison, R., A. Weat, and C.D. Wright (2000), 'Integrating machine design and control', *Journal of Computer Integrated Manufacturing*, vol. 13, pp. 498–516.

Isermann, R. (2000), 'Mechatronic system concept and applications', *Transactions on Instrumentation, Measurement and Control*, vol. 22, pp. 29–49.

Kaelsch, J.R. (1994), 'A new look to transfer lines', *Journal of Manufacturing Engineering*, vol. 5, pp. 73–78.

Kattan, I.A. (1997), 'Design and scheduling of hybrid multicell manufacturing systems', *Journal of Production Research*, vol. 36, pp. 1239–57.

Knotts, W. (1995), 'Ball clay skeleton using experimental design techniques', *Journal of Ceramic Engineering and Science*, vol. 16, pp. 123–26.

Lacher, R.C., S.I. Hruska and S.C. Kumnciskyc (1992), 'back propagation learning in expert networks', *IEEE Transaction on Neural Networks*, vol. 3, pp. 67–72.

Lin, C.M., T.K. Chu, and T.P. Chang (1996), 'Cement roller mill control by fuzzy logic controller', *Journal of Control Systems and Technology*, vol. 4, pp. 133–38.

Martinez, S.E., A.E. Smith, and B. Bidanda (1994), 'Reducing waste in casting with a predictive neural model', *Journal of Intelligent Manufacturing*, vol. 5, pp. 277–86.

Muth, E.J. (1975), 'Modeling and analysis of multi-station closed loop conveyer', *Journal of Production Research*, vol. 13, pp. 559–66.

Noaker, P.M. (1992), 'Down the road with DNC', *Journal of Manufacturing Engineering*, vol. 11, pp. 35–38.

Ogorek, M.(1985), 'CNC standard formats', *Journal of Manufacturing Engineering*, vol. 3, pp. 42–45.

Ohnishi, K., M. Shibato, and T. Musanaki (1996), 'Motor control for advanced mechatronics', *IEEE Transactions on Mechatronics*, vol. 1, pp. 56–67.

Putty, M.W. and S.C. Clang (1989), 'Process integration for active poly-silicon resonant microstructures', *Journal of Sensors and Actuators*, vol. 20, pp. 145–57.

Ralstonl, P.A.S. and T.L. Ward (1990), 'Fuzzy logic control of machining and manufacturing: a review', *IEEE Transactions on Neural Networks*, vol. 3, pp. 147–54.

Readey, M.J. (1993), 'Optimized processing of advanced ceramics: a case study in slip casting', *Journal of Ceramics Engineering and Science*, vol. 14, pp. 288–97.

Rullan, A. (1997), 'Programmable logic controllers versus personal computers for process control', *Journal of Computer and Industrial Engineering*, vol. 33, pp. 421–24.

Ramar, K. and V. Gowrishankar (1975), 'Algorithm for closed loop pole assignment with constant gain output feedback', *Proceedings of IEE*, vol. 122, pp. 574–75.

Somnd, H. Huay and Hang Zhang (1996), 'Neural networks in manufacturing: a survey', *IEEE Transactions on Neural Networks*, vol. 4, pp. 853–65.

Seraji, H. (1978), 'A new method for pole placement using output feedback', *International Journal of Control*, vol. 28, pp. 147–49.

Sridhar, B. and D.P. Linroff (1973), 'Pole placement with constant gain output feedback', *International Journal of Control*, vol. 18, pp. 993–1003.

Thomson, G. (2000), 'Mechatronics in manufacturing', *Journal of Mechatronics*, vol. 10, pp. 419–23.

Tilbury, T. and Khargonekar (2001), 'Challenges and opportunities in logic control for manufacturing systems', *Control System Magazine*, vol. 321, pp. 105–08.

Twomey, J.M. and A.E. Smith (1998), 'Bias and variance of validation methods for function approximation neural networks under conditions of spares data', *IEEE Transactions on Systems, Man and Cybernetics*, vol. 28, pp. 417–30.

Welten, J.R. (1989), 'Designing for manufacture and assembly', *Journal: Industry Week*, vol. 9, pp. 79–82.

Werbos, Paul (1990), 'Back propagation through time: what it does and how to do it?', *IEEE Transactions on Neural Networks*, vol. 78, pp. 1850–60.

Wise, K.O. (1991), 'Integrated micro-electro-mechanical systems: a perspective on MEMS in the 90s', *Journal of IEEE on MEMS*, vol. 27, pp. 33–38.

Witt, C.E. (1999), 'Palletizing unit loads: many options', *Journal of Material Handling Engineering*, vol. 13, pp. 99–106.

Wohlem, T.I. (1991), 'Make fiction fact and fast', *Journal of Manufacturing Engineering*, vol. 15, pp. 44–49.

Wonham, W.M. (1967), 'On pole assignment in multi-input controllable linear systems', *IEEE Transactions on Automatic Control*, vol. 12, pp. 660–65.

Wright, W.E. and J.C. Hall, 'Advanced aircraft gas turbine engine control', *ASME Transactions of Gas Turbine and Power*, vol. 112, pp. 561–68.

Index

accuracy 2, 186
actuation 6
    electrical 6
    hydraulic 6
    mechanical 6
    pneumatic 6, 160
adaptive control 220
    constraints (ACC) 225
    optimization (ACO) 225
address bus bar 91
air bag deployment system 305
aircraft engine control 314
air traffic control 314
alarms 316
algorithm 146
alphanumeric terminals 100
analog to digital conversion 89
antilock devices 304
artificial neural network 254
assembly language 90
automatic camera system 270
automatic controller 132
    integral 132
    PI 132
    proportional 132
autotronics 291
bathroom scale system 279
bearings 30, 46
    collar 46
    journal 30
    magnetic 47
    molecular 48
    pivot 46
    roller 47
bellows 160
Bernoulli's law 112
bimetallic device 293
binary number 105
    decimal 105
    hexadecimal 105
biostructure 303
biotechnology 307
bit 91
bus 91
    address 91
    control 91
    data 91

cam follower   49
capacitive sensor   61
capacitors   61
car park barrier   298
cathode ray tube (CRT)   101
central processing unit (CPU)   91
centrifugal clutch   48
clutches   48
communication   2
   direct   2
   indirect   2
compensation   231
computer intergrated manufacturing (CIM)   5
computer numerical control (CNC)   217
conductors   57
constraint   152
control system   110
   adaptive   110
   closed-loop   110, 111
   critically damped   123
   direct digital   142
   discrete data   144
   feedback   110
   open-loop   110, 111
   sequential   141
   time response of a   120
   under-damped   124
counter   77
cylinders   154
   double-acting   154
   hydraulic   155
   single acting   155
dead band   186
decimal equivalent   105

degree of freedom   151
designing process   5
diaphragms   160
difference formula   144
digital computer   90
digital control systems   141
digital speedometer   291
digital to analog converter   88
diode   67
   tunnel   69
   zener   68
discrete data signal   144
dot matrix printer   103
double lever mechanism   46
drifts   186
drum printer   102
dynamics   3

eddy current sensor   197
electrical relay   161
electroplating   39
engine management system   288
EPRON   99
error   186

fine technology   1
fits and tolerances   39
flexible manufacturing system   17
flip-flops   74
   D   74
   JK   74
   SR   74
floppy disk   98
force   27
force sensors   27
four-bar chain   45
frequency response   134

friction   27
fuzzy logic control   260

gauge factor   208
gear   50
   bevel   50
   helical   50
   ratio   52
   trains   50
   worm   50
genetic algorithm   257
Geneva mechanism   51
glucose detection   303
guideways   42

Hall effect sensors   193
heat treatment   38
hexadecimal equivalent   105
hydraulic servo system   158
hysteresis   186

impedance   64
inductor   62
inkjet printer   272
integral control   133
integrated circuit   71
interpolation   229

kinematics   3

Laplace transform   115
laser printer   273
lever   44
light-emitting diode (LED)   69
linear system   113
logic gates   72
   AND   72
   EXOR   74

NAND   73
NOR   7, 72
OR   72
lubrication   27

machine control unit (MCU)   227
machine language   90
machine structure   41
machine vision   202
magnetic disk   98
manufacturing   13
manufacturing automatic protocol (MAP)   20
materials   38
mechanism   44
   four-bar   45
   lever   44
mechatronics   3
   applications   5
   approach   3
   definition   3
   designs   3
memory   98
   bubble   98
   EEPROM   99
   EPROM   99
   PROM   99
   RAM   98
   ROM   99
microactuation   173
microcomputer   90
microcontroller   94
microgripper   174
micromotor   175
microprocessor   91
   8085   91

architecture   91
microsensors   207
microsystems   8
models   114
   logical   114
   mathematical   114
   physical   114
Moire's fringes   195
motor   164
   AC   165
   DC   165
   stepper   165
multiplexer   76

operational amplifier   83
optical encoder   192
   incremental   192
   normal binary   193
optical sensors   212
optics   8
optoisolator   69
overdamped response   123
overshoot   126

photodiodes   69
PLC (programmable logic controller)   95
pneumatic
   actuator   159
   air receiver   159
pole placement   140
potentiometer   189
   cement   190
   plastic   190
   wire wound   190
pulses   171
   bipolar   171

   unipolar   171
rack and pinion   52
range finder   197
ratchet and powel   51
rectifier   68
register   75
repeatability   186
resistor   59
resolution   186
resolver   191
rise time   125
robotics   6
robot, pick-and-place   282

scraping   40
semiconductors   57
sensitivity   186
sensors   185
   internal   189
   external   195
   proximity   196
   tactile   199
settling time   28
shift register   75
silicon-controlled rectifier/
thyristor   162
slip casting process   277
software   5
standard test signals   118
static performance characteristics   186
steady state error   129
step input   119
strain gauges   207
stress-strain behaviour   33
string   259

switch
  mechanical   161
  micro-   162, 293
  reed   162

tachogenerator   169
tachometer   191
technical office protocol (TOP)   21
thermocouple   209
timer   85
time response   120
torsion   35
transducer   185
transfer function   115
transformer   64

transistor   69
triac   163
trusses and frames   36
tuning fork   313

valves   153
  direction control   153, 160
  flow control   153
  non-return   153
  pressure   153
vehicle suspension system   295
vibration and noise control   7

washing machine   10, 269
wedge mechanism   50
windscreen wiper mechanism   289